Landscapes on Fire

Special Publications 80

LANDSCAPES ON FIRE

Impacts on Uplands, Rivers, and Communities

Ellen Wohl

This Work is a co-publication of the American Geophysical Union and John Wiley and Sons, Inc.

AGU
ADVANCING EARTH AND SPACE SCIENCES

WILEY

Published under the aegis of the AGU Publications Committee

Matthew Giampoala, Vice President, Publications
Steven A. Hauck, II, Chair, Publications Committee

For details about the American Geophysical Union visit us at www.agu.org.

Registered Office
John Wiley & Sons, Inc., 111 River Street, Hoboken, NJ 07030, USA

Editorial Office
111 River Street, Hoboken, NJ 07030, USA

For details of our global editorial offices, customer services, and more information about Wiley products visit us at www.wiley.com.

Wiley also publishes its books in a variety of electronic formats and by print-on-demand. Some content that appears in standard print versions of this book may not be available in other formats.

Library of Congress Cataloging-in-Publication Data Applied for:
Hardback: 978-1-394-23513-1

Cover Image: © Vladimir Melnikov
Adobe Stock Photos
Cover Design: Wiley

Set in 10/12pt TimesNewRomanMTStd by Straive, Pondicherry, India

CONTENTS

PREFACE

I live in Colorado at the base of the Southern Rockies, where gently undulating terrain of the Great Plains abruptly meets the dramatic topography of the mountains. The plains are semiarid steppe, and the mountains are nearly as dry. Annual precipitation increases gradually with elevation, falling mostly as snow in the upper mountains.

This is a landscape of wildfires. I have lived here for more than 30 years, and those years have been punctuated by large fires: Buffalo Creek in 1996, Bobcat Gulch in 2000, Hayman in 2002, Hewlett Gulch in 2012, High Park in 2012, Waldo Canyon in 2012, Fern Lake in 2012, Cameron Peak in 2020, and East Troublesome in 2020. The most recent large fires in my immediate area—High Park and Cameron Peak—directly affected my life for weeks as smoke hazed the air, leaving my eyes stinging and my throat scratchy. I swept ash from the sidewalks around my home and anxiously scanned the online map of the fire perimeter. These fires continue to influence my research on streams in the burned areas—as well as my quality of life. Most of the areas in which I hike and snowshoe are burned and will never again during my lifetime appear as they did before the fires.

Wildfire affects your life also, even if you live in a wet climate and have never directly witnessed a wildfire. The global scope of wildfire smoke creates far-reaching impacts that may be visible when orange or brown haze smudges the skyline or invisible as fire pumps more carbon into the global atmosphere.

I wrote this book to generate a more inclusive understanding of fires. My research focuses on how water and sediment move from hillslopes into and through stream corridors after fire and how these fluxes then affect the distribution of large wood, channel and floodplain morphology, hyporheic exchange, nitrate uptake, and so on. The focus of my research keeps expanding, and I keep adding more collaborators because fire and its aftereffects do not stop at traditional scholarly boundaries. Like wind-borne firebrands that cause a fire to dance across the most rugged terrain and drainage divides, comprehending fire requires moving across disciplinary divides and synthesizing the expertise of diverse scholars. That is my objective in writing this book. I think about fire and write as a geomorphologist, but I strive to draw on knowledge from other disciplines as a means to greater insight.

With that context, I stress that this book is written by a geomorphologist. I have tried to distill relevant information from other disciplines, but readers seeking a detailed discussion of the mechanics of wildfire, the interactions between weather and fire, or the effects of fire on biotic or human communities can find more detailed discussions in other sources. Given my disciplinary background, the book focuses on the movements of water, solutes, sediment, and organic material across uplands and into and through river networks. I review the effects of wildfire on process and form in upland and freshwater environments, as well as the interactions between living organisms, wildfire, and physical environments. Because the book spans multiple disciplines, I have included a glossary of terms.

The discussion of all topics covered in this book benefited greatly by thorough reviews from Stefan Doerr, Amy East, Joan Florsheim, Francis Rengers, and two anonymous reviewers. I very much appreciate their time and efforts.

Ellen Wohl
Colorado State University, USA

GLOSSARY

active layer: within permafrost regions, describes the upper layer of the ground that thaws each summer and freezes in the winter

aerosol: a colloidal suspension of particles dispersed in air

albedo: the proportion of incident sunlight or solar radiation that is reflected by a surface

alluvial fan: a fan-shaped (in planform) alluvial deposit, typically at the junction of a smaller river with a larger river or at the exit of a river from a mountain front

anthropogenic: human-caused

beaver meadow: a beaver-modified portion of river corridor that includes multiple dams and ponds of varying age; despite the name, typically includes abundant woody vegetation such as willows (*Salix* spp.)

benthic macroinvertebrates: live on or within the stream bed for at least some portion of their lifecycle and are visible to the naked eye (e.g., mussels, clams, worms, crayfish, snails, and a diverse array of insects)

biofilm: a matrix-enclosed and surface-attached microbial community

biomass: the total quantity or weight of living organisms in a given area or volume

biome: a major ecological community type, such as desert or grassland

bioturbation: the physical disturbance of sediment deposits by living organisms (e.g., by burrowing)

boreal: an ecosystem or river catchment in the northern hemisphere, between approximately 50 and 70° N latitude

channel head: the upstream boundary of concentrated water flow and sediment transport between definable banks

coarse woody debris: downed, dead wood outside of stream corridors

colluvial deposits: sediment deposited after transport by processes other than eolian or fluvial (e.g., rockfall, dry ravel, debris flow)

colluvial hollows: concave or convergent areas of uplands

conduction: transfer of heat via physical contact between a heat source and fuel or mineral body

connectivity: the movement of materials and organisms between components in a system (e.g., between uplands and river corridor, between active channel and floodplain, or between surface and subsurface)

convection: heat transfer by mixing of air masses

crown fire: a wildfire that burns within the forest canopy

debris flow: mass of poorly sorted, saturated sediment that moves downslope; typically has 40–77% sediment concentration by volume

drylands: designated based on the scarcity of water; defined by the United Nations Environment Program (UNEP) as tropical and temperate areas with an aridity index of less than 0.65; numerous aridity indices have been proposed – UNEP uses an index equal to ($100 \times R/LP$), where R is mean annual net radiation, P is mean annual precipitation, and L is latent heat of vaporization for water

dry ravel: involves rapid, dry, particle-to-particle rolling, sliding, and bouncing of sediment and some organic matter under the force of gravity and without the presence of water

dust: silt and clay-sized particles transported by wind

ecoregions: areas where ecosystems are generally similar

ENSO atmospheric circulation pattern: the El Niño-Southern Oscillation, a recurring climate pattern involving changes in the temperature of waters in the central and eastern tropical Pacific Ocean

fallout radionuclides: residual radioactive material propelled into the upper atmosphere following a nuclear blast and subsequently deposited on Earth's surface

fire-dependent ecosystems: those in which most species have evolved in the presence of fire

fire-independent ecosystems: naturally lack sufficient fuel or ignition sources to support fire as an evolutionary force; examples include desert and tundra

fire-sensitive ecosystems: those in which most species have not evolved in the presence of fire

fire hazard: in the context of wildfire, hazard typically refers to the state of the fuel and is quantified with respect to the likelihood and intensity of wildfire

fire intensity: rate of heat energy release

fire risk: in the context of technical risk assessments, risk includes the probability of an event, as well as values and expected losses; i.e., the probability of an event with harmful consequences, which can be thought of as the combination of hazard and vulnerability; in the context of fire, however, *risk* may refer only to the probability of natural or human-caused ignition, or it may be used to refer to hazard and the susceptibility of values-at-risk

fire severity: the effects of the fire on plant communities, or the loss of or change in above- and below-ground organic matter

fire vulnerability: vulnerability typically describes exposure and susceptibility, which can be differentiated as asset vulnerability and human vulnerability

flammability: reflects multiple metrics that quantify the ease of ignition and the behavior of the fire once ignited

floodplain: a low-relief sedimentary surface adjacent to the active channel that is constructed by fluvial processes and frequently inundated

forest canopy: the upper layer of forest formed by mature tree crowns

forest duff: the layer of organic soil below litter but above the mineral soil; composed of partly decomposed organic material

forest litter: the upper, organic soil layer of dead, fallen plant material; includes recognizable plant parts

forest understory: plant life growing beneath the forest canopy without penetrating the canopy to any great extent, but above the forest floor

fuel bed: a generic description of the complex of fuels occupying a given area

fuel component: subdivides the fuel type as a function of size (e.g., fine woody fuels) or physiological condition (e.g., dead wood)

fuel layer: vertical stratification of the fuel bed into ground, surface, and canopy fuels

fuel load: mass per unit area

fuel moisture: measured as the ratio between the mass of water in the fuel and the mass of dry fuel

fuel particle: individual fuel elements (e.g., twig, needle) that form the fuel complex at coarser scales

fuel type: the dominant fuels in the fuel bed (e.g., forest litter or grass)

gelisols: soils formed in very cold climates that contain permafrost within 2 m of the surface

grassland: an area in which the vegetation is dominated by grasses

ground cover: the lowest layer of aboveground vegetation (living and dead) and biological crusts that are in contact with the soil surface

headcut: a vertical down-step in the streambed that is migrating upstream over time

headwaters: first- and second-order streams

heat-induced spalling: wildfire heat results in lensoid-shaped rock flakes up to a few mm in thickness detaching from the surface of boulders or bedrock outcrops

herbaceous plants: vascular plants that have no persistent woody stem above ground

histosols: soils that are dominantly organic, commonly associated with bogs, moors, or peatlands

Hortonian overland flow: runoff that remains on the ground surface without infiltrating

hydrophobic soils: water-repellent soils that can form during wildfire

hyperconcentrated flow: a flowing mixture of water and sediment in a channel, with properties intermediate between fluvial flow and debris flow

hyporheic zone: subsurface areas in a river corridor containing water that originates from and returns to the channel

ignition: the initiation of combustion

indigenous: people inhabiting a region from the earliest times or before the arrival of European colonists

infiltration capacity: the maximum rate at which soil or regolith is capable of absorbing water under given conditions

landslide: mass movement with relatively rigid motion and deformation localized along persistent slip surfaces or shear zones

large wood: downed, dead wood with dimensions ≥ 10 cm in diameter and 1 m in length

machine learning: computer systems that are able to learn and adapt without following explicit instructions; these systems use algorithms and statistical models to analyze and draw inferences from patterns in data

mass movements: downslope movements of multiple grains simultaneously; types include debris flow, landslide, and soil creep

natural range of variability: the range of specified ecosystem or physical system properties (e.g., river discharge) over a particular time period in the absence of human alteration

nutrients: substances used by organisms to survive, grow, and reproduce

organic matter: remains of formerly living organisms, primarily plants; subdivided into fine particulate organic matter (>0.45 µm–1 mm), coarse particulate organic matter (>1 mm), and small wood (<10 cm in diameter and 1 m in length)

overland flow: water flowing across the surface as either diffuse sheet flow or concentrated flow in rills and gullies

peatlands: terrestrial wetland ecosystems in which waterlogged conditions prevent plant material from fully decomposing

periphyton: synonymous with biofilm

permafrost: permanently frozen ground; found at high altitudes or high latitudes

prescribed burns: fires deliberately set to reduce fuel load or to create desired ecological effects (also known as *controlled burns*)

$PM_{2.5}$: fine particulate matter with an aerodynamic diameter $\leq 2.5\,\mu m$

polycyclic aromatic hydrocarbons (PAHs): a class of chemicals composed of multiple aromatic rings; they occur naturally in coal, crude oil, and gasoline and result from burning coal, oil, gas, wood, garbage, and tobacco; they are carcinogens

Pyrocene: a time when humanity's firepower could interact in ways that promoted conditions favoring further fire

pyroconvection: convection caused or intensified by fire

pyrocumulonimbus: smoke-infused clouds and thunderstorms started or augmented by fires

pyrodiversity: fire-derived heterogeneity of habitats

pyrogenic carbon: a high-C and low-N byproduct of the incomplete combustion of organic matter; includes charred fine residue, but mostly takes the form of woody charcoal

pyrogeography: the study of human-fire relationships across space and through time

pyrolysis: the thermal degradation of solid fuel into gases under the influence of heat

pyrophilous: fire-loving (e.g., applied to types of fungi that appear after fire)

pyrophytes: fire-resistant species (e.g., plants)

pyroregions: areas with similar temporal patterns of fire activity

pyrotornadogenesis: when fires generate atmospheric vortices and tornadoes

radiation: transfer of heat via electromagnetic wave motion

rainsplash: occurs when the force of falling raindrops dislodges individual particles up to sand size

recalcitrant humic substances: humic substances that result from the decomposition of plant and animal residues and are composed of complex heterogeneous mixtures of organic compounds; they tend to be recalcitrant or resistant to biodegradation

regolith: loose, unconsolidated sediment; can include soil

resilience: describes the ability of a system (e.g., ecosystem, river corridor, human community) to return to a state or conditions similar to those before the disturbance

rills: small, linear erosional features on hillslopes, sometimes distinguished from gullies as being ≤ 0.3 m deep; may be transient or persistent features

riparian zone: encompasses the stream channel between low- and high-water marks and that portion of the terrestrial landscape from the high-water mark toward the uplands where vegetation may be influenced by elevated water tables or flooding and by the ability of the soils to hold water

river/stream: although a river is generally understood to be larger than a stream, there is no formal definition of the difference in size, and I use both words interchangeably in this book

river corridor: includes the active channel(s), the floodplain and riparian zone, and the underlying hyporheic zone

saturation overland flow: surface runoff that infiltrates to shallow depths and then returns to the surface

savanna: a grassy woodland in which trees are sufficiently widely spaced that the canopy does not close

sediment cascade: describes the sequential transport of sediment through the landscape or some portion of the landscape (e.g., uplands to river network); the most comprehensive version covers sediment movement from source (weathering of bedrock to regolith) through to sink (e.g., oceanic depositional environments)

sheetwash/sheetflow: surface runoff that forms a continuous layer across the ground surface, submerging small irregularities such as pebbles

shrubland: areas with vegetation dominated by shrubs or short-statured trees, generally <5 m tall, and commonly in a single canopy layer

snag: in a forest context, standing dead wood

soil creep: a slow, seasonal downslope movement of sediment facilitated by freezing and thawing or wetting and drying

solastalgia: psychological distress from losing a valued landscape, as occurs after wildfire

solutes: materials dissolved in water

spotting: the mass transport of solid hot particles carried by winds ahead of the fire front

stream order: a numbering system that reflects spatial position within a river network; first-order streams have no tributaries, second-order streams have two or more first-order streams tributary to them, third-order streams have two or more second-order streams tributary to them, and so on

subcritical cracking: fractures in rock that grow slowly and steadily at stresses much lower than critical stresses

supply-limited sediment: the amount of sediment moved is limited by the supply of weathered material

taiga: a type of forest of cold, subarctic regions; mostly consists of pines, spruces, and larches

talik: unfrozen patches within permafrost terrain

thermokarst: subsidence and slope failure caused by melting of ice disseminated through or concentrated within permafrost

throughflow: water moving below the ground surface but above the groundwater table; can move as diffuse matrix flow or within concentrated zones such as pipes

transpiration (within plants): a plant's loss of water, mainly through the stomata (minute pores) of leaves

transport-limited sediment: the amount of sediment moved is limited by the energy of the transport mechanism

tropical: areas within the tropics of Cancer and Capricorn, or within 23°26' north and south of the equator

tundra: type of biome where tree growth is hindered by cold temperatures and short growing seasons

vapor pressure deficit: the difference between the amount of moisture in the air and the amount that air could hold at saturation

WHI: wildland–human interface

WUI: wildland–urban interface

1

A FIRE-PRONE WORLD

Abstract

Wildfires have been common for nearly 400 million years. They have shaped landscapes and ecosystems from the boreal regions to the tropics. Wildfires have varied in spatial extent, frequency, and intensity through time, but they are currently becoming more widespread, frequent, and intense in regions including Europe's Mediterranean region, the boreal forests of North America, and the drylands of Australia and the western interior of North America. Many of these patterns reflect changing climate. Wildfire can create immediate, direct hazards for human and biotic communities through destructive heat and combustion and air pollution from smoke and other particulates. Wildfire also creates secondary effects by changing fluxes of water, solutes, sediment, and organic matter in uplands and freshwater environments. These secondary effects can shape landscapes and alter ecosystems in ways that persist for years to decades or longer. Most of the studies synthesized in this book come from forested environments in western North America, Australia, and the Mediterranean portions of Europe, even though most wildfires occur in grassland and savanna environments. The geographic distribution of information on wildfire effects reflects the bias in English-language scientific literature on wildfires.

Landscapes on Fire: Impacts on Uplands, Rivers, and Communities, Special Publications 80,
First Edition. Ellen Wohl.

DOI: 10.1002/9781394235186.ch1

Wildfire is likely to have been more or less common since the late Devonian (Schmidt & Noack, 2000), approximately 380 million years ago. Long-term records from diverse locations indicate that the location, extent, and severity of fire have varied through time. Global atmospheric transport and deposition of the smallest particles produced by fire have created records far from the actual burned areas. Isotopic records in sediment cores from the Ross Sea off Antarctica, for example, suggest large, frequent wildfires in the grasslands of the Southern Hemisphere from circa 15,000 to 4,200 years ago (Ren et al., 2022).

In recent decades, increasing frequency, extent, and severity of wildfires have made fires the hot topic in much of the world. Regional and global summaries tell the story:

- Globally, several hundred million hectares burn each year (FAO, 2001), equating to an estimated 8 billion Mg of biomass (Levine, 2000). Increased wildfire activity is observed in diverse regions and ecosystems around the world, from Mediterranean portions of Europe (Jiménez-Morillo et al., 2020; Seidl et al., 2014; Shakesby, 2011) to tropical rainforests (Cochrane, 2003) to Australian savanna (Fairman et al., 2015).
- Grassland and forest wildfires in the United States have increased in frequency, size, and severity since the mid-1980s, especially in the western United States (Joint Fire Service Program, 2004; Paul et al., 2022).
- In Arctic tundra, wildfire is projected to increase 4- to 11-fold by 2100 (Abbott et al., 2016; Hu et al., 2015).
- Fires historically occurred during hot and/or dry seasons in fire-prone ecosystems (Miller et al., 2019). Fire season has increased by nearly 20% across Earth's vegetated surface (Jolly et al., 2015). Parts of western North America, Brazil, and East Africa now have fire seasons more than a month longer than they were 35 years ago (Abatzoglou & Williams, 2016; Jolly et al., 2015). The greatest changes in fire season have occurred where temperature, changes in humidity, length of rain-free intervals, and wind speeds are greatest (Jolly et al., 2015).
- Large and frequent fires in boreal and tropical forests have the potential to cause terrestrial carbon stores to become major sources of greenhouse gases, amplifying climate change (Bowman et al., 2020). Global mean carbon emissions from fire during the period 1997–2016 were about 22% of global annual carbon emissions from fossil fuel combustion (Van der Werf et al., 2017).

Some of these patterns of wildfire reflect changing land use. Increasing wildfire frequency and spatial extent in Portugal, for example, are influenced by land abandonment and subsequent shrub encroachment, along with afforestation of former agricultural land, both of which increase fuel accumulation (Pinto et al., 2008).

Many of the patterns of wildfire reflect the changing climate. Increasing fire in the western United States corresponds to drier conditions, earlier snowmelt, and severe droughts (Paul et al., 2022). Increased tundra fires reflect increased temperature, evapotranspiration, ignition sources, and vegetation shifts (Abbott et al., 2021). Tropical fires become more frequent in years with extended dry seasons (Van der Werf et al., 2008). The area burned annually by wildfires is expected to continue to increase globally due to climate change (Sankey et al., 2017).

Charcoal records indicate a global monotonic increase in wildfire since the last glacial maximum 21,000 years ago, with increased spatial heterogeneity in the last 12,000 years (Power et al., 2008). Fire suppression and changing land use during the first half of the 20th century caused the global average area burned to decrease from 535 to 500 Mha per year (Mouillot & Field, 2005), but global burned area increased to an estimated 608 Mha per year during the second half of the 20th century (Flannigan et al., 2009). This latter change includes increasing savanna and grassland fires in the tropics and temperate regions, with an exponential increase in tropical forest fires driven by deforestation and agricultural development (Carmona-Moreno et al., 2005; Flannigan et al., 2009). During the period 2001-2019, burned area increased by 50% or more in some extratropical forest regions including the Pacific U.S. and high-latitude forests, but decreased by 27% globally because of declines in burned area in African savannas (Jones et al., 2022).

Beyond the immediate hazards of the fire itself—destructive heat and combustion, air pollution from smoke, and other particulates—the secondary effects from wildfire can substantially alter landscapes and ecosystems in ways that persist for years to decades. Burned areas increase rates of runoff and soil erosion, as well as downstream sedimentation in rivers and reservoirs. Fluxes of nutrients such as nitrogen and phosphorus can increase to levels at which they cause harmful algal blooms in receiving waters. Increased water and sediment inputs to river corridors change channel and floodplain form and function, influencing processes such as flood attenuation, surface-subsurface water exchange, biogeochemical cycling, and aquatic and riparian biotic communities.

Some species are resistant to fire, especially lower intensity fires, surviving underground or within thick protective layers such as tree bark. Others are resilient, using cues from heat and smoke to germinate, or moving into a burned area quickly to take advantage of specific resources created by fire. When the frequency and intensity of fire increase beyond levels to which species can adapt, however, or when fires start to occur in ecosystems without an evolutionary history of adaptation to fire, individual species can undergo local extinction and the community of species present can change.

As fires increase in frequency, extent, and severity, these secondary effects also increase: Modeling conditions in the western United States in 2050, for example, Sankey et al. (2017) predict that post-fire sedimentation will increase by >10% for nearly 90% of watersheds and by >100% for one third of the watersheds, which will stress the aquatic ecosystems that are on the receiving end of extra sediment. Increased fire frequency in Mediterranean drylands can be responsible for accelerated soil erosion and reduced soil nutrient content, which then causes a transition from conifer forests to shrublands (Caon et al., 2014). The persistent changes in the geographic range of individual species and in the associations of species that constitute a biotic community and ecosystem are increasingly being documented in ecosystems with a history of fire, as well as those without such a history, creating examples of what have been called novel ecosystems (Jones et al., 2015).

A global map of the fire weather index, a key indicator of extreme fire behavior potential, shows that fire hazards are more common in the western United States, Australia, the southern and central parts of Africa, the southern and northeastern portions of South America, and central Asia (Figure 1.1). Fires are generally absent poleward of 70 °N and 70 °S in regions characterized by tundra and ice, progressively more frequent toward the tropics, and then drop sharply at the equator (Mouillot & Field, 2005). To some extent, the literature reviewed in this book represents this geographic distribution. The great majority of the studies synthesized here come from western North America, Australia, and the Mediterranean portions of Europe. The portions of Africa, South America, and central Asia prone to wildfires have been poorly represented in English-language peer-reviewed scientific literature, although the number of published studies focused on these regions is increasing rapidly.

The majority of studies synthesized in this book also represent forested environments, even though the majority of wildfires occur in grassland and savanna environments (Flannigan et al., 2009). Again, this reflects a bias in the proportion of published studies examining forests versus other biomes. (Biomes are ecosystems with the same dominant plant life-forms; Grimm et al., 2013.) This may reflect the greater severity and duration of forest wildfires relative to the relatively brief duration of grassland and savanna fires. Or perhaps the imbalance reflects the obvious changes in a forest following intense wildfire and the decades required for a mature forest to re-grow. Grasslands, in contrast, survive fire by concentrating plant resources underground and grasses can typically re-sprout rapidly after fire (e.g., Vilà et al., 2001). Whatever the causes underlying the disproportionate focus on forests in wildfire literature, I have sought to include relevant references addressing wildfire in grasslands, savannas, and Arctic and tropical peatlands wherever possible throughout this book.

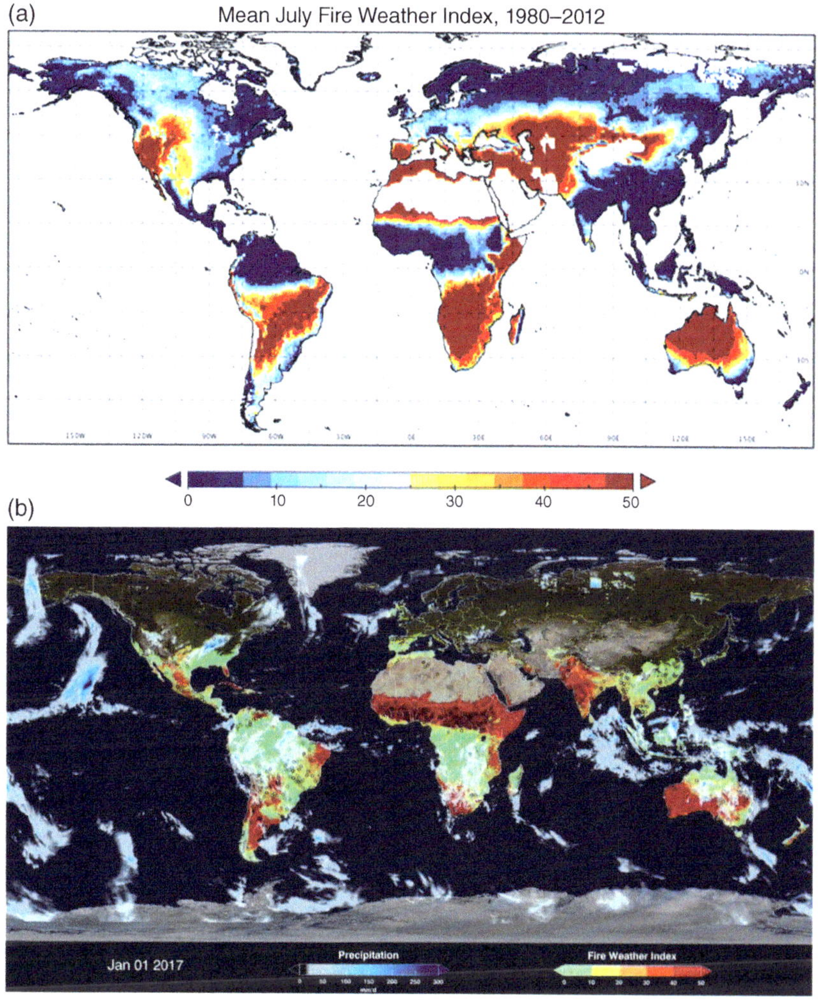

Figure 1.1 (a) Global map of fire weather index based on mean July conditions—summer in the northern hemisphere. (b) A contrasting map during summer in the southern hemisphere, here for 1 January 2017. (NASA / Public domain.)

This book focuses on wildfires. Wildfires here include fires outside of human-built environments started by processes not directly associated with humans (e.g., lightning strikes) or by deliberate or accidental human-generated ignitions of fire. Unless otherwise specified, any subsequent reference to fire refers only to wildfires. This book starts by exploring the details of fire. An introduction to the characteristics of fire, including the concept of a fire regime and fire recurrence intervals, is followed by a brief review of human alteration of fires, and the consequences of different types of fire regimes.

The succeeding two chapters are organized from the viewpoint of a geomorphologist—someone who studies the physical processes that create and maintain landscapes. Gravity is the great driver of physical processes following wildfire, aided and abetted by water. Consequently, I have organized the third chapter around the effects of wildfire in uplands—the unchanneled parts of a river catchment—that most people would simply think of as the terrestrial environment. This chapter proceeds from discussion of fire-induced changes in upland fluxes of water, solutes, sediment, and organic matter, to a review of how these changed fluxes influence landforms and biotic communities. It is worth emphasizing that the description of upland fluxes in this chapter is heavily biased toward the effects of severe wildfires in moderate- to high-relief environments. This bias mirrors the scientific literature, which primarily describes these scenarios. I presume this reflects the tendency to write about substantial change: it is more difficult to have a paper on the effects of wildfire accepted by peer reviewers if the primary conclusion is along the lines of 'nothing much happened.'

The fourth chapter focuses on post-fire changes in fluxes, morphology, and biotic communities within freshwater environments. The great majority of studies examining fire effects in freshwater environments focus on river networks, as reflected in the emphasis in the third chapter. This chapter starts with headwater streams and then moves downstream to higher-order streams, discusses stratigraphic records of long-term fire history in depositional environments such as alluvial fans and lakes, and concludes with a broader discussion of lake ecosystems.

Again, chapters three and four include a disproportionate focus on high-severity wildfires in moderate- to high-relief environments. As with the literature on forest rather than grassland fires, this coverage of severely burned, steep landscapes reflects the preponderance of literature discussing the effects of wildfire on fluxes and landscape forms. Low-relief landscapes also burn in wildfires and low-severity wildfires occur, but neither of these scenarios creates the dramatic landscape changes and post-fire hazards of floods and debris flows that attract scientists to investigate and publish their findings.

The fifth chapter addresses wildfire and human communities, with an emphasis on predicting and mitigating diverse types of fire-related hazards. The book concludes with an overview of fires in the future, including changing fire regimes, shifts in species and biomes, and human responses.

Several important compilations of fire-related research have been published within the past few years and I have used these compilations in developing the summary presented in this book. Important references include *Fire Phenomena and the Earth System: An Interdisciplinary Guide to Fire Science* (Belcher, 2013); *Fire Science: From Chemistry to Landscape Management* (Rego et al., 2021); *Spreading Like Wildfire: The Rising Threat of Extraordinary Landscape Fires* (Sullivan et al., 2022); and *Fire in the Earth System*, a special collection spanning 10 journals of the American Geophysical Union and including more than 100 individual papers (East et al., 2023).

In addition to a wealth of disciplinary and interdisciplinary studies of diverse aspects of wildfire, a specialized vocabulary related to wildfire has developed. Nearly any word can have *pyro*, from the Greek word for fire, added as a prefix to indicate that the word is being used in the context of wildfires. Text Box 1.1 defines some of the pyro vocabulary that I encountered most frequently while writing this book.

Text Box 1.1 Pyro Vocabulary: Examples of Words Incorporating *Pyro* (Fire in Greek) as a Prefix

Pyrogenic carbon: created by fire, includes charred fine residue & woody charcoal
Pyrocene: a time when humanity's firepower could interact in ways that promoted conditions that favored further fire
Pyroconvection: convection caused or intensified by fire
Pyrocumulonimbus: smoke-infused clouds & thunderstorms started or augmented by fires
Pyrodiversity: fire-derived heterogeneity of habitats
Pyrogeography: the study of human-fire relationships across space & through time
Pyrolysis: the thermal degradation of solid fuel into gases under the influence of heat
Pyrophilous: fire-loving (e.g., applied to types of fungi that appear after fire)
Pyrophytes: fire-resistant species (e.g., plants)
Pyroregions: areas with similar temporal patterns of fire activity

References

Abatzoglou, J. T., & Williams, A. P. (2016). Impact of anthropogenic climate change on wildfire across western US forests. *Proceedings of the National Academy of Science*, *113*, 11770–11775.

Abbott, B. W., Jones, J. B., Schuur, E. A. G., Chapin, F. S., Bowden, W. B., Bret-Harte, M. S., et al. (2016). Biomass offsets little or none of permafrost carbon release from soils, streams, and wildfire: An expert assessment. *Environmental Research Letters*, *11*, 034014.

Abbott, B. W., Rocha, A. V., Shogren, A., Zarnetske, J. P., Iannucci, F., Bowden, W. B., et al. (2021). Tundra wildfire triggers sustained lateral nutrient loss in Alaskan Arctic. *Global Change Biology*, *27*, 1408–1430.

Belcher, C. M. (Ed.) (2013). *Fire phenomena and the earth system: An interdisciplinary guide to fire science*. Chichester, UK: Wiley-Blackwell.

Bowman, D. M. J. S., Kolden, C. A., Abatzoglou, J. T., Johnstons, F. H., van der Werf, G. R., & Flannigan, M. (2020). Vegetation fires in the Anthropocene. *Nature Reviews Earth and Environment*, *1*, 500–515.

Caon, L., Vallejo, V. R., Ritsema, C. J., & Geissen, V. (2014). Effects of wildfire on soil nutrients in Mediterranean ecosystems. *Earth-Science Reviews*, *139*, 47–58. http://dx.doi.org/10.1016/j.earscirev.2014.09.001

Carmona-Moreno, C., Belward, A., Malingreau, J.-P., Hartley, A., Garcia-Alegre, M., Antonovskiy, M., et al. (2005). Characterizing interannual variations in global fire calendar using data from Earth observing satellites. *Global Change Biology*, *11*, 1537–1555. http://dx.doi.org/10.1111/J.1365-2486.2005.01003.X

Cochrane, M. (2003). Fire science for rainforests. *Nature*, *421*, 913–919.

East, A., AghaKouchak, A., Caprarelli, G., Filipetti, G., Florindo, F., Luce, C., et al. (2023). Fire in the Earth system: Introduction to the special collection. *Journal of Geophysical Research: Earth Surface*, *128*, e2023JF007184. http://dx.doi.org/10.1029/2023JF007184

Fairman, T. A., Nitschke, C. R., & Bennett, L. T. (2015). Too much, too soon? A review of the effects of increasing wildfire frequency on tree mortality and regeneration in temperate eucalypt forests. *International Journal of Wildland Fire*, *25*, 831–848.

FAO. (2001). Global Forest Fire Assessment 1990–2000. Food and Agriculture Organization of the United Nations, Forestry Department, Forest Resources Assessment Programme Working Paper 55, Rome.

Flannigan, M. D., Krawchuk, M. A., de Groot, W. J., Wotton, B. M., & Gowman, L. M. (2009). Implications of changing climate for global wildland fire. *International Journal of Wildland Fire*, *18*, 483–507. http://dx.doi.org/10.1071/WF08187

Grimm, N. B., Chapin, F. S., Bierwagen, B., Gonzalez, P., Groffman, P. M., Luo, Y., et al. (2013). The impacts of climate change on ecosystem structure and function. *Frontiers in Ecology and Environment*, *11*, 474–482. http://dx.doi.org/10.1890/120282

Hu, F. S., Higuera, P. E., Duffy, P., Chipman, M. L., Rocha, A. V., Young, A. M., et al. (2015). Arctic tundra fires: Natural variability and responses to climate change. *Frontiers in Ecology and Environment*, *13*, 369–377.

Jiménez-Morillo, N. T., Almendros, G., De la Rosa, J. M., Jordán, Z. L. M., Granged, A. J. P., & González-Pérez, J. A. (2020). Effect of a wildfire and of post-fire restoration actions in the organic matter structure in soil fractions. *Science of the Total Environment*, *728*, 138715.

Joint Fire Service Program. (2004). *Joint fire service program 2003 business summary*. Seattle, WA: US Department of Agriculture, Forest Service, Pacific Northwest Research Station.

Jolly, W. M., Cochrane, M. A., Freeborn, P. H., Holden, Z. A., Brown, T. J., Williamson, G. J., & Bowman, D. M. J. S. (2015). Climate-induced variations in global wildfire danger from 1979 to 2013. *Nature Communications*, *6*, 7537. https://doi.org/10.1038%2Fncomms8537

Jones, T. A., Monaco, T. A., & Rigby, C. W. (2015). The potential of novel native plant materials for the restoration of novel ecosystems. *Elementa*, *3*, 000047. https://doi.org/10.12952/journal.elementa.000047

Levine, J. S. (2000). Global biomass burning: A case study of the gaseous and particulate emissions released to the atmosphere during the 1997 fires in Kalimantan and Sumatra, Indonesia. In J. L. Innes, M. Beniston, & M. M. Verstraete (Eds.), *Biomass burning and its inter-relationships with the climate system. Advances in global change research* (Vol. 3, pp. 15–31). Dordrecht: Springer.

Miller, R. G., Tangney, R., Enright, N. J., Fontaine, J. B., Merritt, D. J., Ooi, M. K. J., et al. (2019). Mechanisms of fire seasonality effects on plant populations. *Trends in Ecology and Evolution*, *34*, 1104–1117.

Mouillot, F., & Field, C. B. (2005). Fire history and the global carbon budget: A 1°× 1° fire history reconstruction for the 20th century. *Global Change Biology*, *11*, 398–420. http://dx.doi.org/10.1111/J.1365-2486.2005.00920.X

Paul, M. J., LeDuc, S. D., Lassiter, M. G., Moorhead, L. C., Noyes, P. D., & Leibowitz, S. G. (2022). Wildfire induces changes in receiving waters: A review with considerations for water quality management. *Water Resources Research*, *58*, e2021WR030699.

Pinto, P., Vaz, P., Robinson, C., & Morais, M. (2008). Wildfire impacts on aquatic ecosystems. In L. M. G. Duarte & P. Pinto (Eds.), *Sustainable Development: Energy, Environment and Natural Disasters* (pp. 25–34). Fundação Luis de Molina.

Power, M., Marlon, J., Ortiz, N., Bartlein, P., Harrison, S., Mayle, F., et al. (2008). Changes in fire regimes since the last glacial maximum: An assessment based on a global synthesis and analysis of charcoal data. *Climate Dynamics*, *30*, 887–907. https://doi.org/10.1007/%20S00382-007-0334-X

Rego, F. C., Morgan, P., Fernandes, P., & Hoffman, C. (2021). *Fire science: From chemistry to landscape management*. Cham, Switzerland: Springer Nature Switzerland.

Ren, P., Luo, C., Zhang, H., Cui, C., Sun, S., Song, H., et al. (2022). Isotopic records of ancient wildfires in C_4 grasses preserved in the sediment of the Ross Sea, Antarctica. *Geophysical Research Letters*, *49*, e2022GL098979. http://dx.doi.org/10.1029/2022GL098979

Sankey, J. B., Kreitler, J., Hawbaker, T. J., McVay, J. L., Miller, M. E., Mueller, E. R., et al. (2017). Climate, wildfire, and erosion ensemble foretells more sediment in western USA watersheds. *Geophysical Research Letters*, *44*, 8884–8892.

Schmidt, M. W., & Noack, A. G. (2000). Black carbon in soils and sediments: Analysis, distribution, implications, and current challenges. *Global Biochemical Cycles*, *14*, 777–793. https://doi.org/10.1029/1999GB001208

Seidl, R., Schelhaas, M. J., Rammer, W., & Verkerk, P. J. (2014). Increasing forest disturbances in Europe and their impact on carbon storage. *Nature Climate Change*, *4*, 806–810.

Shakesby, R. A. (2011). Post-wildfire soil erosion in the Mediterranean: Review and future research directions. *Earth-Science Reviews*, *105*, 71–100.

Sullivan, A., Baker, E., & Kurvits, T. (Eds.) (2022). *Spreading like wildfire: The rising threat of extraordinary landscape fires*. United Nations Environment Programme.

Van der Werf, G., Dempewolf, J., Trigg, S., Randerson, J., Kasibhatla, P., Giglio, L., et al. (2008). Climate regulation of fire emissions and deforestation in equatorial Asia. *Proceedings of the National Academy of Sciences*, *105*, 20,350–20,355. http://dx.doi.org/10.1073/pnas.0803375105

Van der Werf, G. R., Randerson, J. T., Giglio, L., Van Leeuwen, T. T., Chen, Y., Rogers, B. M., et al. (2017). Global fire emissions estimates during 1997-2016. *Earth System Science Data*, *9*, 697–720.

Vilà, M., Lloret, F., Ogheri, E., & Terradas, J. (2001). Positive fire-grass feedback in Mediterranean Basin woodlands. *Forest Ecology and Management*, *147*, 3–14. http://dx.doi.org/10.1016/S0378-1127(00)00435-7

2

FIRES: WHAT, WHEN, WHERE, AND HOW

Abstract

Heat is transferred from wildfires via radiation, conduction, convection, and spotting when wind transports solid hot particles ahead of the fire front. Fire intensity describes the energy release rate. Fire severity describes the effects of the fire on plant communities or the loss or change in above- and belowground organic matter. Ground fires burn the organic layer overlying mineral soil. Surface fires burn ground fuels. Crown fires burn through the forest canopy. Although most wildfires result from human activities in the coterminous United States and some other parts of the world, most natural wildfires are ignited by lightning strikes. Fire weather occurs under weather conditions conducive to the ignition and spread of wildfires, which typically means hot, dry, and windy. Fire weather indices that incorporate daily weather variables related to fuel moisture and fire behavior are used to characterize fire weather. A fire regime describes the historical sequence of fires with some stable, recurrent, or cyclic characteristics or properties affecting a specified spatial and temporal window. Human activities have forced fire regimes outside of the natural range of variability in most of the world. Fire-prone ecosystems cover 40% of Earth's land surface and include grasslands, savannas, shrublands, and boreal forests.

Landscapes on Fire: Impacts on Uplands, Rivers, and Communities, Special Publications 80,
First Edition. Ellen Wohl.

DOI: 10.1002/9781394235186.ch2

2.1. Introduction

This chapter summarizes the basic characteristics of wildfires and is intended mainly to provide basic background information for readers who are not fire scientists. *What* involves the chemical and physical processes that create combustion, the parameters used to describe wildfires, and the phases of a fire. *How* refers to mechanisms of fire ignition and spread. *When* involves fire weather, fire regimes, and fire frequency as reflected in historical, botanical, and sedimentary records. *Where* refers to the geographic distribution of fires.

Once ignition occurs and a wildfire starts, the controls on the behavior of the fire vary across spatial scales (Figure 2.1). Wildfire science provides a reasonably good understanding of what controls fire behavior and the effects of these controls at different scales, but the interactions between scales that can create emergent properties are less well understood (Moritz et al., 2005).

2.2. What Is Fire?

Fundamentally, fire is a chemical reaction, known as combustion, between a fuel and oxygen (Drysdale, 2011). This reaction results in the release of energy and a variety of chemical products known as emissions (Rego et al., 2021). Combustion occurs when the fuel is heated to its ignition temperature and mixed with an oxidant (e.g., oxygen in the atmosphere) at a sufficient concentration. Heat released during combustion then sustains the combustion reaction.

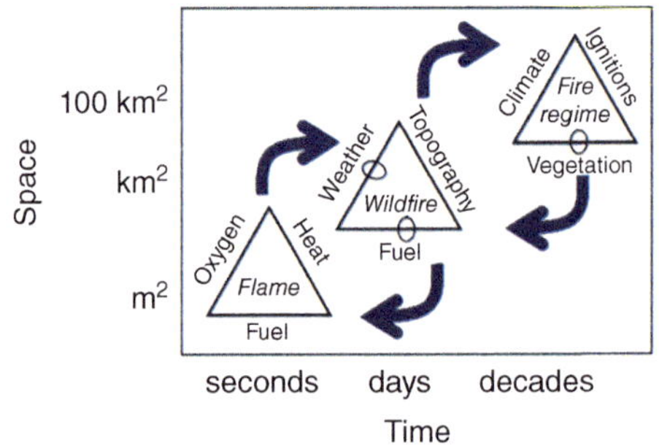

Figure 2.1 Schematic illustration of controls on wildfire at different time and space scales, showing the dominant factors that influence fire at the scale of a flame, a single wildfire, and a fire regime. The small loops indicate feedbacks that wildfire has on the controls, and the large arrows indicate feedbacks between processes at different scales. (Moritz et al., 2005/Proceedings of the National Academy of Sciences.)

Before the vegetation fuels burned in wildfires can ignite, they must be dried and sufficiently heated to create the thermal degradation known as pyrolysis (Rego et al., 2021). During this process, volatiles formed from the thermal degradation of celluloses and hemicelluloses in the fuel become flammable gases, which burn when there is sufficient oxygen available. Consequently, both the types of fuel (e.g., volatile organic compound content within the fuel) and the environmental conditions (dryness of the fuel) influence how readily fuels ignite (Rego et al., 2021).

Heat transfer from fires occurs via radiation, conduction, convection, and spotting (Chandler et al., 1983; Rego et al., 2021). Radiation is the transfer of heat via electromagnetic wave motion. Conduction, which occurs via physical contact between a heat source and fuel or mineral body, is a primary mode of heat transfer in heavier fuels such as forest duff and organic soils. Convection is heat transfer by mixing of air masses. Radiation and convection create the greatest heat transfer from light fuels such as grass and foliage. Spotting is the mass transport of solid hot particles (firebrands) carried by winds ahead of the fire front.

2.3. Describing Wildfire

An individual wildfire can be described with respect to the spatial extent, locations within the fire, the intensity or severity of the fire, the type of fire, the fire danger, diverse aspects of the fuel available, the behavior of the fire, or the cause of the fire (Text Box 2.1).

The minimum area that constitutes a large fire varies between regions. Area thresholds for designating a fire as large vary from 200 ha in Russia and Canada to 2,000 ha in the United States Intermountain West (Gill & Allan, 2008). The largest fires have extended over a million hectares in Australia, Canada, and parts of Africa (Gill & Allan, 2008). Whatever the reported area of a fire, that area always includes unburned portions where winds, terrain, and fuel supply combine to prevent the burning of vegetation.

Analogously, the phrase "extreme fire" is used in different ways, including fires with significant social, economic, or ecological impacts; especially large fires; and fires exhibiting unusual behavior, including spread other than steady surface spread (Rego et al., 2021). "Other than steady surface spread" refers to fires that spread with tremendous energy and speed via processes such as fire-generated vortices that are upright columns of "vigorously rotating ash, smoke, and often flame" (Lareau et al., 2018, p. 13107). Vigorous might be an understatement: a fire vortex during the 2018 Carr Fire in California, USA included surface wind speeds within the vortex estimated to exceed 64 m/s, equivalent to an EF-3 tornado, and

Text Box 2.1 Fire Descriptors

Fuels

Fuel-bed: a generic description of the complex of fuels occupying a given area

Fuel component: subdivides the fuel type as a function of size (e.g., fine woody fuels) or physiological condition (e.g., dead wood)

Fuel layer: vertical stratification of the fuel-bed into ground, surface, and canopy fuels

Fuel load: mass per unit area

Fuel moisture: measured as the ratio between the mass of water in the fuel and the mass of dry fuel

Fuel particle: individual fuel elements (e.g., twig, needle) that form the fuel complex at coarser scales

Fuel type: the dominant fuels in the fuel-bed (e.g., forest litter or grass)

Fire behavior

Flammability: reflects multiple metrics that quantify the ease of ignition and the behavior of the fire once ignited

Ignitability: the ability of a fuel to ignite and produce flames; reflects the temperature at which a substance or mixture may ignite when exposed to an ignition source; and the relative concentration of fuel and oxygen

Rate of spread: linear advance of flaming fire front per unit of time

Residence time: duration of flaming combustion, or length of time for flame front to pass a given point

Reaction time: total duration of combustion, or time for all fuel fractions to burn out

Flame shape: height, angle (angle between fire front and unburned fuel-bed), tilt angle (angle between vertical and fire front), depth, length

Reaction intensity: heat release per unit area within the flaming front

Fireline intensity: heat release within the flaming front per unit time and unit length of the fire front

Power of the fire: heat release per unit time within the flaming zone

(From Rego et al., 2021/Springer Nature)

the convective plume grew by kilometers in height within 15 minutes (Lareau et al., 2018).

Extreme fires are sometimes referred to as megafires or firestorms. They burn larger areas, create greater economic damages, and have more impacts than the combined effects of numerous average fires, not least because features such as fire-generated vortices can quickly overwhelm firefighting resources and evacuation capabilities. During the 2003 Canberra Firestorm, for example, the fire gave rise to a 20-km-long F2-intensity tornado, a phenomenon referred to as pyrotornadogenesis (McRae et al., 2013).

Fire intensity describes the energy release rate during a wildfire. Fire severity describes the effects of the fire on plant communities, or the loss of or change in, above- and below-ground organic matter (Keeley, 2009). A low-intensity fire, for example, has minimal effects on overstory trees, which mostly survive the fire. Although fuel in grasslands and shrublands can be very dry, these biomes have less fuel compared to forests, so fire intensity can be relatively low, but fires spread quickly (Stavi, 2019). Approximately 80–86% of the global area burned occurs in grassland and savannas, primarily in Africa and Australia, but also in South Asia and South America (Flannigan et al., 2009).

Wildfire types include ground fires that burn the organic layer overlying mineral soil and may be of low temperature but last for hours, days, or even years (Scott, 2000). Surface fires burn ground fuels such as foliage, limbs, logs, brush, grass, duff, and roots. Surface fires are typically of low temperature, but in places they can reach 600 °C and remain high for several hours, a situation that can produce large quantities of charcoal (Pyne et al., 1996). Crown fires that burn through the forest canopy feed on aerial fuels of foliage, branches, snags, and moss. These fires tend to be hotter than surface fires (Scott, 2000).

Fire danger is used to quantify burning conditions in which a wildfire may occur using a relative scale. Fire danger results from the factors affecting the start, spread, and difficulty of controlling a fire (Rego et al., 2021) and is described via rating classes from low to extreme in systems such as U.S. National Fire Danger Rating System (Cohen & Deeming, 1985), the Canadian Forest Fire Weather Index System (Van Wagner, 1974), and the long-used Australian Forest Fire Danger Index (McArthur, 1967) and its replacement (the Australian Fire Danger Rating System; McCoy & Field, 2022). All of these rating systems use basic weather variables observed over consecutive days, along with some measure of available fuels.

Estimates of fuels depend on detailed mapping of vegetation spatial extent, type, and condition (e.g., moisture). Satellite-derived mapping is increasingly used for this purpose but can be limited because of the logistics and high costs involved in the acquisition of some types of data, such as hyperspectral remote sensing. Development of methods that can be used to simulate hyperspectral images is consequently an area of active research (e.g., Badola et al., 2022). Soil moisture conditions also inform fire danger rating systems, especially in grasslands.

Combining *in situ*, remotely sensed, and modelled soil moisture information can improve estimates of soil moisture and associated fire danger (Krueger et al., 2023).

2.4. Where There's Smoke, There's Fire

A wildfire includes four phases (Pyne et al., 1996). During the pre-ignition phase, the temperature of the fuel is raised by endothermic reactions in which reactants absorb heat energy from the surroundings and form products. These reactions evaporate water and release volatiles. The volatile gases, which are combustible, are produced by pyrolysis and volatilization of organic compounds such as plant waxes and oils (Pyne et al., 1996; Rego et al., 2021) (Text Box 2.2).

The ignition phase of a wildfire is a transition from pre-ignition to combustion. This transition produces a strong exothermic reaction during

Text Box 2.2 How Do We Know? Mechanics and Details of Fire

Fire spread and other aspects of fire behavior are studied in laboratory experiments and in deliberately set fires that represent field experiments (Rego et al., 2021). These experiments provide the empirical data used to calibrate predictive relationships. Small-scale laboratory experiments expose vegetation samples, such as a few pine needles, to constant heat flux and note the time until ignition. Slightly larger experiments use a fuel-bed (such as dead pine needles) a few meters in size with a slope gradient, a wind source, and ignition from a point source to measure fire spread. Prepared fuel-beds are also used to study the effect of fuel moisture content on the heat of pre-ignition and fire spread rate. Wind tunnels are used to measure the terminal velocity of burning firebrands. Experimental fires in field plots or prescribed burn areas can also supply measurements of fire parameters that help to calibrate predictive models. A spectacular example comes from the International Crown Fire Modeling Experiment carried out during 1995–2001 in Canada's Northwest Territories. This series of 18 experimental high-intensity and highly instrumented crown fires was used to quantify parameters essential to modeling the initiation and spread of crown fires (Stocks et al., 2004).

which energy is released in the form of heat, typically at around 325 °C. The reaction must continue over an extended period for combustion to occur and spread.

The third fire phase, combustion, can occur as flaming and smoldering or glowing (Rego et al., 2021), which involve different processes and differ in appearance (Pyne et al., 1996). Flaming combustion dominates during the startup phase but the proportion of different types of combustion can change during a fire. Flaming results from a rapid oxidation reduction reaction at high temperatures with an abundant oxygen supply (Rego et al., 2021). Flames occur at 425–480 °C, but can exceed 1,000 °C (Pyne et al., 1996). Smoldering is a non-flaming form of combustion that occurs at lower temperatures around 300 °C and produces abundant smoke with large amounts of particulate matter (Rego et al., 2021).

Smoldering occurs at a slower rate than flaming combustion and reflects limited oxygen supply in fuel arrays that are more tightly packed than those that support flaming (Pyne et al., 1996). Smoldering in organic soil horizons can persist for months or years (Pyne et al., 1996). Smoldering is referred to as glowing combustion when it heats the solid fuel to a sufficiently high temperature to radiate in the visible spectrum (Rego et al., 2021).

The final phase, extinction, occurs when the fuel is exhausted or the amount of heat produced per unit volume is reduced (Scott, 2000). At the most basic level, a wildfire can be limited by the presence of fuels or by the condition of those fuels (i.e., flammability).

Wildfires can also be described with respect to location within the fire and different locations can have different behaviors (Rego et al., 2021). The ignition location is where the fire originates. Where fuels, wind, and topography are homogeneous, fires spread in a circular shape. Where wind or slope exerts an influence, fires are more likely to spread in an elliptical shape. The portion of fire spreading upslope and with the wind is the headfire or fire front, whereas the portion spreading (usually more slowly) against the wind and/or against the slope is the backfire or heel. Flankfire moves perpendicular to the wind on the sides of the fire area. When parts of the fire front spread more rapidly in long, narrow strips, these are known as fingers. Unburned embayments along the fire perimeter are pockets and unburned areas within the fire perimeter are islands.

Combustion may not be complete in wildfires (Rego et al., 2021). Incomplete combustion products include gases (carbon monoxide, methane, other hydrocarbons) and particles such as soot (Novakov et al., 1997), which can combine to form smoke. Solid residues include condensed hydrocarbons, tars, soot, ash, charcoal, and partially charred plants (Scott, 2000).

Complete combustion of organic material forms white ash, whereas partial combustion forms gray or black ash, of which 30–90% by weight is organic material (Ranalli, 2004). The details of ash matter because of the effects on soil properties, water quality, animal and human health, and distance over which different types of ash can be dispersed, as detailed in subsequent chapters.

Ash can also be subdivided into large and fine ash particles. Ash particles greater than about 10 μm in diameter are mostly deposited on surfaces near the fire source due to their rapid gravitational settling (Bodí et al., 2014). Fine ash particles with smaller diameters are composed mainly of Ca and Mg carbonates (Adachi et al., 2022). These particles can remain suspended in the air for several days or more and are transported >100 km from the fire source. The fine particles can affect cloud condensation and act as ice nuclei (Adachi et al., 2022).

2.5. Catching Fire

In some parts of the world, a substantial portion of wildfire ignitions result from human activities. In the coterminous United States, for example, human activities accounted for 84% of all wildfires and 44% of the total area burned from 1992 to 2012 (Balch et al., 2017). The great majority of wildfires that start from natural rather than human causes are ignited by lightning strikes, although of course not every strike starts a fire. Both field observations and lab experiments suggest that long-continuing-current lightning flashes may be the main igniters (Pérez-Invernón et al., 2023). Long-continuing-currents have continuing electrical currents flowing in the lightning for multiple tens of milli-seconds.

Data from boreal regions suggest that short-term fuel drying associated with fire weather is the main driver of lightning ignition efficiency (Hessilt et al., 2022). Sometimes lightning-generated fires can last for decades to centuries if they are maintained along slow-burning layers of coal. In 1789, Alexander Mackenzie described burning coal seams along the Canadian river that now bears his name and these smoldering coal layers can still be observed today (Wohl, 2011).

Pyne et al. (1996) estimated that there are more than 8 million lightning strikes each day worldwide. More recently, Cecil et al. (2014) documented a mean global lightning flash rate of 46 flashes per second, varying from ~35 flashes per second in austral summer to 60 flashes per second in boreal summer. The global champion is Lake Maracaibo, Venezuela, with an average of 233 flashes per square kilometer per year (Albrecht et al., 2016).

Prior to human influence, lightning strikes were the main cause of fire. By the late 20th century, lightning fires contributed only an estimated 10%

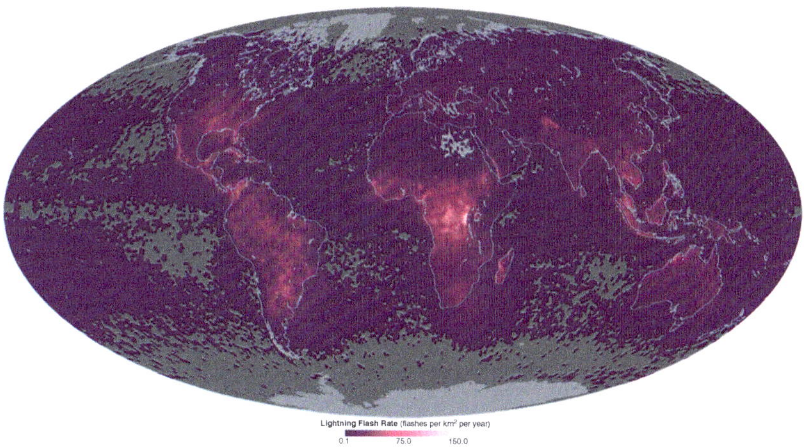

Figure 2.2 Average yearly counts of lightning flashes per square kilometer based on data collected by NASA satellites during 1995–2013. (NASA / public domain / https://earthobservatory.nasa.gov/images/85600/global-lightning-activity accessed at 24 May 2024.)

of global biomass burning as human-started fires associated with changes in land cover became more common and widespread (Lobert & Warnatz, 1993). In drier and less populated regions such as the western United States, lightning may still account for the majority of total area burned (Prestemon et al., 2013), although human-started wildfires accounted for 84% of all wildfires and most of the area burned across the entire United States during 1992–2012 (Balch et al., 2017). A map of annual global average lightning activity (flashes/km^2/year) reveals the greatest terrestrial frequencies at equatorial to subtropical latitudes between about 30° N and S (Figure 2.2) (Collier et al., 2010; Albrecht et al., 2016).

Lightning strikes and wildfire can also interact to enhance lightning activity. Lightning originates from electrification in convective cloud systems. Added aerosols in dry or moist conditions can create more small ice crystals and invigorate the convective strength, which results in more lightning flashes (Wang et al., 2018). Wildfires release aerosols into the atmosphere, which can then trigger additional lightning flashes (Liu et al., 2021). In a worst case scenario, this enhanced lightning in the region of a wildfire could ignite more fires, or human-started fires could enhance lightning that ignites additional fires (Altaratz et al., 2010). Other forms of atmospheric pollution can also enhance lightning strikes, as revealed in comparisons of lightning flash density over relatively polluted and clear oceanic locations (Liu et al., 2020).

Spontaneous combustion can also ignite wildfires. Carbon-rich soils such as histosols or gelisols cover more than 3% of Earth's surface, mostly in tropical and boreal regions. These soils can ignite by self-heating if the moisture conditions are sufficiently low (Yuan et al., 2021), although the prevalence of this form of fire ignition is poorly constrained. Self-heating describes the ability of a porous solid fuel to undergo spontaneous reactions in the presence of oxygen (Drysdale, 2011; Restuccia et al., 2017). These reactions generate sufficient heat to create combustion (Restuccia et al., 2017). Coal, biomass, and shale are among the natural materials in which slow oxidation can generate more heat than the material can lose, leading to combustion. Wildfires can likely also be ignited by volcanic eruptions, but there are few published descriptions of this process (Scott, 2000).

Once wildfire starts, it grows from the point of ignition. Rate of spread increases with a "tail wind" and up topographic slopes (Rego et al., 2021) and can be especially fast in grassland fires, which have been measured over multi-hour periods with sustained rates of spread in excess of 6 km/hour, or 100 m/minute (Cruz et al., 2022), and can attain 27.3 km/hour (Noble, 1991). Measurements during laboratory and field experiments, and during unplanned wildfires, are used to develop equations for predicting the rate and direction of spread and other fire characteristics. For example, Cruz et al. (2022) used a dataset of nearly 60 high-intensity grassland wildfire observations from southern Australia to develop a relationship between wind speed and fire spread rate. Use of data from past fires to inform future predictions is increasingly likely to employ machine learning methods (Khanmohmmadi et al., 2022), especially in tropical environments where access to climate data can be limited (Dahan et al., 2023).

Rego et al. (2021) includes a detailed discussion of the variables used in predictive equations for fire spread and the resulting numerical models.

2.6. Fire Weather

Fire weather occurs under weather conditions conducive to the ignition and spread of wildfires (Jones et al., 2022). This typically means hot, dry, and windy. Extreme fire weather is typically evaluated using fire weather indices that incorporate daily weather variables related to fuel moisture and fire behavior (Jain et al., 2022), including surface air temperature, relative humidity, wind speed, vapor pressure deficit (the difference between the amount of moisture in the air and the amount that air could hold at saturation), and precipitation. Fire weather indices also model the moisture content of downed, dead wood of different diameter classes (for forests) and use these indicators to predict the rate of fire spread and fuel available for combustion, as well as the general fire danger.

Fire weather indices detect weather conditions conducive to uncontrollable fire but do not model the probability of ignition and fire behavior. Numerous regional studies, however, indicate strong correlations between actual occurrence of wildfires and fire weather indices (e.g., Carvalho et al., 2008; Barbero et al., 2020; Trnka et al., 2021).

The Canadian Fire Weather Index (Van Wagner, 1974; Van Wagner & Pickett, 1985) is the most widely used fire-danger rating system globally (Field et al., 2015). Other indices also exist, such as the McArthur Forest Fire Danger Index developed in Australia (Clarke & Evans, 2019).

The typical approach is to use a fire weather index with numerical weather predictions to predict fire danger. Time is of the essence when preparing for and responding to fires because advance warning can aid in creating management and containment strategies. Using observed weather data can facilitate daily monitoring of fire conditions. Adding satellite data can provide four to six hours of warning time and using advanced numerical weather models can extend warning times to one to two weeks (Di Giuseppe et al., 2020). These advances are incorporated into international initiatives such as the Global Fire Early Warning System.

One limitation of fire weather indices may be that the same result can be produced by different combinations of weather variables. Analyzing fire weather components separately can reveal the relative importance of each factor, as well as specific interactions between factors that trigger specific types of fire scenarios (Rodrigues et al., 2020). Resco de Dios et al. (2022) used fire weather and patterns of burned area to characterize pyroregions—areas with similar temporal patterns of fire activity—in the Mediterranean zone of the European Union. Investigating fire weather patterns at the very local scale, such as a single river catchment, can also reveal spatial variations in fire risk associated with factors such as elevation-related snowpack (Spittlehouse & Dymond, 2022).

Fire weather indices are also used to understand the occurrence and behavior of past fires. The Global Fire WEather Database (GFWED), for example, includes daily fire weather index calculations starting in 1980 (Field et al., 2015). Records of the McArthur Forest Fire Danger Index from 1973 to 2017 suggest that the ENSO atmospheric circulation pattern is the primary driver of interannual variability for fire weather for most of Australia, although upward linear trends through time largely reflect the higher mean temperatures caused by anthropogenic climate change (Harris & Lucas, 2019; Wang et al., 2022).

Use of a fire weather index can focus on all conditions conducive to wildfire or on extreme fire weather. In regions such as Canada, most area is burned during a small proportion of the time with extreme fire weather (Wang et al., 2015). These are the times that also create the greatest challenge for fire suppression and that can leave fewer refugia for biota during and after the fire (Meigs et al., 2020). The correlation between extreme fire weather and burned area extends to many other regions (Jain et al., 2022).

2.7. Fire Regimes

The history of fires within a region can be used to describe a fire regime. The phrase fire regime is common, although it can have different meanings. A simple definition for a fire regime is a sequence of fire events with some stable, recurrent, or cyclic characteristics or properties affecting a specified spatial and temporal window (Krebs et al., 2010). General fire regimes are primarily driven by climate (Westerling et al., 2006; Taylor et al., 2008) and fire frequency increases during periods of drought (Clark, 1988). Fire frequency, size, intensity, severity, seasonality, type (surface, crown, ground), and ignition cause are characteristics typically included in a fire regime (Erni et al., 2020).

Fire regimes can be described for the Quaternary Period of the past 2 million years (Ruan et al., 2020), the Holocene Epoch covering the past 10,000 years (Marlon et al., 2013), the past century (Li et al., 2022), or the past decade, and fire regimes can be forecast for the future. They can be described for an ecosystem, a vegetation type, or a geographic region. They can include natural fires, anthropogenic fires (Guyette et al., 2002), or both. Fire regimes can include fire intensity, the conditions of fuel type and weather that affect fire occurrence, and the severity of the fire. In other words, the definition of a fire regime is flexible and varies among users (Figure 2.3).

Krebs et al. (2010) traced the use of 'fire regime' to French-speaking scientists working in France's African colonies during the 1920s. Fire regime came into use in the United States during the early 1960s as ecologists and natural resources managers accepted the idea of disturbance by wildfire as a basic natural force shaping ecosystems (Leopold et al., 1963; Krebs et al., 2010). Underlying the continued use of the phrase is the understanding that the characteristics of a fire regime exert an important influence on ecosystems and human communities (Gill & Allan, 2008). Examining the interactions between fire regime and ecosystems over millennia, Whitlock et al. (2010) suggested the use of super-fire regimes to describe the variability of fire characteristics over the lifespan and spatial extent of a biome.

Zubkova et al. (2022) introduced the concept of fire regions, analogous to the pyroregions of Resco de Dios et al. (2022). In this framework, the variables that promote fuel accumulation and desiccation are used to

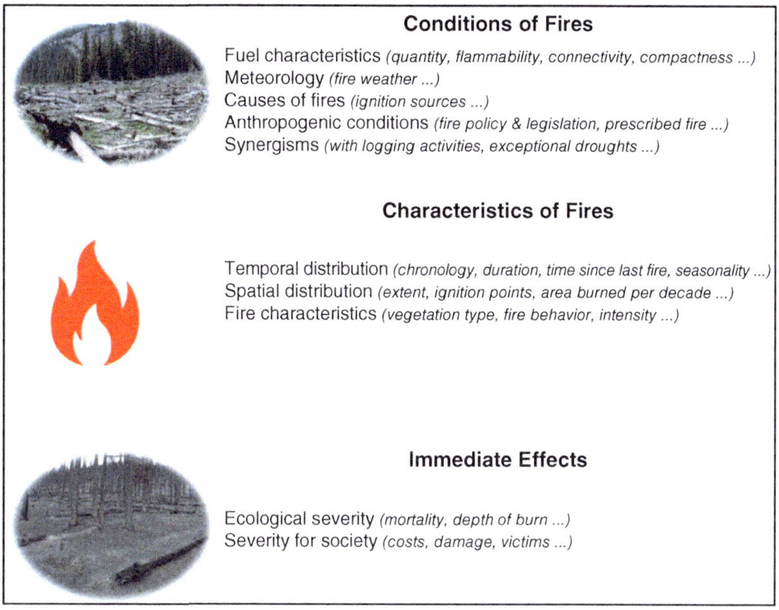

Figure 2.3 Schematic illustration of the components of a fire regime. In the most restrictive definition, a fire regime is a description of parameters of when, where, and which fires occur (characteristics of fire). Used more broadly, the fire regime can include parameters that refer to conditions of fires and immediate effects. (Modified from Krebs et al., 2010, Figure 5.)

define an environmental space at the continental level, which is then divided into fire regions with distinct fire potential. Africa has nine fire regions, for example, and Australia has eight, and these regions have statistically different fire activity.

Fire regime is now used as a descriptive term globally. However, spatial and temporal variability (random fluctuations) and heterogeneity (meaningful patterns over time and space) make it challenging to precisely describe and model any particular fire regime (Krebs et al., 2010). Comprehensive, consistent, and comparable fire data are not available worldwide (Shlisky et al., 2009).

The recent historical characteristics of a fire regime also may not adequately represent future characteristics because of continuing changes in global and regional climates that influence fire ignition, intensity and severity, and other fire characteristics. Nonetheless, in regions with relatively detailed fire histories for the later twentieth century, investigators have used frequency-area power-law relationships to calculate return intervals for fires equaling or exceeding a specified burned area within an ecoregion. This provides a means of classifying fire regimes for probabilistic hazard estimations of the type employed for earthquakes (Malamud et al., 2005).

Despite the long history of thinking about fire regimes, Bowman et al. (2020) emphasized the need for basic research that describes regional fire regimes by quantifying the seasonality, frequency, spatial extent, and severity of fires, as well as determining how these patterns are shaped by climate, terrain, and vegetation. More detailed quantification of fire regimes in turn requires improved data acquisition and database assembly that will go beyond the relatively coarse-scale mapping possible with existing satellite records (Bowman et al., 2020).

2.8. The Long History of Fires

The difficulty of estimating mean fire recurrence intervals illustrates the challenges of diverse types of data that are not consistently available for different regions. Fire recurrence interval can be estimated from multiple sources, primarily historical records, burn scars and tree-ring records, and charcoal preserved in sedimentary records.

Maps and dates of past fires (Malamud et al., 2005) may date back more than a century (e.g., Guyette et al., 2006), but commonly cover only a relatively short time span of decades. Archived information can include maps, databases, and historical accounts. Maps of daily or hourly perimeters are now commonly created during the management of large fires. Information on fire extent at a point in time comes from ground-based systems, aerial surveys using aircraft or unmanned aerial systems, or satellite imagery. This information is used to generate fire perimeter maps or fire atlases. Satellite imagery can be particularly effective in covering large spatial areas or places where it is dangerous or difficult to access, but the fire needs to be either large relative to the sensor's pixel size or, if smaller, burning intensely enough to saturate the sensor detectors (Roy et al., 2013; Rego et al., 2021).

Burn scars and tree-ring records (Guyette et al., 2006; Falk et al., 2011; Marlon et al., 2012) preserved in living trees or cross-dated for dead trees record the occurrence of fires of diverse severities. These indicators can be used to reconstruct past frequency, severity, spatial variability, and extent of wildfire. Burn scars and tree-ring records can focus on individual fires and stand-level fire regimes (Moritz et al., 2009; Swetnam et al., 2011; Daniels et al., 2017). Combining evidence of stand-level fire severity with fire extent allows reconstruction of spatial variability in past fire regimes and comparison with modern fire regimes (Daniels et al., 2017).

Different types of dendrological indicators are used for different fire regimes. High-frequency and low-severity fire regimes cause low levels of canopy tree mortality but produce characteristic fire scars on individual trees (Swetnam et al., 1999). Fire scars can be distinguished from other sources of damage that scar trees based on their triangular shape at the base of the tree and the presence of charred bark or evidence of healing over the scar (Smith & Sutherland, 2001). Targeting large trees with multiple scars can maximize the fire record (Swetnam et al., 1999).

The crown fires of low-frequency and high-severity fire regimes cause high levels of tree mortality that result in forest patches with distinct boundaries visible on air photographs and in the field (Daniels et al., 2017), even in fire-adapted forests such as those of *Eucalyptus* spp. (Fairman et al., 2015; Etchells et al., 2020). Age sampling of distinct forest patches can be combined with analysis of fire scars formed at polygon boundaries to refine estimates of fire recurrence interval (Sibold et al., 2006).

Tree stand ages and fire scars are both used to simultaneously characterize the timing and severity of individual fires and the spatio-temporal variations among sites for areas with mixed-severity fire regimes (Baker et al., 2007).

The most extensive regional tree-ring chronologies and associated inferences about past fire characteristics come from western North America, where networks of the fire-scar records spanning centuries to millennia now include hundreds to thousands of trees sampled across hundreds to many thousands of hectares (Falk et al., 2011). The North American tree-ring fire-scar network (NAFSN) contains 2,562 sites and >37,000 fire-scarred trees and covers large parts of North America (Margolis et al., 2022). Regional tree-ring records have also been developed and used to understand fire in other regions, including Australia (Allen et al., 2022) and South America, especially the temperate forests along the slopes of the Southern Andes (Boninsegna et al., 2009).

The International Tree Ring Data Bank provides access to diverse parameters from more than 3,000 tree-ring sites across the globe. Parameters in the database include ring widths, wood density, stable isotope ratios, wood anatomical features, and the abundance of specific chemical compounds (Babst et al., 2014). These data are applied to many questions, including understanding the history of wildfire.

Charcoal preserved in lake, bog, or alluvial fan sediments (Meyer et al., 1992; Pierce & Meyer, 2008; Marlon et al., 2013; Vannière et al., 2016; Remy et al., 2023) can also record the occurrence of wildfire within the contributing catchment. As with tree rings, the largest number of records come from fire-prone regions of western North America, but records are rapidly being developed for other regions such as northern China (Ji et al., 2021). Millennial-scale records of forest fire inferred from sedimentary deposits are particularly important in regions with too few recent fires to describe the natural range of variability (Gavin et al., 2007).

Natural or historical range of variability describes the range between the minimum and maximum values of selected ecosystem characteristics, such as fire frequency and spatial extent, through time. This range can be inferred from diverse records of past fires (e.g., Veblen et al., 2008; Cyr et al., 2009; Hu et al., 2015). An important aspect of natural range of variability is that this range is not static through time but has varied in relation to climatic shifts and pre-industrial human activities, such as indigenous use of fire (e.g., Twidwell et al., 2021).

The Global Charcoal Database (now Global Paleofire Database; https://www.paleofire.org/index.php) provides a global perspective on past fire based on more than 650 sedimentary charcoal records from around the globe that together span the past 21,000 years (Power, 2013). This span of time includes the disappearance of northern hemisphere ice sheets, changes in the land surface related to isostatic rebound and rising sea level, and changes in atmospheric and oceanic conditions that affect weather patterns and fire. Together, the charcoal records indicate low levels of fire activity until circa 14,000 years ago. Prior to that time, cooler and drier climate, lower atmospheric CO_2 levels, and less terrestrial vegetation all contributed to low global fire activity (Power, 2013). As the ice sheets retreated, atmospheric CO_2 concentrations increased and terrestrial vegetation expanded into high latitudes, causing steady increases in global fire activity, despite increasing spatial heterogeneity in charcoal records (Power, 2013).

Studies using both tree-ring records and sediment records can enhance the data available to infer past fire characteristics in a region (e.g., Marlon et al., 2012; Brossier et al., 2014). Such studies are more common in regions where lakes are present and fire frequencies are low. Combined records are rare where high-frequency surface fires dominate and sedimentary deposits are primarily bogs and wetlands (Allen et al., 2008).

Fire scars and alluvial stratigraphy can cover hundreds to thousands of years but may contain spatial or temporal gaps and can blur local heterogeneities. Charcoal can contain more information than simply indicating fire occurrence, however. The chemical signatures of charcoal provide a fingerprint for burning conditions, allowing differentiation of grass, shrub, and forest, as well as domestic fires based on burn temperatures, charring duration, and fuel (Wolf et al., 2013).

Other sources of information on past fires include speleothems and chemical signals in ice cores. Speleothems are naturally occurring cave deposits. They are dateable using radiometric techniques and can record past fires based on a subset of typically measured variables such as stable isotopes, vertical extension rates, and trace and minor element concentrations (Campbell et al., 2023). Chemical signals in ice cores that can reflect fires include biomarkers such as polyaromatic hydrocarbons, aerosols such as black carbon, isotopes, and trace elements (Grieman et al., 2018; Marlon, 2020).

Data inferred from natural proxies such as tree rings and sediment cores are stored in the public archive of the International Multiproxy Paleofire Database (NCEI, 2020), maintained by the U.S. National Oceanic and Atmospheric Administration.

2.9. People Altering Wildfire

People alter wildfire regimes deliberately and inadvertently. Deliberate alterations include fire suppression and arson or the use of fires to remove, alter, or maintain native land cover. During 1979–2009, for example, anthropogenic burning in grasslands and savannas dominated global fire emissions (O'Connor et al., 2011). Inadvertent alterations include changes in land cover (e.g., afforestation with flammable species such as *Eucalyptus* spp.) and increased activities that can ignite fires (e.g., use of machinery).

Changes in land cover can also be direct or indirect. Clearing of native land cover for agriculture is an example of a direct change in land cover. An indirect change in land cover is a secondary effect of some other human alteration. Extinction of large wild herbivores by over-hunting, significant population increases in large wild herbivores as a result of human over-hunting of predators, diseases of domesticated herbivores that reduce grazing intensity, and the spread of exotic and invasive plants are all examples of indirect changes in land cover that may alter fire regime (Pausas & Keeley, 2014). Deliberate or inadvertent alterations can then lead to management such as forest fuels reduction or prescribed burns (Knapp et al., 2009) that are designed to reduce hazards to ecosystems and human communities created by the altered fire regime.

People seem unable to resist altering fire regimes. The use of fire for managing plant and animal populations has a long history (e.g., Vannière et al., 2016). Fire became a regular technology for many hominin populations between 400,000 and 300,000 years ago (Roebroeks & Villa, 2011). Many areas in the Americas and Oceania that were regarded as pristine wilderness at the time of European settlement were strongly shaped by centuries to millennia of deliberate use of fire by indigenous populations (e.g., Preece, 2002; Knight et al., 2022).

Increasing global population has only intensified the manipulation of fires. Observations of the conversion of fire-limiting landscapes to burnable landscapes because of diverse human activities has given rise to the term pyrogeography for the study of human-fire relationships across space and through time (O'Connor et al., 2011; Roos et al., 2014).

The Global Fire Partnership (GFP) classifies intact, degraded, and very degraded fire regimes (Shlisky et al., 2007). Intact fire regimes have characteristics such as fire frequency, severity, extent, and season that are within their natural range of variability. Degraded fire regimes are outside of their range of natural variation but considered to be restorable via prescribed burns or manipulation of vegetation. Very degraded fire regimes are so far from their natural range of variability that it may not be possible to restore them. Only an estimated 25% of the global terrestrial area—northern regions of Canada and Asia and scattered areas in South America, Africa, and Australasia—has an intact fire regime (Shlisky et al., 2007).

Although natural range of variability can inform whether a fire regime is degraded, it may be less useful as a management paradigm when conditions are expected to be very different in future. Hessburg et al. (2015) recommended using both the natural range of variability and the future range of variability to inform fire-related management. Future range of variability, as generated from simulation models or approximated from other parts of a landscape, represents the range of predicted future characteristics.

How can a fire regime become so altered that it is impossible to restore it and past conditions are no guide to the future? Primarily as a result of vegetation shifts toward fire-dependent species. These species then support a fire regime that precludes fire-sensitive species that did not largely evolve in the presence of fire. The spread of fire-adapted, invasive cheatgrass (*Bromus tectorum*) and buffelgrass (*Cenchrus ciliaris*) across semiarid rangelands and deserts in the western United States, for example, has facilitated the replacement of a low-frequency, mixed-severity shrubland fire regime dominated by sagebrush (*Artemisia* spp.) with a high-frequency, high-severity grassland fire regime (Fusco et al., 2019). This change limits the survival of both sagebrush and animals such as the sage grouse (*Centrocercus urophasianus*) that depend on sage-dominated shrubland (Rowland et al., 2006). Analogously, the introduction of gamba grass (*Andropogon gayanus*) to tropical eucalypt savanna in northern Australia has led to the replacement of high-frequency, low-intensity fires with high-intensity fires, which appears to be progressively changing the savanna toward a treeless state (Setterfield et al., 2010).

At least 20% of global ecosystems are composed of fire-sensitive species and more than 70% of these fire-sensitive habitats have altered fire regimes (Shlisky et al., 2009).

2.10. Consequences of Different Fire Regimes

The details of a fire regime exert a critically important influence on landscape processes, biotic communities, human communities, and even weather in areas directly or indirectly affected by wildfire. Directly affected areas are those that burn at some level of intensity. Indirectly affected areas are downslope or downwind and receive gases and solid residues, as well as increased yields of water and sediment, from burned areas.

The subsequent sections of this book address the direct and indirect influences of wildfires in greater detail, but fire regimes characterized by relatively frequent surface fires, for example, create very different ecosystem responses than fire regimes characterized by infrequent crown fires. Frequent, low intensity fires result in a plant community adapted to fire (fire-dependent ecosystem) and fires may not generate the significantly larger yields of water and sediment from burned areas that commonly occur following high intensity fires. In contrast, infrequent crown fires may substantially change the community composition of vegetation for decades after the fire and lead to sediment yields from uplands that are at least an order of magnitude higher than "background" weathering and erosion rates (e.g., Orem & Pelletier, 2016). Changes in fire regime may also force a significant shift in plant community composition, such as may be occurring now in the fire-adapted giant sequoia (*Sequoiadendron giganteum*) forests of California, USA (Shive et al., 2022).

Wildfires can also alter local to regional weather. Organic compounds released by fires can serve as ice-nucleating particles that influence cloud radiative properties and precipitation development (Barry et al., 2021). Intense heating by fires can generate pyrocumulonimbus (pyroCb), which are smoke-infused thunderstorms started or augmented by fires (Fromm et al., 2010). These storms can release a large quantity of smoke particles above the cruising altitudes of jet aircraft—about 10 km (Peterson et al., 2018)—and can change the temperature, humidity, and height of the atmospheric boundary layer (Castellnou et al., 2022). They can also create a situation in which the convective plume collapses, resulting in strong outward winds near the ground that can exceed 100 km/h and spread out in all directions, causing erratic fire behavior (Rego et al., 2021).

On one day in August 2017, five intense pyroCb systems occurring nearly simultaneously in western North America injected into the lower stratosphere a mass of smoke aerosol particles comparable to those from a moderate volcanic eruption (Peterson et al., 2018). The resulting stratospheric plume encircled the northern hemisphere and remained evident for four months. During their early development, wildfire plumes can also cause substantial warming of up to 10 K in the lower stratosphere (Stocker et al., 2021). Clearly, wildfires start locally but can act globally.

2.11. Geography of Wildfires

Fire-prone ecosystems cover 40% of Earth's land surface and include major global biomes such as grasslands, savannas, shrublands, and boreal forests (Machete & Dintwe, 2023). Together, these fire-prone ecosystems account for more than 85% of global fire (Hao & Liu, 1994; Tansey et al., 2004). Grassland fires account for roughly 80% of global annual fires (Glassman et al., 2023); however, because grassland fire emissions are lower per area burned than forests, savannas are responsible for 65% of the gross global mean fire emissions (Machete & Dintwe, 2023). Part of the variation in estimates of grassland versus savanna fire area and emissions may be the overlap in definitions of grasslands (vegetation dominated by grasses, although other vegetation categories such as sedges and legumes can also be present) and savannas (mixed grassland-woodland ecosystems with trees sufficiently widely spaced that the canopy does not close); a grassland can have some trees, and a savanna has a great deal of grass.

Grass-dominated biomes, including grasslands, steppes, and savanna shrublands and woodlands, account for more than 40% of Earth's vegetative cover and occur naturally on all continents except Antarctica (Neary & Leonard, 2020). Fire, along with grazing and cropping, is key in maintaining these grassland areas (Cruz et al., 2022) as well as some types of shrublands (Van Etten et al., 2021). The global distribution of grasslands and savannas certainly overlaps substantially with the global distribution of wildfires (Figure 2.4).

Africa has the largest extent of savannas, which cover about 15 million km^2, or 40–50% of Africa's land surface, with a concentration in southern Africa (Machete & Dintwe, 2023). South America has another 2.7 million km^2 and northern Australia has about 2 million km^2 of savanna (Chen et al., 2003). Together, these regions equate to 20–22% of the global savanna biome (Machete & Dintwe, 2023) and all of them are prone to wildfires. Examining global fire emissions during 1997–2016, Van der Werf et al. (2017) estimated that tropical savannas are the most fire-prone biomes on Earth and emit the greatest annual volume of greenhouse gases.

Fires are also widespread in relatively dry forests such as those of southern Australia (Fairman et al., 2015) and interior western North America (Schoennagel et al., 2017; Coop et al., 2020); the extensive boreal forests of North America (Whitman et al., 2019) and Eurasia (Whitman et al., 2022); and some boreal (Bourgeau-Chavez et al., 2022)

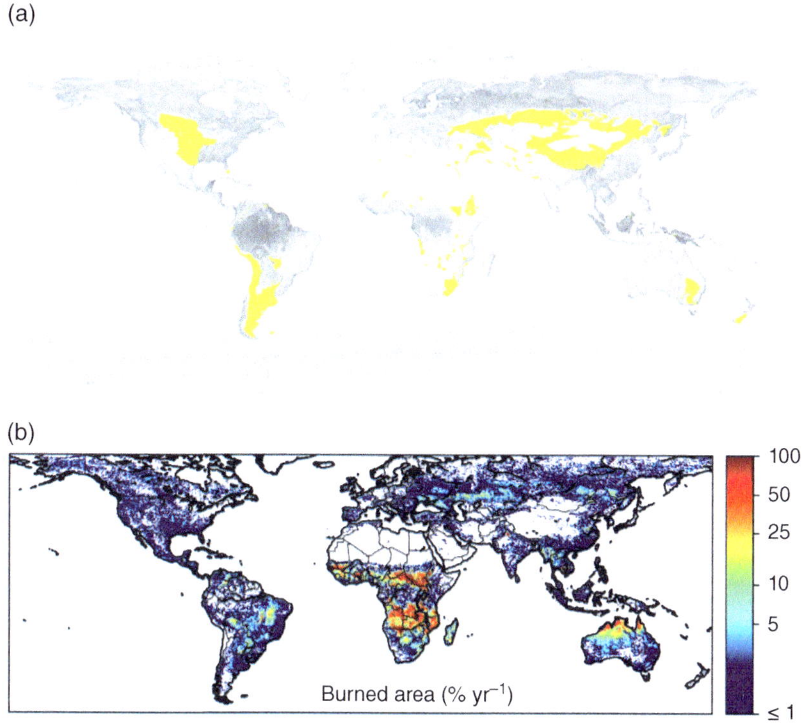

Figure 2.4 Global distribution of grasslands (a) and average global burned area over the period 2003–2016 (b). (Grasslands map from NASA Earth Observatory, https://earthobservatory.nasa.gov/biome/biograssland.php, and burned area map Andela et al., 2019 / Copernicus Publications / CC BY 4.0.)

and tropical (Wooster et al., 2018; Hayasaka, 2023) peatlands. The occurrence of wildfires in these environments, which is increasing with time, at least partly reflects climate and land-use changes.

Major wildfires in tropical peatlands, for example, were relatively rare in the past but expansion of clearcutting and large-scale land drainage has made these areas increasingly vulnerable to fire (Sullivan et al., 2022). This is of concern for many reasons, including the fact that peatlands cover only 3% of the world's surface but store 30% of global total soil carbon (Dargie et al., 2017) and contain more carbon than all other vegetation types in the world combined (Turetsky et al., 2015; Sullivan et al., 2022). Nearly half the peatlands are in the Arctic, and these contain the majority of peatland carbon and nitrogen stocks (Hugelius et al., 2020).

Wildfires can also occur in desert environments, despite the sparse, discontinuous vegetation and low fuel loads common in deserts

(Van Etten et al., 2022). Studies from deserts and sagebrush (*Artemisia* spp.) steppe in the southwestern United States indicate recent increases in the frequency, spatial extent, and severity of wildfires (e.g., Moloney et al., 2019; Wilder et al., 2021). These fires are tied to years with above-average precipitation and the spread of exotic, invasive grasses (Moloney et al., 2021; Brunelle, 2022). Fire can kill the native woody vegetation (e.g., *Artemisia*) and cacti such as saguaro (*Carnegiea gigantea*), whereas the invasive grasses are resilient to fire and some of the grasses (e.g., cheatgrass, *Bromus tectorum*) increase in response to burn severity, which can cause the vegetation community to transition to an ecologically novel community (Rodhouse et al., 2020).

2.12. Summary

Fire is a rapid chemical reaction between a fuel and an oxidant. The resulting combustion releases heat that is transferred via radiation, conduction, convection, and spotting. A fire includes the four phases of pre-ignition, ignition, combustion, and extinction. Most naturally caused wildfires are ignited by lightning strikes, which are most frequent at equatorial to subtropical latitudes.

Fire intensity describes the rate of heat energy release during a wildfire and fire severity describes the effects of the fire on plant communities or changes in organic matter. Complete combustion creates white ash and partial combustion forms gray or black ash, as well as gases. Wildfires include ground fires, surface fires, and crown fires.

Fire weather refers to weather conditions conducive to the ignition and spread of wildfires. Extreme fire weather is typically evaluated using fire weather indices that incorporate daily weather variables related to fuel moisture and fire behavior.

The spatial and temporal patterns of fire in a specified geographic region are used to describe a fire regime, although the recent historical characteristics of wildfire may not adequately represent future characteristics because of continuing changes in global and regional climates and land use. Temporal patterns of wildfire can be estimated from multiple sources, primarily historical records, burn scars and tree-ring records, and charcoal preserved in sedimentary records.

People alter fire regimes deliberately and inadvertently and have been doing so for millennia in many parts of the world. Consequently, only an estimated 25% of the global terrestrial area now has a fire regime that remains largely unaltered by humans.

Fire-prone ecosystems cover 40% of Earth's land surface and grassland fires account for roughly 80% of global annual area burned.

References

Adachi, K., Dibb, J. E., Scheuer, E., Katich, J. M., Schwarz, J. P., Perring, A. E., et al. (2022). Fine ash-bearing particles as a major aerosol component in biomass burning smoke. *Journal of Geophysical Research: Atmospheres*, *127*, e2021JD035657. https://doi.org/10.1029/2021JD035657

Albrecht, R. I., Goodman, S. J., Buechler, D. E., Blakeslee, R. J., & Christian, H. J. (2016). Where are the lightning hotspots on Earth? *Bulletin of the American Meteorological Society*, *97*, 2051–2068. https://doi.org/10.1175/BAMS-D-14-00193.1

Allen, C. D., Anderson, R. S., Jass, R. B., Toney, J. L., & Baisan, C. H. (2008). Paired charcoal and tree-ring records of high-frequency Holocene fire from two New Mexico bog sites. *International Journal of Wildland Fire*, *17*, 115–130. https://doi.org/10.1071/WF07165

Allen, K., Stewart, S. B., Tozer, C., Richardson, D., Nitschken, C., Risbey, J., et al. (2022). Reconstructing the seasonal fire danger in southeastern Australia using tree rings. *International Journal of Wildland Fire*, *31*, 559–571. https://doi.org/10.1071/WF21072

Altaratz, O., Koren, I., Yair, Y., & Price, C. (2010). Lightning response to smoke from Amazonian fires. *Geophysical Research Letters*, *37*, L07801. https://doi.org/10.1029/2010GL042679

Andela, N., Morton, D. C., Giglio, L., Paugam, R., Chen, Y., Hantson, S., et al. (2019). The Global Fire Atlas of individual fire size, duration, speed and direction. *Earth System Science Data*, *11*, 529–552. http://dx.doi.org/10.5194/essd-11-529-2019

Babst, F., Alexander, M. R., Szejner, P., Bouriaud, O., Klesse, S., Roden, J., et al. (2014). A tree-ring perspective on the terrestrial carbon cycle. *Oecologia*, *176*, 307–322. http://dx.doi.org/10.1007/s00442-014-3031-6

Badola, A., Panda, S. K., Roberts, D. A., Waigl, C. F., & Jandt, R. R. (2022). A novel method to simulate AVIRIS-NG hyperspectral image from Sentinel-2 image for improved vegetation/wildfire fuel mapping, boreal Alaska. *International Journal of Applied Earth Observations and Geoinformation*, *112*, 102891. http://dx.doi.org/10.1016/j.jag.2022.102891

Baker, W. L., Veblen, T. T., & Sherriff, R. L. (2007). Fire, fuels and restoration of ponderosa pine-Douglas fir forests in the Rocky Mountains, USA. *Journal of Biogeography*, *34*, 251–269. http://dx.doi.org/10.1111/j.1365-2699.2006.01592.x

Balch, J. K., Bradley, B. A., Abatzoglou, J. T., & Mahood, A. L. (2017). Human-started wildfires expand the fire niche across the United States. *Proceedings National Academy of Sciences*, *114*, 2946–2951. http://dx.doi.org/10.1073/pnas.1617394114

Barbero, R., Abatzoglou, J. T., Pimont, F., Ruffault, J., & Curt, T. (2020). Attributing increases in fire weather to anthropogenic climate change over France. *Frontiers in Earth Science*, *8*, 104. https://doi.org/10.3389/feart.2020.00104

Barry, K. R., Hill, T. C. J., Levin, E. J. T., Twohy, C. H., Moore, K. A., Weller, Z. D., et al. (2021). Observations of ice nucleating particles in the free troposphere from western US wildfires. *Journal of Geophysical Research: Atmospheres*, *126*, e2020JD033752. http://dx.doi.org/10.1029/2020JD033752

Bodí, M. B., Martin, D. A., Balfour, V. N., Santín, C., Doerr, S. H., Pereira, P., et al. (2014). Wildland fire ash: Production, composition, and eco-hydro-geomorphic effects. *Earth-Science Reviews*, *130*, 103–127. http://dx.doi.org/10.1016/j.earscirev.2013.12.007

Boninsegna, J. A., Argollo, J., Aravena, J. C., Barichivich, J., Christie, D., Ferrero, M. E., et al. (2009). Dendroclimatological reconstructions in South America: a review. *Palaeogeography, Palaeoclimatology, Palaeoecology*, *281*, 210–228. http://dx.doi.org/10.1016/j.palaeo.2009.07.020

Bourgeau-Chavez, L. L., Graham, J. A., Vander Bilt, D. J. L., & Battaglia, M. J. (2022). Assessing the broadscale effects of wildfire under extreme drought conditions to boreal peatlands. *Frontiers in Forests and Global Change*, *5*, 965605. https://doi.org/10.3389/ffgc.2022.965605

Bowman, D. M. J. S., Kolden, C. A., Abatzoglou, J. T., Johnstons, F. H., van der Werf, G. R., & Flannigan, M. (2020). Vegetation fires in the Anthropocene. *Nature Reviews Earth and Environment*, *1*, 500–515. http://dx.doi.org/10.1038/s43017-020-0085-3

Brossier, B., Oris, F., Finsinger, W., Asselin, H., Bergeron, Y., & Ali, A. A. (2014). Using tree-ring records to calibrate peak detection in fire reconstruction based on sedimentary charcoal records. *The Holocene*, *24*, 635–645. http://dx.doi.org/10.1177/0959683614526902

Brunelle, A. (2022). Interactions among the fire, vegetation, the North American monsoon and the El Niño-Southern Oscillation in the North American Desert Southwest. *Frontiers in Ecology and Evolution*, *10*, 656462. http://dx.doi.org/10.3389/fevo.2022.656462

Campbell, M., McDonough, L., Treble, P. C., Baker, A., Kosarac, N., Coleborn, K., et al. (2023). A review of speleothems as archives for paleofire proxies, with Australian case studies. *Reviews of Geophysics*, *61*, e2022RG000790. http://dx.doi.org/10.1029/2022RG000790

Carvalho, A., Flannigan, M. D., Logan, K., Miranda, A. I., & Borrego, C. (2008). Fire activity in Portugal and its relationship to weather and the Canadian Fire Weather Index System. *International Journal of Wildland Fire*, *17*, 328–338. http://dx.doi.org/10.1071/WF07014

Castellnou, M., Bachfischer, M., Miralles, M., Ruiz, B., Stoof, C. R., & de Arellano, J. V. G. (2022). Pyroconvection classification based on atmospheric vertical profiling correlation with extreme fire spread observations. *Journal of Geophysical Research: Atmospheres*, *127*, e2022JD036920. http://dx.doi.org/10.1029/2022JD036920

Cecil, D. J., Buechler, D. E., & Blakeslee, R. J. (2014). Gridded lightning climatology from TRMM-LS and OTD: Dataset description. *Atmospheric Research*, *135-135*, 404–414. http://dx.doi.org/10.1016/j.atmosres.2012.06.028

Chandler, C., Cheney, P., Thomas, P., Trabaud, L., & Williams, D. (1983). *Fire in forestry* (Vol. 1, Forest Fire Behavior and Effects.). New York: John Wiley & Sons.

Chen, X., Hutley, L. B., & Eamus, D. (2003). Carbon balance of a tropical savanna of northern Australia. *Oecologia*, *137*, 405–416. http://dx.doi.org/10.1007/s00442-003-1358-5

Clark, J. S. (1988). Effect of climate change on fire regimes in northwestern Minnesota. *Nature*, *334*, 233–235. http://dx.doi.org/10.1038/334233a0

Clarke, H., & Evans, J. P. (2019). Exploring the future change space for fire weather in southeast Australia. *Theoretical and Applied Climatology*, *136*, 513–527. http://dx.doi.org/10.1007/s00704-018-2507-4

Cohen, J. D, & Deeming, J. E. (1985). The national fire-danger rating system: Basic equations. *General Technical Report PSW-GTR-82*. Berkeley, CA: USDA Forest Service Pacific Southwest Forest and Range Experiment Station.

Collier, A. B., Bremner, S., Lichtenberger, J., Downs, J. R., Rodger, C. J., Steinbach, P., & McDowell, G. (2010). Global lightning distribution and whistlers observed at Dunedin, New Zealand. *Annales Geophysicae*, *28*, 499–513.

Coop, J. D., Parks, S. A., Stevens-Rumann, C. S., Crausbay, S. D., Higuera, P. E., Hurteau, M. D., et al. (2020). Wildfire-driven forest conversion in western North American landscapes. *BioScience*, *70*, 659–673. http://dx.doi.org/10.1093/biosci/biaa061

Cruz, M. G., Alexander, M. E., & Kilinc, M. (2022). Wildfire rates of spread in grasslands under critical burning conditions. *Fire*, *5*, 55. http://dx.doi.org/10.3390/fire5020055

Cyr, D., Gauthier, S., Bergeron, Y., & Carcaillet, C. (2009). Forest management is driving the eastern North American boreal forest outside its natural range of variability. *Frontiers in Ecology and the Environment*, *7*, 519–524. http://dx.doi.org/10.1890/080088

Dahan, K. S., Kasei, R. A., Husseini, R., Said, M. Y., & Rahman, M. M. (2023). Towards understanding the environmental and climatic changes and its contribution to the spread of wildfires in Ghana using remote sensing tools and machine learning (Google Earth Engine). *International Journal of Digital Earth*, *16*, 1300–1331. http://dx.doi.org/10.1080/17538947.2023.2197263

Daniels, L. D., Yocom Kent, L. L., Sherriff, R. L., & Heyerdahl, E. K. (2017). Deciphering the complexity of historical fire regimes: diversity among forests of western North America. In M. M. Amoroso, et al. (Eds.), *Dendroecology*, Ecological Studies (Vol. 231, pp. 185–210). Springer. https://doi.org/10.1007/978-3-319-61669-8_8

Dargie, G. C., Lewis, S. L., Lawson, I. T., Mitchard, E. T. A., Page, S. E., Bocko, Y. E., et al. (2017). Age, extent and carbon storage of the central Congo Basin peatland complex. *Nature*, *542*, 86–90. http://dx.doi.org/10.1038/nature21048

Di Giuseppe, F., Vitolo, C., Krzeminski, B., Barnard, C., Maciel, P., & San-Miguel, J. (2020). Fire weather index: the skill provided by the European Centre for medium-range weather forecasts ensemble prediction system. *Natural Hazards and Earth System Sciences*, *20*, 2365–2378. http://dx.doi.org/10.5194/nhess-20-2365-2020

Drysdale, A. (2011). *An introduction to fire dynamics*. Chichester, UK: John Wiley & Sons.

Erni, S., Wang, X., Taylor, S., Boulanger, Y., Swystun, T., Flannigan, M., & Parisien, M. A. (2020). Developing a two-level fire regime zonation system for Canada. *Canadian Journal of Forest Research*, *50*, 259–273. http://dx.doi.org/10.1139/cjfr-2019-0191

Etchells, H., O'Donnell, A. J., McCaw, W. L., & Grierson, P. F. (2020). Fire severity impacts on tree mortality and post-fire recruitment in tall eucalypt forests of southwest Australia. *Forest Ecology and Management*, *459*, 117850. http://dx.doi.org/10.1016/j.foreco.2019.117850

Fairman, T. A., Nitschke, C. R., & Bennett, L. T. (2015). Too much, too soon? A review of the effects of increasing wildfire frequency on tree mortality and regeneration in temperate eucalypt forests. *International Journal of Wildland Fire*, *25*, 831–848. http://dx.doi.org/10.1071/WF15010

Falk, D. A., Heyerdahl, E. K., Brown, P. M., Farris, C., Fule, P. Z., McKenzie, D., et al. (2011). Multi-scale controls of historical forest-fire regimes: new insights from fire-scar networks. *Frontiers in Ecology and Environment*, *9*, 446–454. http://dx.doi.org/10.1890/100052

Field, R. D., Spessa, A. C., Aziz, N. A., Camia, A., Cantin, A., Carr, R., et al. (2015). Development of a global fire weather database. *Natural Hazards and Earth System Sciences*, *15*, 1407–1423. http://dx.doi.org/10.5194/nhess-15-1407-2015

Flannigan, M. D., Krawchuk, M. A., de Groot, W. J., Wotton, B. M., & Gowman, L. M. (2009). Implications of changing climate for global wildland fire. *International Journal of Wildland Fire*, *18*, 483–507. http://dx.doi.org/10.1071/WF08187

Fromm, M., Lindsey, D. T., Sevranckx, R., Yue, G., Trickl, T., Sica, R., et al. (2010). The untold story of pyrocumulonimbus. *Bulletin of the American Meteorological Society*, *91*, 1193–1210. http://dx.doi.org/10.1175/2010BAMS3004.1

Fusco, E. J., Finn, J. T., Balch, J. K., Nagy, R. C., & Bradley, B. A. (2019). Invasive grasses increase fire occurrence and frequency across US ecoregions. *Proceedings National Academy of Sciences*, *116*, 23594–23599. http://dx.doi.org/10.1073/pnas.1908253116

Gavin, D. G., Hallett, D. J., Hu, F. S., Lertzman, K. P., Prichard, S. J., Brown, K. J., et al. (2007). Forest fire and climate change in western North America: insights from sediment charcoal records. *Frontiers in Ecology and Environment*, *5*, 499–506. http://dx.doi.org/10.1890/060161

Gill, A. M., & Allan, G. (2008). Large fires, fire effects and the fire-regime concept. *International Journal of Wildland Fire*, *17*, 688–695. http://dx.doi.org/10.1071/WF07145

Glassman, S. I., Randolph, J. W. J., Saroa, S. S., Capocchi, J. K., Walters, K. E., & Pulido-Chavez, L. L. (2023). Prescribed versus wildfire impacts on exotic plants and soil microbes in California grasslands. *Applied Soil Ecology*, *185*, 104795. http://dx.doi.org/10.1016/j.apsoil.2022.104795

Grieman, M. M., Aydin, M., Isaksson, E., Schwikowski, M., & Saltzman, E. S. (2018). Aromatic acids in an arctic ice core from Svalbard: A proxy record of biomass burning. *Climate of the Past*, *14*, 637–651. http://dx.doi.org/10.5194/cp-14-637-2018

Guyette, R. P., Muzika, R. M., & Dey, D. C. (2002). Dynamics of an anthropogenic fire regime. *Ecosystems*, *5*, 472–486. http://dx.doi.org/10.1007/s10021-002-0115-7

Guyette, R. P., Dey, D. C., Stambaugh, M. C., & Muzika, R. M. (2006). Fire scars reveal variability and dynamics of eastern fire regimes. In M. B. Dickinson (Ed.), *Fire in Eastern Oak Forests: Delivering Science to Land Managers*. Newtown Square, PA: USDA Forest Service General Technical Report NRS-P-1.

Hao, W., & Liu, M. H. (1994). Spatial and temporal distribution of tropical biomass burning. *Global Biogeochemical Cycles*, *8*, 495–503. http://dx.doi.org/10.1029/94GB02086

Harris, S., & Lucas, C. (2019). Understanding the variability of Australian fire weather between 1973 and 2017. *PLoS ONE*, *14*, e0222328. http://dx.doi.org/10.1371/journal.pone.0222328

Hayasaka, K. (2023). Peatland fire weather conditions in Sumatra, Indonesia. *Climate*, *11*, 92. http://dx.doi.org/10.3390/cli11050092

Hessburg, P. F., Larson, A. J., Churchill, D. J., Haugo, R. D., Miller, C., Spies, T. A., et al. (2015). Restoring fire-prone forest landscapes: Seven core principles. *Landscape Ecology*, *30*, 1805–1835. http://dx.doi.org/10.1007/s10980-015-0218-0

Hessilt, T. D., Abatzoglou, J. T., Chen, Y., Randerson, J. T., Scholten, R. C., van der Werf, G., & Veraverbeke, S. (2022). Future increases in lightning ignition efficiency and wildfire occurrence expected from drier fuels in boreal forest ecosystems of western North America. *Environmental Research Letters*, *17*, 054008. http://dx.doi.org/10.1088/1748-9326/ac6311

Hu, F. S., Higuera, P. E., Duffy, P., Chipman, M. L., Rocha, A. V., Young, A. M., et al. (2015). Arctic tundra fires: natural variability and responses to climate change. *Frontiers in Ecology and the Environment*, *13*, 369–377. http://dx.doi.org/10.1890/150063

Hugelius, G., Loisel, J., Chadburn, S., Jackson, R. B., Jones, M., MacDonald, G., et al. (2020). Large stocks of peatland carbon and nitrogen are vulnerable to permafrost thaw. *Proceedings National Academy of Sciences*, *17*, 20438–20446. http://dx.doi.org/10.1073/pnas.1916387117

Jain, P., Castellanos-Acuna, D., Coogan, S. C. P., Abatzoglou, J. T., & Flannigan, M. D. (2022). Observed increases in extreme fire weather driven by atmospheric humidity and temperature. *Nature Climate Change*, *12*, 63–70. http://dx.doi.org/10.1038/s41558-021-01224-1

Ji, P., Chen, J., Zhou, A., Ma, R., Chen, R., Chen, S., et al. (2021). Biofuels reserve controlled wildfire regimes since the last deglaciation: A record from Gonghai Lake, North China. *Geophysical Research Letters*, *48*, e2021GL094042. http://dx.doi.org/10.1029/2021GL094042

Jones, M. W., Abatzoglou, J. T., Ververbeke, S., Andela, N., Lasslop, G., Forkel, M., et al. (2022). Global and regional trends under climate change. *Reviews of Geophysics*, *60*, e2020RG000726. https://doi.org/10.1029/2020RG000726

Keeley, J. E. (2009). Fire intensity, fire severity and burn severity: a brief review and suggested usage. *International Journal of Wildland Fire*, *18*, 116–126. http://dx.doi.org/10.1071/WF07049

Khanmohmmadi, S., Arashpour, M., Golafshani, E. M., Cruz, M. G., Rajabifard, A., & Bai, Y. (2022). Prediction of wildfire rate of spread in grasslands using machine learning methods. *Environmental Modelling and Software, 156*, 105507. http://dx.doi.org/10.1016/j.envsoft.2022.105507

Knapp, E. E, Estes, B. L., & Skinner, C. N. (2009). Ecological Effects of Prescribed Fire Season: A Literature Review and Synthesis for Managers. *USDA Forest Service General Technical Report PSW-GTR-224*. Albany, CA.

Knight, C. A., Anderson, L., Bunting, M. J., Champagne, M., Clayburn, R. M., Crawford, J. N., et al. (2022). Land management explains major trends in forest structure and composition over the last millennium in California's Klamath Mountains. *Proceedings National Academy of Sciences, 119*, e2116264119. http://dx.doi.org/10.1073/pnas.2116264119

Krebs, P., Pezzatti, G. B., Mazzoleni, S., Talbot, L. M., & Conedra, M. (2010). Fire regime: history and definition of a key concept in disturbance ecology. *Theory in Biosciences, 129*, 53–69. http://dx.doi.org/10.1007/s12064-010-0082-z

Krueger, E. S., Levi, M. R., Achieng, K. O., Bolten, J. D., Carlson, J. D., Coops, N. C., et al. (2023). Using soil moisture information to better understand and predict wildfire danger: a review of recent developments and outstanding questions. *International Journal of Wildland Fire, 32*, 111–132. http://dx.doi.org/10.1071/WF22056

Lareau, N. P., Nauslar, N. J., & Abatzoglou, J. T. (2018). The Carr fire vortex: A case of pyrotornadogenesis? *Geophysical Research Letters, 45*, 13107–13115. http://dx.doi.org/10.1029/2018GL080667

Leopold, A. S., Cain, S. A., Cottam, C. M., Gabrielson, I. N., & Kimball, T. L. (1963). Study of wildlife problems in national parks: wildlife management in the national parks. In J. B. Trefethen (Ed.), *Transactions of the 28th North American wildlife and natural resources conference* (pp. 28–45). Washington: Wildlife Management Institute.

Li, C., Ma, X., Dowdy, A., Wang, X., Ding, M., Du, Z., et al. (2022). Changes in refractory black carbon (rBC) deposition to coastal eastern Antarctica during the past century. *Global Biogeochemical Cycles, 36*, e2021GB007223. http://dx.doi.org/10.1029/2021GB007223

Liu, Y., Guha, A., Said, R., Williams, E., Lapierre, J., Stock, M., & Heckman, S. (2020). Aerosol effects on lightning characteristics: A comparison of polluted and clean regimes. *Geophysical Research Letters, 47*, e2019GL086825. http://dx.doi.org/10.1029/2019GL086825

Liu, Y., Williams, E., Li, Z., Guha, A., Lapierre, J., Stock, M., et al. (2021). Lightning enhancement in moist convection with smoke-laden air advected from Australian wildfires. *Geophysical Research Letters, 48*, e2020GL092355. http://dx.doi.org/10.1029/2020GL092355

Lobert, J. M., & Warnatz, J. (1993). Emissions from the combustion process in vegetation. In P. J. Crutzen & J. G. Goldammer (Eds.), *Fire in the environment: The ecological, atmospheric and climatic importance of vegetation fires* (pp. 15–37). Chichester: Wiley.

Machete, R. L., & Dintwe, K. (2023). Cyclic trends of wildfires over sub-Saharan Africa. *Fire*, *6*, 71. http://dx.doi.org/10.3390/fire6020071

Malamud, B. D., Millington, J. D. A., & Perry, G. L. W. (2005). Characterizing wildfire regimes in the United States. *Proceedings of the National Academy of Sciences*, *102*, 4694–4699. http://dx.doi.org/10.1073/pnas.0500880102

Margolis, E. Q., Guiterman, C. H., Chavardes, R. D., Coop, J. D., Copes-Gerbitz, K., Dawe, D. A., et al. (2022). The North American tree-ring fire-scar network. *Ecosphere*, *13*, e4159. http://dx.doi.org/10.1002/ecs2.4159

Marlon, J. R. (2020). What the past can say about the present and future of fire. *Quaternary Research*, *96*, 66–87. http://dx.doi.org/10.1017/qua.2020.48

Marlon, J. R., Bartlein, P. J., Daniau, A. L., Harrison, S. P., Maezumi, S. Y., Power, M. J., et al. (2013). Global biomass burning: a synthesis and review of Holocene paleofire records and their controls. *Quaternary Science Reviews*, *65*, 5–25. http://dx.doi.org/10.1016/j.quascirev.2012.11.029

Marlon, J. R., Bartlein, P. J., Gavin, D. G., Long, C. J., Anderson, R. S., Briles, C. E., et al. (2012). Long-term perspective on wildfires in the western USA. *Proceedings of the National Academy of Sciences*, *109*, E535–E543. http://dx.doi.org/10.1073/pnas.1112839109

McArthur, A. G. (1967). *Fire behaviour in eucalypt forests* (p. 107). Leaflet No: Forestry and Timber Bureau.

McCoy, L., & Field, D. (2022). Lessons from the NSW RFS trial of the Australian fire danger rating system. *The Australian Journal of Emergency Management*, *37*, 55–58. https://doi.org/10.3316/informit.738269919163057

McRae, R. H., Sharples, J. J., Wilkes, S. R., & Walker, A. (2013). An Australian pyro-tornadogenesis event. *Natural Hazards*, *65*, 1801–1811. http://dx.doi.org/10.1007/s11069-012-0443-7

Meigs, G. W., Dunn, C. J., Parks, S. A., & Krawchuk. (2020). Influence of topography and fuels on fire refugia probability under varying fire weather conditions in forests of the Pacific Northwest, USA. *Canadian Journal of Forest Research*, *50*, 636–647. http://dx.doi.org/10.1139/cjfr-2019-0406

Meyer, G. A., Wells, S. G., Balling, R. C., & Jull, A. J. T. (1992). Response of alluvial systems to fire and climate change in Yellowstone National Park. *Nature*, *357*, 147–150. http://dx.doi.org/10.1038/357147a0

Moloney, K. A., Fuentes-Ramirez, A., & Holzapfel, C. (2021). Climate impacts on fire risk in desert shrublands: A modeling study. *Frontiers in Ecology and Evolution*, *9*, 601877. http://dx.doi.org/10.3389/fevo.2021.601877

Moloney, K. A., Mudrak, E. L., Fuentes-Ramirez, A., Parag, H., Schat, M., & Holzapfel, C. (2019). Increased fire risk in Mojave and Sonoran shrublands due to exotic species and extreme rainfall events. *Ecosphere*, *10*, e02592. http://dx.doi.org/10.1002/ecs2.2592

Moritz, M. A., Moody, T. J., Miles, L. J., Smith, M. M., & de Valpine, P. (2009). The fire frequency analysis branch of the pyrostatistics tree: sampling decisions and censoring in fire interval data. *Environmental and Ecological Statistics*, *16*, 271–289. http://dx.doi.org/10.1007/s10651-007-0088-y

Moritz, M. A., Morais, M. E., Summerell, L. A., Carlson, J. M., & Doyle, D. (2005). Wildfires, complexity, and highly optimized tolerance. *Proceedings National Academy of Sciences*, *102*, 17912–17917. https://doi.org/10.1073/pnas.0508985102

NCEI (National Centers for Environmental Information). (2020). International Multiproxy Paleofire Database, Asheville. https://www.ncdc.noaa.gov/data-access/paleoclimatology-data/datasets/fire-history.

Neary, D. G., & Leonard, J. M. (2020). Effects of fire on grassland soils and water: A review. In V. M. Kindomihou (Ed.), *Grasses and grassland aspects* (pp. 1–22). Norderstedt, Germany: Books on Demand. https://www.intechopen.com/books/grasses-and-grassland-aspects/effects-of-fire-on-grassland-soils-and-water-a-review

Noble, J. C. (1991). Behaviour of a very fast grassland wildfire on the Riverine Plain of southeastern Australia. *International Journal of Wildland Fire*, *1*, 189–196. http://dx.doi.org/10.1071/WF9910189

Novakov, T., Cachier, H., Clark, J. S., Gaudichet, A., Macko, S., & Masclet, P. (1997). Characterization of particulate products of biomass combustion. In J. S. Clark, H. Cachier, J. G. Goldammer, & B. Stocks (Eds.), *Sediment records of biomass burning and global change*. NATO ASI Series v. 51, (pp. 117–143). Berlin: Springer.

O'Connor, C. D., Garfin, G. M., Falk, D. A., & Swetnam, T. W. (2011). Human pyrogeography: A new synergy of fire, climate and people is reshaping ecosystems across the globe. *Geography Compass*, *5*(6), 329–350. http://dx.doi.org/10.1111/j.1749-8198.2011.00428.x

Orem, C. A., & Pelletier, J. D. (2016). The predominance of post-wildfire erosion in the long-term denudation of the Valles Caldera, New Mexico. *Journal of Geophysical Research Earth Surface*, *121*, 843–864. http://dx.doi.org/10.1002/2015JF003663

Pausas, J. G., & Keeley, J. E. (2014). Abrupt climate-independent fire regime changes. *Ecosystems*, *17*, 1109–1120. http://dx.doi.org/10.1007/s10021-014-9773-5

Pérez-Invernón, F. J., Gordillo-Vázquez, F. J., Huntrieser, H., & Jöckel, P. (2023). Variation of lightning-ignited wildfire patterns under climate change. *Nature Communications*, *4*, 739. http://dx.doi.org/10.1038/s41467-023-36500-5

Peterson, D. A., Campbell, J. R., Hyer, E. J., Fromm, M. D., Kablick, G. P., Cossuth, J. H., & DeLand, M. T. (2018). Wildfire-driven thunderstorms cause a volcano-like stratospheric injection of smoke. *Climate and Atmospheric Science*, *1*, 30. http://dx.doi.org/10.1038/s41612-018-0039-3

Pierce, J., & Meyer, G. (2008). Long-term fire history from alluvial fan sediments: the role of drought and climate variability, and implications for management of Rocky Mountain forests. *International Journal of Wildland Fire*, *17*, 84–95. http://dx.doi.org/10.1071/WF07027

Power, M. J. (2013). A 21,000-year history of fire. In C. M. Belcher (Ed.), *Fire phenomena and the Earth system: An interdisciplinary guide to fire science* (pp. 207–228). Chichester, UK: Wiley-Blackwell.

Preece, N. (2002). Aboriginal fires in monsoonal Australia from historical accounts. *Journal of Biogeography*, *29*, 321–336. http://dx.doi.org/10.1046/j.1365-2699.2002.00677.x

Prestemon, J. P., Hawbaker, T. J., Bowden, M., Carpenter, J., Brooks, M. T., Abt, K. L. et al. (2013). Wildfire ignitions: A review of the science and recommendations for empirical modeling. *USDA Forest Service, General Technical Report SRS-171*. Boise, ID.

Pyne, S. J., Andrews, P. L., & Laven, R. D. (1996). *Introduction to wildland fire*. NY: Wiley.

Ranalli, A. J. (2004). A summary of the scientific literature on the effects of fire on the concentration of nutrients in surface water. *US Geological Survey Open-File Report No. 2004–1296*. Reston, VA.

Rego, F. C., Morgan, P., Fernandes, P., & Hoffman, C. (2021). *Fire science: From chemistry to landscape management*. Cham, Switzerland: Springer Nature.

Remy, C. C., Magne, G., Stivrins, N., Aakalas, T., Asselin, H., Seppa, H., et al. (2023). Climatic and vegetational controls of Holocene wildfire regimes in the boreal forest of northern Fennoscandia. *Journal of Ecology*, *111*, 845–860. http://dx.doi.org/10.1111/1365-2745.14065

Resco de Dios, V., Camprubi, A. C., Perez-Zanon, N., Pena, J. C., Martinez del Castillo, E., Rodrigues, M., et al. (2022). Convergence in critical fuel moisture and fire weather thresholds associated with fire activity in the pyroregions of Mediterranean Europe. *Environment*, *806*, 151462. http://dx.doi.org/10.1016/j.scitotenv.2021.151462

Restuccia, F., Huang, X., & Rein, G. (2017). Self-ignition of natural fuels: can wildfires of carbon-rich soil start by self-heating? *Fire Safety Journal*, *91*, 828–834. http://dx.doi.org/10.1016/j.firesaf.2017.03.052

Rodhouse, T. J., Irvine, K. M., & Bowersock, L. (2020). Post-fire vegetation response in a repeatedly burned low-elevation sagebrush steppe protected area provides insights about resilience and invasion resistance. *Frontiers in Ecology and Evolution*, *9*, 584726. http://dx.doi.org/10.3389/fevo.2020.584726

Rodrigues, M., Trigo, R. M., Vega-Garcia, C., & Cardil, A. (2020). Identifying large fire weather typologies in the Iberian Peninsula. *Agricultural and Forest Meteorology*, *280*, 107789. http://dx.doi.org/10.1016/j.agrformet.2019.107789

Roebroeks, W., & Villa, P. (2011). On the earliest evidence for habitual use of fire in Europe. *Proceedings of the National Academy of Sciences USA*, *108*, 5209–5214. http://dx.doi.org/10.1073/pnas.1018116108

Roos, C. I., Bowman, D. M. J. S., Balch, J. K., Artaxo, P., Bond, W. J., Cochrane, M., et al. (2014). Pyrogeography, historical ecology, and the human dimensions of fire regimes. *Journal of Biogeography*, *41*, 833–836. http://dx.doi.org/10.1111/jbi.12285

Rowland, M. M., Wisdom, M. J., Suring, L. H., & Meinke, C. W. (2006). Greater sage-grouse as an umbrella species for sagebrush-associated vertebrates. *Biological Conservation*, *129*, 323–335. http://dx.doi.org/10.1016/j.biocon.2005.10.048

Roy, D. P., Boschetti, L., & Smith, A. M. (2013). Satellite remote sensing of fires. In C. M. Belcher (Ed.), *Fire phenomena and the Earth system: An interdisciplinary guide to fire science* (pp. 77–90). London: Wiley.

Ruan, Y., Mohtadi, M., Dupont, L. M., Hebbeln, D., van der Kaars, S., Hopmans, E. C., et al. (2020). Interaction of fire, vegetation, and climate in tropical ecosystems: A multiproxy study over the past 22,000 years. *Global Biogeochemical Cycles*, *34*, e2020BG006677. http://dx.doi.org/10.1029/2020GB006677

Schoennagel, T., Balch, J. K., Brenkert-Smith, H., & Whitlock, C. (2017). Adapt to more wildfire in western North American forests as climate changes. *Proceedings of the National Academy of Sciences*, *114*, 4582–4590. http://dx.doi.org/10.1073/pnas.1617464114

Scott, A. C. (2000). The pre-Quaternary history of fire. *Palaeogeography, Palaeoclimatology, Palaeoecology*, *164*, 281–329. http://dx.doi.org/10.1016/S0031-0182(00)00192-9

Setterfield, S. A., Rossiter-Rachor, N. A., Hutley, L. B., Douglas, M. M., & Williams, R. J. (2010). Biodiversity research: turning up the heat: the impacts of *Andropogon gayanus* (gamba grass) invasion on fire behaviour in northern Australian savannas. *Diversity and Distributions*, *16*, 854–861. http://dx.doi.org/10.1111/j.1472-4642.2010.00688.x

Shive, K. L., Wuenschel, A., Hardlund, L. J., Morris, S., Meyer, M. D., & Hood, S. M. (2022). Ancient trees and modern wildfires: Declining resilience to wildfire in the highly fire-adapted giant sequoia. *Forest Ecology and Management*, *511*, 120110. http://dx.doi.org/10.1016/j.foreco.2022.120110

Shlisky, A., Alencar, A., Manta, M., & Curran, L. M. (2009). Overview: Global fire regime conditions, threats, and opportunities for fire management in the tropics. In M. A. Cochrane (Ed.), *Tropical fire ecology* (pp. 65–83). Chichester, UK: Praxis.

Shlisky, A., Waugh, J., Gonzalez, P., Gonzalez, M., Manta, M., Santoso, H., et al. (2007). Fire, Ecosystems and People: Threats and Strategies for Global Biodiversity Conservation. *Global Fire Initiative Technical Report 2007–2*. Arlington, VA: The Nature Conservancy.

Sibold, J. S., Veblen, T. T., & Gonzalez, M. E. (2006). Spatial and temporal variation in historic fire regimes in subalpine forests across the Colorado Front Range in Rocky Mountain National Park, Colorado, USA. *Journal of Biogeography*, *33*, 631–647. http://dx.doi.org/10.1111/j.1365-2699.2005.01404.x

Smith, K. T., & Sutherland, E. K. (2001). Terminology and biology of fire scars in selected central hardwoods. *Tree-Ring Bulletin*, *57*, 141–147.

Spittlehouse, D. L., & Dymond, C. C. (2022). Interaction of elevation and climate change on fire weather risk. *Canadian Journal of Forest Research*, *52*, 237–249. http://dx.doi.org/10.1139/cjfr-2021-0137

Stavi, I. (2019). Wildfires in grasslands and shrublands: A review of impacts on vegetation, soil, hydrology, and geomorphology. *Water*, *11*, 1042. http://dx.doi.org/10.3390/w11051042

Stocker, M., Ladstadter, F., & Steiner, A. K. (2021). Observing the climate impact of large wildfires on stratospheric temperature. *Scientific Reports*, *11*, 22994. http://dx.doi.org/10.1038/s41598-021-02335-7

Stocks, B. J., Alexander, M. E., & Lanoville, R. A. (2004). Overview of the International Crown Fire Modelling Experiment (ICFME). *Canadian Journal of Forest Research*, *34*, 1543–1547. http://dx.doi.org/10.1139/x04-905

Sullivan, A., Baker, E., & Kurvits, T. (Eds.) (2022). *Spreading like wildfire: The rising threat of extraordinary landscape fires.* United Nations Environment Programme.

Swetnam, T., Allen, C., & Betancourt, J. (1999). Applied historical ecology: using the past to manage for the future. *Ecological Applications*, *9*, 1189–1206. http://dx.doi.org/10.1890/1051-0761(1999)009%5b1189:AHEUTP%5d2.0.CO;2

Swetnam, T. L., Falk, D. A., Hessl, A., & Farris, C. (2011). Reconstructing landscape patterns of historic fires and fire regimes. In D. McKenzie, C. Miller, & D. A. Falk (Eds.), *Landscape ecology of fire*. Dordrecht, The Netherlands: Springer.

Tansey, K., Gregoire, J. M., Binaghi, E., Boschetti, L., Ershov, D., Flasse, S., et al. (2004). A global inventory of burned areas at 1 km resolution for the year 2000 derived from Spot vegetation data. *Climatic Change*, *67*, 345–377. https://doi.org/10.1007/s10584-004-2800-3

Taylor, A. H., Trouet, V., & Skinner, C. N. (2008). Climatic influences on fire regimes in montane forests of the southern Cascades, California, USA. *International Journal of Wildland Fire*, *17*, 60–71. http://dx.doi.org/10.1071/WF07033

Trnka, M., Mozny, M., Jurecka, F., Balek, J., Semeradov, D., Hlavinka, P., et al. (2021). Observed and estimated consequences of climate change for the fire weather regime in the moist-temperate climate of the Czech Republic. *Agricultural and Forest Meteorology*, *310*, 108583. http://dx.doi.org/10.1016/j.agrformet.2021.108583

Turetsky, M. R., Benscoter, B., Page, S., Rein, G., Van der Werf, G. R., & Watts, A. (2015). Global vulnerability of peatlands to fire and carbon loss. *Nature Geoscience*, *8*, 11–14. https://doi.org/10.1038/ngeo2325

Twidwell, D., Bielski, C. H., Scholtz, R., & Fuhlendorf, S. D. (2021). Advancing fire ecology in 21st century rangelands. *Rangeland Ecology and Management*, *78*, 201–212. https://doi.org/10.1016/j.rama.2020.01.008

Van der Werf, G. R., Randerson, J. T., Giglio, L., Van Leeuwen, T. T., Chen, Y., Rogers, B. M., et al. (2017). Global fire emissions estimates during 1997–2016. *Earth System Science Data*, *9*, 697–720. https://doi.org/10.5194/essd-9-697-2017

Van Etten, E. J. B., Brooks, M. L., Greenville, A. C., & Wardle, G. M. (2022). Editorial: Fire regimes in desert ecosystems: Drivers, impacts and changes. *Frontiers in Ecology and Evolution*, *10*, 968031. http://dx.doi.org/10.3389/fevo.2022.968031

Van Etten, E. J. B., Davis, R. A., & Doherty, T. S. (2021). Fire in semi-arid shrublands and woodlands: spatial and temporal patterns in an Australian landscape. *Frontiers in Ecology and Evolution*, *9*, 653870. http://dx.doi.org/10.3389/fevo.2021.653870

Vannière, B., Blarquez, O., Rius, D., Doyen, E., Brucher, T., Colombaroli, D., et al. (2016). 7000-year human legacy of elevation-dependent European fire regimes. *Quaternary Science Reviews*, *132*, 206–212. http://dx.doi.org/10.1016/j.quascirev.2015.11.012

Van Wagner, C. E. (1974). *Structure of the Canadian forest fire weather index*. Ontario, Canada: Environment Canada, Departmental Publication 1333.

Van Wagner, C. E., & Pickett, T. L. (1985). Equations and FORTRAN program for the Canadian forest fire weather index system. In *Canadian Forestry Service* (Vol. 33). Ottawa, Canada.

Veblen, T. T., Kitzberger, T., Raffaele, E., Mermoz, M., Gonzalez, M. E., Sibold, J. E., & Holz, A. (2008). The historical range of variability of fires in the Andean-Patagonian *Nothofagus* forest region. *International Journal of Wildland Fire*, *17*, 724–741. http://dx.doi.org/10.1071/WF07152

Wang, B., Spessa, A. C., Feng, P., Hou, X., Yue, C., Luo, J. J., et al. (2022). Extreme fire weather is the major driver of severe bushfires in southeast Australia. *Science Bulletin*, *67*, 655–664. http://dx.doi.org/10.1016/j.scib.2021.10.001

Wang, Q., Li, Z., Guo, J., Zhao, C., & Cribb, M. (2018). The climate impact of aerosols on the lightning flash rate: is it detectable from long-term measurements? *Atmospheric Chemistry and Physics*, *18*, 12797–12816. http://dx.doi.org/10.5194/acp-18-12797-2018

Wang, X., Thompson, D. K., Marshall, G. A., Tymstra, C., Carr, R., & Flannigan, M. D. (2015). Increasing frequency of extreme fire weather in Canada with climate change. *Climatic Change*, *130*, 573–586. http://dx.doi.org/10.1007/s10584-015-1375-5

Westerling, A. L., Hidalgo, H. G., Cayan, D. R., & Swetnam, T. W. (2006). Warming and earlier spring increases western U.S. forest wildfire activity. *Science*, *313*, 940–943. http://dx.doi.org/10.1126/science.1128834

Whitlock, C., Higuera, P. E., McWethy, D. B., & Briles, C. E. (2010). Paleoecological perspectives on fire ecology: revisiting the fire-regime concept. *The Open Ecology Journal*, *3*, 6–23. http://dx.doi.org/10.2174/1874213001003020006

Whitman, E., Parisien, M. A., Thompson, D. K., & Flannigan, M. D. (2019). Short-interval wildfire and drought overwhelm boreal forest resilience. *Scientific Reports*, *9*, 18796. http://dx.doi.org/10.1038/s41598-019-55036-7

Whitman, T., Woolet, J., Sikora, M., Johnson, D. B., & Whitman, E. (2022). Resilience in soil bacterial communities of the boreal forest from one to five years after wildfire across a severity gradient. *Soil Biology and Biochemistry*, *172*, 108755. http://dx.doi.org/10.1016/j.soilbio.2022.108755

Wilder, B. T., Jarnevich, C. S., Baldwin, E., Black, J. S., Franklin, K. A., Grissom, P., et al. (2021). Grassification and fast-evolving fire connectivity and risk in the Sonoran Desert, United States. *Frontiers in Ecology and Evolution*, *9*, 655561. http://dx.doi.org/10.3389/fevo.2021.655561

Wohl, E. (2011). *A world of rivers: Environmental change on ten of the world's great rivers*. Chicago, IL: University of Chicago Press.

Wolf, M., Lehndorff, E., Wiesenberg, G. L. B., Stockhausen, M., Schwark, L., & Amelung, W. (2013). Towards reconstruction of past fire regimes from geochemical analysis of charcoal. *Organic Geochemistry*, *55*, 11–21. http://dx.doi.org/10.1016/j.orggeochem.2012.11.002

Wooster, M. J., Gaveau, D. L. A., Salim, M. A., Zhang, T., Xu, W., Green, D. C., et al. (2018). New tropical peatland gas and particulate emissions factors indicate 2015 Indonesian fires released far more particulate matter (but less methane) than current inventories imply. *Remote Sensing*, *10*, 495. http://dx.doi.org/10.3390/rs10040495

Yuan, H., Restuccia, F., & Rein, G. (2021). Spontaneous ignition of soils: a multi-step reaction scheme to simulate self-heating ignition of smouldering peat fires. *International Journal of Wildland Fire*, *30*, 440–453. http://dx.doi.org/10.1071/WF19128

Zubkova, M., Boschetti, L., Abatzoglou, J. T., & Giglio, L. (2022). Fire regions as environmental niches: a new paradigm to define potential fire regimes in Africa and Australia. Journal of Geophysical Research. *Biogeosciences*, *127*, e2021JG006694. http://dx.doi.org/10.1029/2021JG006694

3

FIRES AND TERRESTRIAL ENVIRONMENTS

Abstract

The primary effects of wildfire in uplands are to (i) reduce infiltration capacity and the stability of regolith, especially in steep terrain with severe fire, and (ii) alter soil microbial communities and aboveground plant biomass. These effects influence biogeochemical processes and, in steep, severely burned environments, increase downslope fluxes of materials for varying timespans. The effects of wildfire on terrestrial environments vary with (i) pre-existing conditions such as topography, regolith thickness, and plant communities, and (ii) fire characteristics such as intensity, severity, and return interval. Although post-fire upland fluxes increase substantially, the morphology and stability of upland landforms are only locally affected. The responses of individual species and biotic communities vary greatly in relation to their evolutionary adaptation to wildfire and the characteristic or atypical nature of the fire. Alterations of microbial communities influence biogeochemical reactions in uplands and fluxes to rivers. The rate and extent of upland vegetation recolonization influence upland stability and fluxes, and both reflect and influence soil microbial communities. The return of animals influences vegetation regrowth by affecting propagule dispersal, pollination, soil stability and

Landscapes on Fire: Impacts on Uplands, Rivers, and Communities, Special Publications 80,
First Edition. Ellen Wohl.

DOI: 10.1002/9781394235186.ch3

nutrient content, herbivory, and parasitism. Grasslands are less likely to exhibit persistent declines in vegetation and may be more resilient than forests.

3.1. Introduction

Few places on Earth exist outside a river catchment. Even very dry environments that do not produce sufficient surface runoff to support a river network that flows to a larger river or an ocean can have an internal river network with channels that flow only occasionally. Consequently, the great majority of Earth's terrestrial environments can be considered part of a river catchment. The catchment includes all the area that ultimately drains downslope into the river channel. Organizing the terrestrial environment into river catchments acknowledges the dominant effects of gravity acting on water and everything that water carries, from dissolved chemicals to mineral sediment to particulate organic matter.

This is a geomorphologist's view of the world. Ecologists may view the world differently—many organisms can freely cross catchment boundaries with the aid of wind or their own mobility. Organisms can be divided, however, into those that live primarily in terrestrial environments versus those that live in (aquatic) or along (riparian) river networks, and the organization of this chapter and the next reflects this division.

Uplands—the portions of a river's catchment that include hillslopes and colluvial hollows without defined, persistent stream channels—constitute most of the surface area within a catchment. It is important to emphasize that the average elevation of uplands may differ little from that of channels in flat landscapes. The presence of a network of channels reflects at least a slight topographic gradient, however, and provides a fundamental division of the landscape into the channel network dominated by fluvial processes and the remainder of the catchment, which is dominated by materials moving under gravity by other processes, such as diffusion.

In the context of wildfire-related changes in terrestrial environments, the temperature at which the fire burns and the duration of heating exert primary controls on fire-induced changes (Ryan & Noste, 1985). Each of these variables reflects the influence of other factors. The temperature at which the fire burns, for example, depends on fuel availability and spatial distribution, moisture content of the fuel, and the type of fuel (e.g., vegetation species) (Santi et al., 2013), as discussed in the preceding chapter.

This chapter reviews and synthesizes information on the increased magnitudes and rates of fluxes of materials that commonly occur following wildfires, especially in steep terrains that experience high-severity fire, and how these fluxes alter the morphology and biota of uplands. The text is subdivided into fluxes, forms, and biota.

3.2. Terrestrial Fluxes

Uplands are the source for materials moved downslope and into river networks. Materials considered here are water at the surface and in the subsurface, solutes, mineral sediment, organic matter, and large wood. Organic matter includes dead plant parts from fine particulate organic matter (>0.45 µm–1 mm) through coarse particulate organic matter (>1 mm) (Tank et al., 2010) to small wood (<10 cm diameter and 1 m length).

Water and solutes move downslope into rivers via surface flow and subsurface flow paths at differing depths below the ground surface (Figure 3.1). Mineral sediment, organic matter, and large wood move downslope as individual particles; within shallow, transient flows of surface water (sheetwash, rills); and as aggregates within mass movements such as landslides and debris flows. This section provides a conceptual understanding of how these materials move down through the landscape over a range of spatial and temporal scales in environments with diverse topographic relief and fire frequency and severity.

Connectivity describes the movement of materials and organisms between components in a system (Wohl et al., 2019). Water and sediment connectivity are most commonly characterized by physical scientists examining connectivity (e.g., Bracken et al., 2013, 2015), whereas ecologists may focus more on connectivity for specific types of organisms (e.g., Komonen & Müller, 2018; Chapman et al., 2020). If a river catchment is the system, components could be uplands and the river network. A completely connected catchment would be a flat, featureless plain in which water, sediment, and other materials were uniformly distributed. Disconnectivity (Fryirs, 2013) allows catchments to have uplands, topographic relief, and discrete river networks by preventing at least some materials from moving to the same, lowest level under gravity. Gravity is always present and creates episodic and local connectivity, but much of the connectivity within river catchments is driven by water (as driven by gravity).

Water falling from the sky reduces the friction between sediment particles once the water reaches the ground surface, and it provides the energy to carry the particles downslope. Water flowing through fractured bedrock dissolves minerals and transports the resulting solutes down pressure gradients. Plants take up water through their roots, move the water upward within their tissues, and release the water back to the

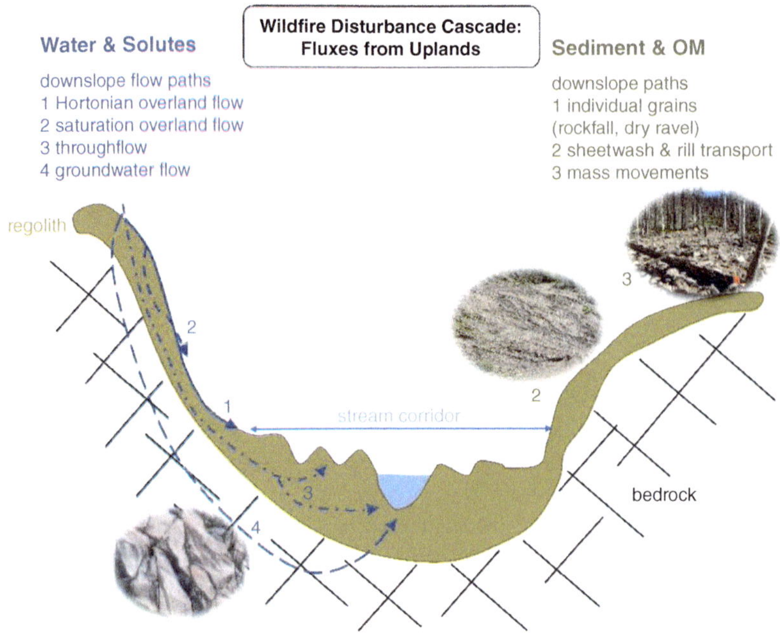

Figure 3.1 Schematic illustration of the first part of the wildfire disturbance cascade, from uplands into the river network. In this idealized cross-sectional view of a river valley and adjacent uplands, dashed lines at left indicate pathways of water. Transport processes for sediment and organic matter are listed at right and illustrated with inset photos of, clockwise from lower left, fractured bedrock, rilled surface, and debris-flow deposit. *OM* refers to organic matter. Hortonian overland flow does not infiltrate, and saturation overland flow infiltrates but then returns to the surface downslope.

atmosphere through transpiration, in the process influencing the water remaining in the subsurface and the ability of plant roots to hold unconsolidated material in the uplands.

The connectivity resulting from moving water is opposed by sources of disconnectivity. These include the absence of flowing water and associated transport capacity, as well as features that dissipate flow energy or increase the frictional resistance of sediment. Connectivity can be limited by above- and belowground portions of vegetation (Estrany et al., 2019) and by irregular topography that creates low-energy depositional areas (Lu et al., 2019). Disconnectivity can also reflect areas with limited sediment mobility because of cohesive sediment grains, very large, heavy sediment particles, or lack of sediment (Neely et al., 2019). This background disconnectivity varies among portions of the landscape and through time. Even a gentle rain can increase connectivity. More severe disturbances, such as intense rain, an earthquake, a volcanic eruption, or a forest blowdown, can substantially alter connectivity.

One of the effects of wildfire is to overcome background disconnectivity for water, sediment, and solutes across the burned area by removing barriers to connectivity, such as vegetation, and by increasing the rapid downslope flow of water as the primary agent of connectivity. The resulting increases in connectivity are known as the "wildfire disturbance cascade" (e.g., Kemter et al., 2021; Ortega-Becerril et al., 2024) or the "sediment cascade" (Fryirs, 2013; Murphy et al., 2019). This cascade moves through a burned catchment from uplands into headwater streams and down through larger channels, increasing fluxes of water, solutes, sediment, and organic matter from the uplands and through the channel network, and changing landforms and biotic communities en route. A sediment cascade is always present, although sediment movement may be largely episodic. A wildfire disturbance cascade is a version of the sediment cascade with enhanced magnitude and rate. The proportion of burned area and fire severity particularly influence connectivity and the enhancement that constitutes a wildfire disturbance cascade (Ortíz-Rodríguez et al., 2019; Santi & Rengers, 2022).

Wildfire has the potential to alter rates and magnitudes of all upland fluxes (Moody et al., 2013; Santi et al., 2013) by reducing (i) interception of precipitation by the forest canopy or other surface vegetation, as well as uptake and transpiration of water by vegetation (Santi et al., 2013; McGuire & Youberg, 2020), (ii) erosional resistance of the ground surface by consuming understory plants (the layer of trees and shrubs between the forest floor and forest canopy) (Robichaud et al., 2016), (iii) uptake of nutrients and carbon by vegetation (Song et al., 2021), (iv) storage of sediment because of roughness associated with vegetation (Lamb et al., 2013), (v) infiltration capacity of the regolith as litter and duff layers are destroyed (litter is the upper layer of dead, fallen plant material and includes recognizable plant parts; duff is below litter but above the mineral soil and is partly decomposed organic material) and a hydrophobic layer forms in the regolith (Ebel & Moody, 2017), and (vi) increasing surface velocities and connectivity of flow paths due to bare ground (Moody et al., 2013). These alterations commonly operate over timespans of days to a few years following a fire, with the duration of alterations partly dependent on whether hydrophobic soil layers form and how long they persist, the severity of the wildfire, and how quickly understory plants and sapling trees regrow in forests or grasses and herbaceous plants regrow in grasslands and savanna. Altered upland conditions and associated fluxes are common in temperate-latitude environments, especially those with moderate to high topographic relief. Accelerated fluxes to river networks from adjacent uplands may be negligible in very flat environments or following low-severity wildfires (e.g., Bonell & Williams, 1987).

High-latitude environments underlain by permafrost exhibit some distinctive responses to the fires that are becoming increasingly common in these areas. Permafrost, or permanently frozen ground, is overlain by an active layer that thaws each summer. Active layer thickness varies from a few centimeters to meters, but the underlying permafrost limits root penetration and restricts vegetation to tundra or black spruce (*Picea mariana*) forests.

Boreal forests are subject to larger and more severe fires as the global climate continues to warm (Holloway et al., 2020). The heat of a fire can increase the thickness of the active layer as the insulating plant, litter, and duff layers are reduced or removed, and albedo changes (Yoshikawa et al., 2002). The active layer may return to pre-fire thickness after several decades; or, increasingly at the southern limit of permafrost extent, the active layer and permafrost may remain altered in a manner that promotes the growth of a different vegetation community.

Both high- and low-relief permafrost regions may also include thermokarst. Melting of ice disseminated through or concentrated within permafrost can trigger subsidence known as "thermokarst" and slope failure in the form of slowly flowing slumps. These failures can be distinguished as retrogressive thaw slumps or deep-seated permafrost landslides (Young et al., 2022). Retrogressive thaw slumps initiate where surface disturbance exposes permafrost and ground ice. The slumps take their name from their upslope expansion, typically at rates of several tens of meters per year, as the ice-rich headwall ablates (Jones et al., 2019). Deep-seated permafrost landslides are a rapid, thaw-induced failure typically occurring at depths of meters to tens of meters below the active layer. These landslides are characterized by rapid translocation of frozen materials downslope along a planar failure surface (Young et al., 2022). Thermokarst and slope failure can occur without wildfire, but the direct heat of wildfire and the changes in surface insulation over a period of weeks to months following the fire can exacerbate subsidence and slumping, although the spatial heterogeneities of the active layer influence the magnitude of these changes (Lewkowicz & Harris, 2005; Yanagiya et al., 2023). Satellite-based remote sensing can be used to map the surface deformation resulting from thermokarst (Yanagiya & Furuya, 2020).

3.2.1. Bedrock to Soil

Although most of this chapter addresses fluxes that occur over timespans of minutes to years, wildfire can also influence processes better characterized by timespans of decades to centuries, including the slow flux of bedrock weathering to soil. Here, soil refers to all unconsolidated material—regolith—rather than formally defined soil with distinct horizons.

A global compilation of soil production rates from diverse climates and lithologies indicates a range of 0.1–594 mm per thousand years (mm/kyr), with mean and median rates of 67 and 27 mm/kyr, respectively (Stockmann et al., 2014). Most studies also indicate that soil production rates decrease exponentially as soil thickness increases (e.g., Heimsath et al., 1997). This suggests that when soil is removed, such as during post-fire erosion, rates of conversion of bedrock to soil increase as newly exposed bedrock is subject to enhanced physical and chemical weathering.

Physical weathering can occur as heat-induced spalling (Figure 3.2) during the fire (e.g., Blackwelder, 1927; Ollier & Ash, 1983; Birkeland, 1999; Buckman et al., 2021) and as freeze–thaw cracking following the fire. Heat-induced rock spalling, in which lensoid-shaped rock flakes up to a few millimeters in thickness become detached, has been recognized for a century (e.g., Blackwelder, 1927). Observers have speculated that heat-induced rock weathering rates associated with wildfire may be one to two

(a) (b)

Figure 3.2 Examples of heat-induced rock spalling of granitic rocks in areas burned during the 2012 High Park Fire, Colorado, USA. (a) Boulder is approximately 0.5 m across. (b) Boulder face is about 2 m tall. Pale gray areas close to the ground were exposed by erosion of the adjacent surface after the fire; pale gray areas above the ground reflect spalling.

magnitudes higher than those from frost action, but quantitative data are lacking (Shakesby & Doerr, 2006).

Subcritical cracking may also be accelerated by wildfire. Subcritical cracking describes fractures in rock that grow slowly and steadily at stresses much lower than critical stresses (Eppes & Keanini, 2017). Confining pressure from overlying soil or rock may serve to suppress subcritical cracking in near-surface environments, so the removal of overlying material and confining pressure during widespread soil erosion and mass movements following wildfire may accelerate subcritical cracking. Subcritical cracking is likely the primary mechanism by which all surface and near-surface rocks break down under natural conditions (Eppes & Keanini, 2017).

Enhanced physical and chemical weathering of bedrock may also occur as vegetation regrows after fire. Non-woody vegetation (herbaceous vegetation and grasses) typically regrows most rapidly after fire, but in forested areas the forest cover returns in stages (Bartels et al., 2016). As trees return to a site after wildfire, they strongly influence the physical and chemical weathering of bedrock (Brantley et al., 2017). Tree roots physically penetrate rock fractures and can help to expand the fractures as the roots grow in diameter and as the stresses of the aboveground portions of the tree, swaying in the wind, are transmitted to the roots. Trees draw water from the subsurface and transpire it into the atmosphere. A tree and its associated microbiota, such as myccorhizal fungi, secrete acids and other compounds into the soil and fractured rock, dissolving nutrients that the tree takes up and subsequently returns to the soil via litterfall or when the tree dies (e.g., Marañón-Jiménez & Castro, 2013). Finally, treefall lifts rock and soil to the surface with the root wad and can create pit and mound topography that influences downslope surface and subsurface pathways of water and thus the weathering environment (Brantley et al., 2017; Valtera & Schaetzl, 2017). All of these tree-induced processes of rock weathering decrease immediately after wildfire but gradually resume importance as a forest regrows.

3.2.2. Surface and Subsurface Water

American poet Emily Dickinson wrote, "Then look out for the little brook in March/When the rivers overflow/And the snows come hurrying from the hills/And the bridges often go." The snows come hurrying from the hills much faster and sooner in the melt season following wildfire in a forest. This acceleration of snowmelt reflects three primary changes caused by wildfire (Burles & Boon, 2011; Gleason et al., 2013; Uecker et al., 2020; Giovando & Niemann, 2022; Hatchett et al., 2023). First, the lack of interception by the now-burned ground vegetation or forest canopy allows more snow to settle and melt on the ground surface. Second, greater solar exposure (landscape snow albedo) due to the absence of shading by vegetation enhances sublimation and melt of the snow. Third, wind-redistributed ash and pyrogenic carbon can darken the snow surface and accelerate melting. An evaluation of snow albedo across a chronosequence of eight burned forests in Wyoming, USA, indicated that post-fire landscape snow albedo can increase by up to 63% for 10 years, with the effects lingering at lesser intensity for at least 18 years after fire (Gersh et al., 2022), despite the fact that deposition of ash and soot on the snow in the year(s) immediately after the fire may darken the snow surface and reduce albedo.

Work in southern California indicates that the net effect of post-fire alterations of uplands seems to favor surface runoff relative to deeper, subsurface flow paths during storms for the first few months to years following fire (Jung et al., 2009) (Text Box 3.1). This is reflected in river flow that becomes flashier after fire, with higher magnitude and shorter duration for a given precipitation input (Shakesby & Doerr, 2006; Paul et al., 2022). Although not all studies report increased post-fire surface runoff, most do, with magnitudes of increase from 100% to 10,000% (Paul et al., 2022). The median duration for these increases is 5 years, although some persist for 15 years (Paul et al., 2022). Increases in surface runoff reflect multiple changes after wildfire, including hydrophobic soils in which a water-repellent layer limits infiltration. Reduced surface roughness because of the removal of ground cover and soil organic layers also allows surface water to move downslope more rapidly, as does reduced interception by vegetation. Finally, clogging of soil pores with ash and fine soil particles limits infiltration. Groundwater levels can also increase, however, because of the lack of transpiration of water by plants. In other words, both surface runoff and infiltration can increase post-fire (Atwood et al., 2023).

Multiple studies document a nonlinear upland response between precipitation and runoff following wildfire. This nonlinearity takes the form of rainfall intensity thresholds above which runoff and streamflow from

Text Box 3.1 How Do We Know? Water Sources

Identifying and quantifying the relative contributions of water from different surface and subsurface sources can be challenging, as can understanding the effects of wildfire on water sources and pathways in a catchment. End-member mixing analysis compares the ratios of different solutes in water from different sources (groundwater, overland flow, hillslope soil water, riparian water) to infer the sources and relative contributions to river water. Bivariate scatter plots of all possible combinations of solutes such as Cl^-, Mg^{2+}, Ca^{2+}, Na^+, K^+, and SO_4^{2-} are used for these comparisons (Jung et al., 2009).

Figure 3.3 Examples of burned roots that left depressions in the ground surface following the 2012 High Park Fire in Colorado, USA. Surface dimensions of the depressions are about 30 × 30 cm.

burned areas increase substantially (Moody et al., 2013). These runoff thresholds may reflect the connectivity of surface and subsurface flow paths. They may also reflect depression storage in burned-out stumps and root holes (Figure 3.3) or pits created by treefall (Gallaway et al., 2009). Depressions could either fill-and-spill or create pathways for infiltration that reduce overland flow (Moody et al., 2013; Valtera & Schaetzl, 2017; McDonnell et al., 2021).

The effects of fire on groundwater are relatively understudied. Effects are likely to involve decreased recharge at shorter timespans but potentially greater recharge over longer periods if infiltration recovers but vegetation responds slowly or has lower rates of evapotranspiration (Paul et al., 2022; Atwood et al., 2023). Even immediately after the fire, though, groundwater continues to contribute to rivers, suggesting some level of recharge (Jung et al., 2009). Post-fire changes to source water (runoff versus groundwater) also appear to be more pronounced in smaller catchments (Jung et al., 2009) (Figure 3.4). A comparison of south- versus north-facing slopes in semiarid regions of the U.S. Southern Rockies suggests that snowmelt-driven groundwater recharge may be larger in burned areas, especially on south-facing slopes, because of earlier soil thaw and longer durations of soil water contents above field capacity (Ebel et al., 2012).

One of the important influences on the downslope movement of water is the formation of a water-repellent, or hydrophobic, soil layer (DeBano et al., 1979; DeBano, 2000; Weirich et al., 2024). As fire combusts waxy or resinous hydrophobic compounds, these compounds move downward to cooler levels of the soil at depths ranging from a few millimeters to several centimeters below the ground surface (Stavi, 2019) and distill into a layer that can inhibit water infiltration (Gustine et al., 2022). High-severity fires create more persistent, spatially extensive, and deeper hydrophobic layers,

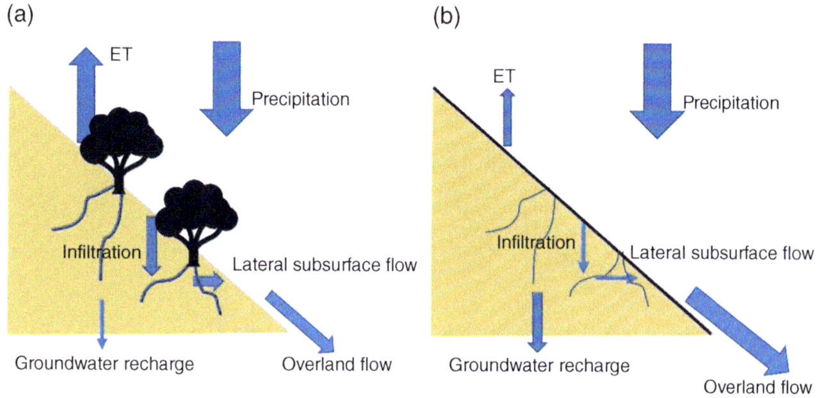

Figure 3.4 Conceptual diagram of expected hydrologic behavior in a semiarid catchment for (a) pre-fire and (b) post-fire conditions. The width of arrows and lines for infiltration indicates the relative magnitude of the water flux. (Adapted from Jung et al., 2009, Figure 1.)

as one might expect, but steeper slopes can also have larger hydrophobic layers. A large proportion of the organic matter in coarse-textured soils is not enveloped within macro-aggregates, making these soils more prone to becoming water-repellent (Stavi, 2019). Water repellency may persist only a few months after a fire or for as long as six years (Shakesby & Doerr, 2006).

Some unburned forest soils also exhibit water repellency in the absence of wildfire, including soils under common coniferous forest types in the western United States (Doerr et al., 2009) and eucalypt forests (Doerr et al., 2000; Santi et al., 2013). And, wildfire can also destroy water repellency, leaving a wettable surface layer after the fire (Doerr et al., 2000; Santi et al., 2013). As with many aspects of post-wildfire fluxes, site-specific details exert considerable influence on the magnitude and rates of processes, creating fundamentally different post-fire changes across diverse environments.

Although hydrophobic layers can inhibit infiltration, the overall effect on upland runoff may be reduced by other features, and infiltration is not always reduced after wildfire (McGuire et al., 2024). Pervasive discontinuities that reflect pre-fire variability in soil organic compounds and that facilitate rapid infiltration after fire in the Oregon Coast Range, USA, for example, can interrupt the spatial continuity and hydrological effects of water-repellent soil layers (Jackson and Roering, 2009). Such discontinuities include tension cracks in the soil and cavities left after roots burn.

Ash deposited on uplands can also exert important influences on downslope movement of water after a fire. Ash incorporated into soil can

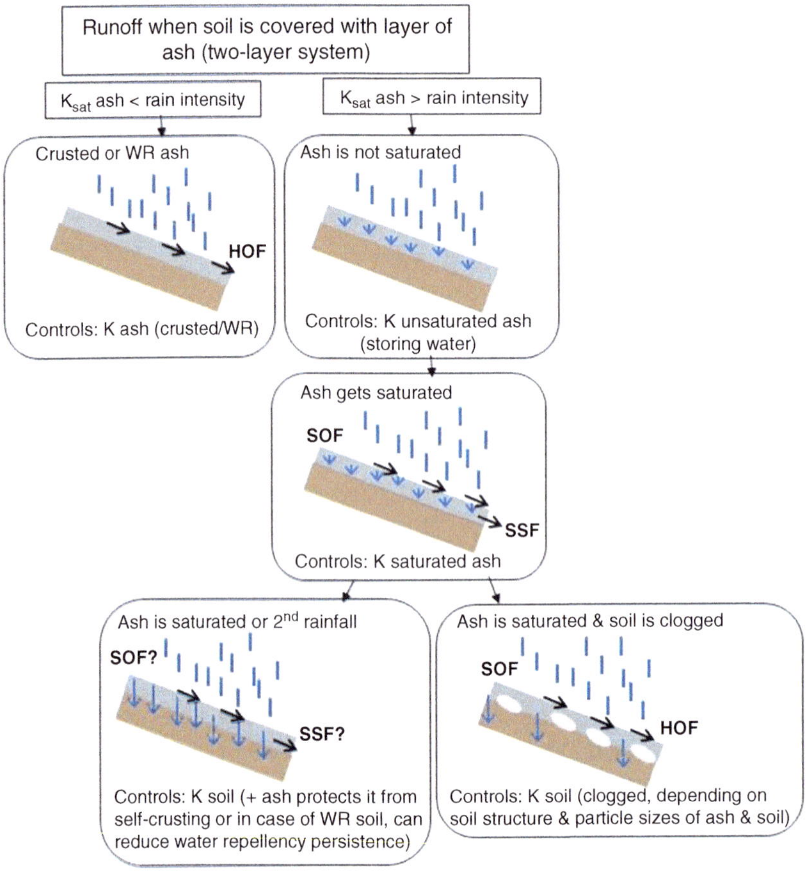

Figure 3.5 Schematic diagram of the different runoff generation mechanisms for the ash-soil two-layer system following wildland fire: K = hydraulic conductivity, HOF = Hortonian overland flow; SOF = saturation overland flow; SSF = subsurface flow; WR = water repellent. (Bodí et al., 2014 / with permission of ELSEVIER.)

either increase or reduce soil water repellency, depending on whether the ash is wettable or water-repellent (Bodí et al., 2014). The permeability of ash tends to be greater than that of the underlying soil, facilitating ponding at the ash-soil interface and delaying overland flow and runoff until the ash becomes saturated. Different scenarios can then occur (Figure 3.5): (i) the infiltration rate of the layered ash-soil system can become similar to or higher than that of bare soil, reducing overland flow; (ii) ash may clog the soil pores, reducing infiltration rates and increasing overland flow

(Ortega-Becerril et al., 2024); or (iii) ash may lose its structure after prolonged or successive rains, although the ash can still protect the soil from compaction, or the ash can be entirely removed from the burned site (Bodí et al., 2014).

Increased post-fire runoff in catchments underlain by permafrost has also been attributed to the decreased water-retention capacity of the surface organic layer (Holloway et al., 2020). Permafrost also has some distinctive longer-term responses. Permafrost terrain typically contains talik, or unfrozen patches. In one scenario, the increased thickness of the active layer and the development of talik create perennial pathways between the surface and groundwater. Groundwater propagating laterally can transfer advective heat from burned to unburned regions, spatially amplifying the effects of the fire. In another scenario, shallow closed taliks in burned peatlands isolate subsurface water, promoting talik and bog expansion. A third scenario can occur in coarse-grained upland areas in which post-fire thaw stops permafrost from creating an impermeable layer, facilitating vertical drainage instead of downslope drainage above the permafrost table (Holloway et al., 2020).

Human activities that alter land cover include deforestation, afforestation (Hou et al., 2023), prescribed burns (Holden et al., 2015), agriculture, and urbanization. These activities, as well as those that alter topography, such as transportation corridors (Cao et al., 2021) or urbanization in upland environments, have the potential to alter post-fire fluxes of water. The magnitude and direction of this alteration (i.e., increased or decreased post-fire water yields) depend on specific site conditions and the type of human alterations to land cover and topography, but human alterations are so ubiquitous in many parts of the temperate latitudes (e.g., Wohl, 2013b) that they need to be considered when examining or predicting post-fire water yields.

In summary, multiple factors influence upland water sources following wildfire. The most important factors in temperate latitudes are changes in vegetation that affect interception and transpiration, the removal of organic soil horizons and declines in infiltration, and the development of soil water repellency. In permafrost terrains, the most important factors include changes in vegetation, organic ground cover, and the active layer. The net effect of these changes in both environments is that more surface runoff is generated from precipitation inputs. In synthesizing individual studies from high-latitude forests in North America, however, Robinne et al. (2020) noted that hydrologic responses were highly variable both across and within studies. This generalization also applies to temperate-latitude uplands. Understanding site- or region-specific processes and feedbacks between factors such as soil water repellency and ash can enhance predictions of the magnitude of post-fire changes in terrestrial hydrology.

3.2.3. Solutes

Wildfire alters the mobility of a diverse array of chemical constituents within uplands, influencing both groundwater and soil water chemistry (e.g., Mansilha et al., 2020) as well as the chemistry of perennial and intermittent rivers (MacNeille et al., 2020). The various forms of nitrogen (N), phosphorus (P), carbon (C), and dissolved organic matter (DOM) receive the greatest attention, but some investigations into post-fire nutrient and solute dynamics also include calcium, chloride, magnesium, manganese, mercury, potassium, sodium, sulfur, heavy metals, and other constituents moving into river networks from uplands (e.g., Blank & Zamudio, 1998; Johnson et al., 2007; Smith et al., 2011; Costa et al., 2014; Swindle et al., 2021).

This review focuses on N, P, C, and DOM. Elevated levels of nitrogen and phosphorus can stimulate harmful algal blooms after fire, which raises concerns due to their effects on water quality. DOM is similarly important because of its potential to discolor water and create an unpleasant taste, necessitating treatment for drinking water. Carbon is also a key area of investigation due to concerns about the effects of wildfire on regional and global carbon cycles.

Nitrogen (N) has received substantial attention after fires because it can fertilize and pollute receiving water bodies, leading to enhanced fluxes to rivers, lakes, and reservoirs following fire. N is an essential nutrient that is taken up by plants and microbial communities but the ability of these organisms to utilize N also depends on other factors such as the availability of carbon and the conditions that allow plants to photosynthesize and grow. Catchments that historically burned under a low-frequency, high-severity fire regime—such as wet forests in the U.S. Pacific Northwest or high-elevation forests in the U.S. Rocky Mountains—can release substantial amounts of N previously stored in aboveground biomass, as can catchments that experience high levels of N deposition (e.g., chaparral ecosystems in southern and central California, or the Colorado Front Range) (Gustine et al., 2022). Although elevated post-fire concentrations of nutrients such as N may be transient, they can persist for more than a decade after severe fires (Gustine et al., 2022). Rates of post-fire N export can vary widely among locations, ranging from 0.3 to 431 times unburned rates (Smith et al., 2011).

Nitrogen also provides an introduction to the complexities in solute dynamics within uplands, as thoroughly reviewed by Gustine et al. (2022).

Nitrogen is distributed among the atmosphere, bedrock, soil (especially the upper, organic-rich layers), living and dead vegetation, and rivers; however, rivers typically provide only small, transient storage compared to other N pools (Gustine et al., 2022). The amount of N stored in these pools, as well as fluxes between them, varies throughout the year due to factors such as uptake by plants and soil microbes, along with precipitation that carries N downslope in runoff or deeper into the subsurface via leaching (Johnson & Turner, 2014). Storage and fluxes also fluctuate based on climate, lithology, and regolith properties (Gustine et al., 2022).

Ecosystems can be N-limited or N-saturated (Aber et al., 1989). When N inputs exceed biotic demands, nitrogen saturation occurs (Fenn et al., 2008), leading to elevated N concentrations in soil and leaching of N during runoff (Stoddard, 1994). Even N-limited systems can leach N, however, especially if seasonal precipitation flushes N during periods when plants are not actively growing, such as during early snowmelt or winter rains in xeric or Mediterranean climates (Homyak et al., 2014). Precipitation after wildfire is particularly effective at leaching N in the absence of vegetation (Hanan et al., 2017).

Nitrogen naturally enters catchments through wet and dry atmospheric deposition, abiotic and biological fixation of nitrogen gas, and the weathering of rock and soil (Gustine et al., 2022). The relative importance of these different sources varies regionally and partly depends on precipitation regimes (Cook et al., 2018). In forests, throughfall deposition moves N from the canopy to the forest floor (Gustine et al., 2022). When N is limiting, plants and microorganisms compete for available N (Gustine et al., 2022).

Soil contains the largest N pool in many catchments, with particularly large amounts of soil N found in old-growth and wetter forests (Moghimian et al., 2020). Downed, dead wood can also contribute additional soil N in old-growth forests (Fisk et al., 2002). Most of this N is biologically unavailable and must be transformed into usable forms (mainly ammonium (NH_4^+) and nitrate (NO_3^-)) through microbially mediated redox reactions in the soil (Gustine et al., 2022).

N also leaves uplands through denitrification when microbes transform nitrate to gaseous N_2, and through runoff and leaching (Parton et al., 2001; Gustine et al., 2022). Rainfall can increase rates of N transformations, but much of this N can be taken up by plants or immobilized by soil microbes, which reduces N runoff and leaching (Kelley et al., 2017; Lin et al., 2019). For catchments in which warmth coincides with rainfall, N is taken up rapidly for plant growth, limiting fluxes to rivers (Skorbiłowicz & Ofman, 2014). Wet winters, however, promote N fluxes to rivers even in N-limited systems because plants are less active at low temperatures (Gustine et al., 2022).

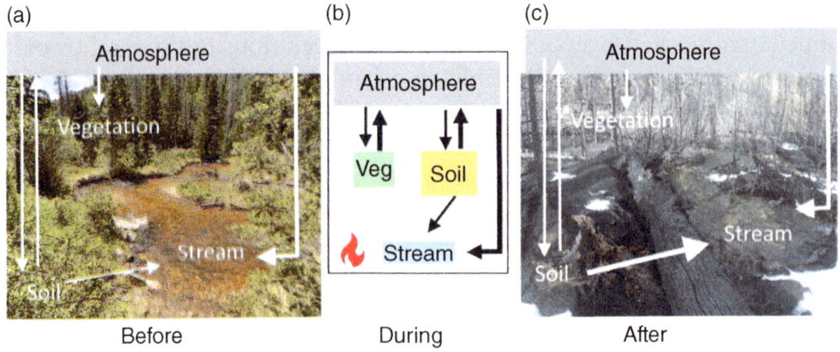

Figure 3.6 Conceptual diagram of N stores and fluxes in a watershed before, during, and after wildfire. (a) Before fire, N is stored in the atmosphere, vegetation, soils (including litter and rocks), and rivers (which are small, often transient stores relative to the others). During this stage, there can be relatively small N fluxes to river water, depending on the N status of the system (N-saturated or N-limited). (b) Fire transforms and moves N through combustion, oxidation, and volatilization back to the atmosphere while simultaneously depositing N with ash on burned hillslopes or in surface waters. Much N also remains in the soil. (c) After fire, N is transported with soil erosion, surface runoff, and leaching to the rivers or taken up by recovering vegetation. (Adapted from Gustine et al., 2022, Figure 1.)

Wildfires alter these interactions in multiple ways (Figure 3.6). N within vegetation, forest litter, and soil can be volatilized by fire or attached to smoke particles and released back to the atmosphere (Caon et al., 2014; Lindaas et al., 2021). N can also be redeposited on soil surfaces with ash (Lindaas et al., 2021). Wildfire always reduces total catchment N content because N volatilizes at relatively low temperatures, but the amount of N volatilized depends on fire temperature and duration, as well as fuel moisture and composition (Johnson & Turner, 2014). Severe fires volatilize more N and usually create more intense and long-duration effects on catchment N dynamics (Neary et al., 1999). Ash increases N availability, but ash is also easily mobilized by wind or runoff (Neris et al., 2021).

Fire can kill soil microorganisms, especially in drier soils that transmit more heat energy, and char soil organic matter, thus affecting N dynamics (Neary et al., 2005). Excess N mobilization can increase as an altered, C-starved microbial community and reduced root uptake due to vegetation mortality result in reduced N demands (Eckdahl et al., 2023).

Fires generally reduce soil microbial and fungal abundance, although this effect may be greater in forests than in grasslands because of the lower fire severity in grasslands (Stavi, 2019). (Tropical grasslands in which fires are more frequent and intense than in tropical moist broadleaf forests may be an exception to this generalization; Ruan et al., 2020.) Lower severity fires may affect primarily the upper soil layers, but a longer and more severe fire may affect N compounds deeper in the soil because of the tendency of N to volatilize (Sharma et al., 2017).

Fire can also make soil N more available by releasing the N from organo-clay minerals (Dijkstra et al., 2017). Multiple studies document increases in soil ammonium and nitrate concentrations after fire (Gustine et al., 2022). Sometimes these conditions last for a few months; sometimes for up to 10 years (Kurth et al., 2014), depending on fire severity and rates of plant regrowth (Gustine et al., 2022).

Many of the processes that govern N cycling and export in an unburned catchment remain present after a fire, but the magnitude and timing usually change post-fire as a result of changes in N availability from ash (e.g., Ferreira et al., 2005), runoff, sediment movement, soil microbial communities, and vegetation demand (Hanan et al., 2016). Soil microbial biomass responds to direct heating during the fire and to soil biogeochemical changes post-fire, making fire effects on microbial biomass highly variable (DeBano et al., 1998; Smith et al., 2008). Soil microbial biomass recovery and associated N dynamics depend on pre-fire N conditions and post-fire litter and carbon inputs (Gustine et al., 2022). Ash inputs can lead to an initial spike in microbial biomass followed by a sharp decline after the first post-fire growing season as ash-related resources become depleted (Hanan et al., 2016). At other sites, microbial biomass can recover within a year after less severe fires and with sufficient C and N (Smith et al., 2008).

The details of fire-generated hydrophobic soil layers can strongly influence soil N dynamics. Although hydrophobic layers increase runoff and erosion, they can reduce N leaching because less water is transported through the soil profile (Fernelius et al., 2017). On the other hand, hydrophobic layers can delay the regrowth of vegetation that might take up mobilized N (Fernelius et al., 2017). Hardwoods and seed-bank shrub species can regrow rapidly, particularly if they are associated with N-fixing bacteria, but burn severity is an important control on vegetation recovery (Shatford et al., 2007). The rate of vegetation regrowth and associated N demand strongly influence post-fire fluxes of N downslope (Gustine et al., 2022).

In regions with substantial atmospheric inputs of N derived from agricultural and urban sources, such as the Colorado Front Range, USA, excess N in a catchment can interact with fire in multiple ways (Heindel et al., 2022): (i) elevated N deposition over many years can increase biomass and the thickness of the forest litter layer, as well as increase vulnerability to drought stress, thus making the ecosystem more susceptible to fire (Grulke et al., 2008); (ii) fires can be large sources of N emissions (e.g., Benedict et al., 2017), and subsequent deposition can further overload receiving ecosystems; (iii) fire can mobilize accumulated N and alter the ecosystem's ability to retain and process subsequent N deposition, causing elevated N export and reduced water quality downstream from burned areas (Mast et al., 2016); and (iv) N deposition before and after the fire can alter nutrient cycling in soils and the species composition of vegetation that regrows after the fire (e.g., Rhoades et al., 2011, 2019; Southon et al., 2012).

In catchments with permafrost, fire can increase the thickness of the active layer and facilitate plant growth that results in the uptake of N, P, and other nutrients previously stored in permafrost (Abbott et al., 2021). Conversely, wildfire can remove a substantial portion of above- and belowground nutrients via combustion, which can decrease organic nutrient sources (Neff et al., 2005). Combustion of surface vegetation and the organic horizon decreases insulation and albedo, which enhances heat transfer into the subsurface and deepens the active layer. This can allow water to enter deeper soil horizons that are richer in solutes, increasing nutrient supply and potentially triggering permafrost collapse and thermal erosion. Such collapse abruptly releases above- and belowground nutrients from terrestrial to aquatic environments, although post-fire regrowth of vegetation could result in greater uptake of nutrients (Abbott et al., 2021).

In other words, prior to fire, there is low nutrient availability in the active layer because of limited decomposition and high demand from microorganisms and plants. There are large stores of N, P, and S in permafrost. During fire, combustion, thermal erosion, and high demand from regrowth exacerbate nutrient limitation on land while increasing erosion and the associated flux of particulate nutrients. After the fire, regrowth stabilizes erosion while plants take up deeper nutrients, enhancing the lateral flux of dissolved organic nutrients (Abbott et al., 2021). As the climate warms, permafrost degrades and the active layer becomes thicker, which is changing nutrient cycling in both burned and unburned terrestrial environments in permafrost regions (Chen et al., 2018; Salmon et al., 2018; Li et al., 2021).

In summary, wildfire typically increases downslope fluxes of N to rivers, but the details of how much more N enters rivers and for how long reflect (i) pre-fire characteristics of N inputs, N pools in soil and plants, and N fluxes between pools; (ii) fire characteristics of severity, duration, and spatial extent, and associated effects on vegetation and soils; and (iii) post-fire presence of a soil hydrophobic layer, characteristics of precipitation, soil microbial biomass, and vegetation regrowth. Much of the significant post-fire N flux to rivers occurs during intense rainfall, although flat topography and limited hydrologic connectivity between uplands and rivers can limit such increased fluxes (e.g., Hauer & Spencer, 1998; Jordan, 2012; Francos et al., 2014). In some burned areas, stemflow (rainfall reaching the ground by draining down the trunks of trees rather than dripping from the canopy) can divert relatively high volumes of precipitation to limited areas of the ground, creating nutrient hotspots surrounding burned tree trunks (White, 2015).

Phosphorus (P), like nitrogen, is a critical nutrient for plants. Despite the presence of P in soils, however, it is commonly a limiting factor for plant growth because it is mainly present in forms that are not available to plants (Caon et al., 2014).

P in soil occurs in dissolved inorganic and organic forms and in particulate inorganic and organic forms. Other compounds, such as iron and aluminum oxides, oxyhydroxides, aluminum minerals, and amorphous oxides, can bind P or release it into solution (Reddy et al., 1999; Records et al., 2016). The ratios of C and N to P, as well as the availability of C and N, also strongly influence biotic immobilization or release of P (Records et al., 2016). The abundance and forms of P in upland soils reflect the concentrations and forms of P and other compounds in the bedrock, the degree of weathering, and, in some locations, atmospheric deposition (Records et al., 2016).

P can move downslope in surface overland flow, in subsurface matrix flow, or in subsurface preferential flow paths such as cracks, animal burrows, and soil pipes. Particulate forms of P adsorb to silt and clay and move in both surface runoff and subsurface preferential flow. Inorganic dissolved P dominates groundwater, and dissolved P generally dominates subsurface transport (Heathwaite & Dils, 2000). Matrix flow appears to move relatively little P relative to the other flow paths (Records et al., 2016).

P fluxes from uplands typically increase during (Hauer & Spencer, 1998) and following wildfire (Bitner et al., 2001; Noske et al., 2010; Sherson et al., 2015; Mishra et al., 2021), although the magnitude of these changes varies widely (Smith et al., 2011). Increased P yields can result from atmospheric and runoff inputs of ash. Combustion and partial combustion of plant litter and vegetation and the subsequent deposition of ash increase P concentrations in surface soil. Nutrients within the ash are susceptible to solution and physical erosion, increasing the potential for the export of dissolved and particulate P (Lane et al., 2008; Blake et al., 2010; Paul et al., 2022). Based on work in Greece (Blake et al., 2010) and Alberta, Canada (Emelko et al., 2016), particulate P can represent the great majority of P yield from burned uplands, and up to 20% of this material is potentially bioavailable.

Increased P yields can also result from soil erosion on hillslopes (Smith et al., 2011), and eroded sediment stored in downslope or downstream depositional areas can protract post-fire P yields (Blake et al., 2010). P tends to be bound to fine mineral sediment such as clays, and the particle-associated fraction of P may dominate exports during stormflow in the first year after a fire. These contributions decline with time as upland vegetation regrows and sediment movement on hillslopes decreases. Burning can decrease P stored in surface organic layers of soil via combustion and volatilization, especially during high-severity fires. Wildfire does not generally directly decrease soil P, however, because the volatilization temperature of P (>550 °C) is much higher than that of N (200 °C) (Smith et al., 2011).

A synthesis of 51 studies indicates that post-fire P yields increase from 0 to 1,000,000% in diverse locations (median increase of nearly 1,000%), and the duration of these effects ranges from 0 to 7 years (median duration 4 years) (Paul et al., 2022). Among all of the post-fire changes in hydrology and water quality reviewed by Paul et al. (2022), P had an especially high median value of increase (along with surface runoff, suspended sediment, cadmium, iron, lead, nickel, and zinc). Simulation modeling can be used to predict these increases and to understand temporal and spatial patterns of change (e.g., Basso et al., 2020, 2021).

Carbon pools in uplands include carbon in bedrock, mineral soil carbon, soil microbial biomass, and carbon within living and dead vegetation (Seedre et al., 2011). Carbon can occur in organic and inorganic forms and in dissolved and particulate forms.

The formation of soil humic materials that constitute carbon in mineral soil has been described as a tortuous pathway (Pingree & DeLuca, 2017). Microbial degradation of dead plants and animals creates humic substances, which are complex heterogeneous mixtures of organic compounds that tend to resist further biodegradation. Plant tissues are largely composed of complex organic compounds including ether-soluble fats, oils, waxes, and resins; water-soluble compounds; cellulose; hemicelluloses; crude protein; and lignin. Individual organic components of plant material decompose at different rates. Lignin is the most resistant, whereas components such as sugars, polysaccharides, and proteins are highly degradable. The proportion of different constituents varies with plant species and the type of plant tissue, so the rate and completeness of microbial degradation to create humic substances vary greatly.

The availability of N also influences C dynamics. C and N dynamics in soil involve three basic steps. The first is the leaching of soluble substances from plant litter. This is followed by accumulation as decomposer microorganisms utilize and retain N, some of which is later converted into recalcitrant humic substances that resist biodegradation. Finally, N is released via net mineralization (decomposition and oxidation). This occurs when N no longer limits microbial growth and activity, allowing N to be released as ammonium (Haynes, 1986). C and N cycling through leaching, accumulation, and release follow diverse pathways as soil microorganisms trade different forms of nutrients. Humus formation can take a few days to thousands of years, but the decomposition contributes to the formation of soil organic matter. Humus is the organic residue that is not readily available as an energy source for heterotrophic soil microflora (organisms that cannot produce their own food) and thus accumulates in soil (Haynes, 1986).

Under near steady-state conditions, net C emissions from ecosystems reflect the balance between net primary production, which sequesters C in living tissue, and heterotrophic respiration, which is primarily decomposition that results in C release. Wildfire creates additional fluxes and changes this balance (Chapin et al., 2009).

Wildfires can result in C loss via the release of C in the gases CO_2, CO, and CH_4 during combustion and via the decay of fire-killed vegetation (Restaino & Peterson, 2013). Wildfires can cause nearly complete loss of organic C stored in surface, organic-rich soil horizons, litter, duff, and ground cover (e.g., Shibata et al., 2003), especially if this C-rich layer is subject to enhanced erosion (Bitner et al., 2001). However, subsequent redeposition of this material elsewhere can create sequestration of the C in other landscape locations such as the toe of a hillslope or an alluvial fan. Deposition of dark ash may return C to the ground surface, but ash is readily erodible by wind and water and hence is rapidly depleted (Reneau et al., 2007).

The completeness of combustion determines the C content of ash. At low combustion completeness ($T < 450\,°C$), ash is primarily organic carbon. At higher temperatures and combustion completeness, most organic C is volatilized, and ash is composed of minerals and inorganic carbonates (Bodí et al., 2014). Increased erosion and inputs of C-rich dark ash can increase particulate organic carbon exports from uplands after fire, but ash deposition also stimulates soil microbial activity and vegetation growth (Bodí et al., 2014).

Soil organic matter may experience only minor effects from the volatilization of some constituents during low-intensity fire and may increase following fire (Babur et al., 2022), but high-intensity fires can create greater charring and the formation of pyrogenic carbon (charred fine residue and charcoal) or complete oxidation at high temperatures. The effects of fire on soil organic carbon vary with fire type, duration, and intensity; soil moisture and type; and vegetation (Caon et al., 2014).

Fire decreases forest soil organic carbon in the short term in both temperate and tropical forests (López-Cruz et al., 2022). Over the longer term, the incorporation of ash, vegetation recovery, and decomposition of partially burned wood fragments can increase soil organic carbon (Caon et al., 2014). Fire in grassland environments may have little or no effect on soil organic carbon, based on a wildfire in a semiarid grassland in Italy (Novara et al., 2013), or may cause a transient increase in this carbon stock, as indicated by the responses of Mediterranean shrublands and grasslands (Martí-Roura et al., 2011), a semiarid restored grassland in China (Liu et al., 2018), and a Mediterranean mountain scrubland in Portugal (Fonseca et al., 2022). Negligible change may reflect the limited increase in soil temperature during the fire and increases may reflect ash deposition, decomposition of incompletely combusted organic material, and new carbon compounds transported from the litter layer into the underlying mineral soil (Fonseca et al., 2022).

Fire can affect the soil microorganisms—archaea, bacteria, fungi, and protists—that influence C dynamics. Fire directly affects these microorganisms via heat—usually fire decreases microbial biomass—and indirectly via altered soil properties such as pH that influence microbial community composition and function (Köster et al., 2021). Increased burn severity, for example, corresponds to the loss of symbiotic fungi that feed on plant roots (Hewitt et al., 2022). These ectomycorrhizal fungi facilitate plant access to limiting nutrients (Nelson et al., 2022). High fire severity that reduces soil fungi may limit conifer seedling regeneration in boreal forests, for example, whereas soil fungi can survive low fire severity, and this favors post-fire regeneration of conifers (Ibáñez et al., 2022).

Carbon losses from the decomposition of plant litter and soil organic matter by microorganisms (heterotrophic soil respiration) may increase CO_2 loss immediately after a fire before the catchment returns to net CO_2 uptake a few years later as vegetation regrows (Granath et al., 2021). The rates of vegetation regrowth and C uptake vary between sites based on fire severity and climate (Seedre et al., 2011), and there is some evidence that increased fire frequency and drought in semiarid regions are limiting forest recovery and will result in less C storage in living biomass (Bowman et al., 2021).

Wildfires can also create pyrogenic carbon, a high-C and low-N byproduct of the incomplete combustion of organic matter. Pyrogenic carbon includes charred fine residue, but most takes the form of woody charcoal. Pyrogenic carbon, in the form of charcoal, is uniquely recalcitrant, or resistant to decomposition, and can persist in soils for hundreds to thousands of years (Santín et al., 2020). Consequently, it represents net ecosystem C sequestration after fire and greatly extends the timespan over which fires influence soil processes (Pingree & DeLuca, 2017).

Pyrogenic carbon has a high surface area that allows adsorption of a broad spectrum of organic compounds that then influence microbial activities after fire. This is the charosphere effect, in which greater microbial activity occurs in the vicinity of pyrogenic carbon (Pingree & DeLuca, 2017). Pyrogenic carbon is of low density and can remain in the upper soil layers, move readily downslope during post-fire overland flow, and be transported in suspension in receiving rivers (Cotrufo et al., 2016). A review of more than 560 empirical pyrogenic carbon measurements in surface and subsurface soils reported in 55 studies suggests that this form of carbon can constitute up to 60% of soil organic matter across forests, grasslands, peat, and urban soils, although the average value is 13.7% of soil organic carbon, and grassland and forest soils tend to have lower pyrogenic carbon content than agricultural land (Reisser et al., 2016; Pingree & DeLuca, 2017).

The net effect of wildfire on soil carbon varies with burn severity and time since fire (Heath et al., 2015). The net effect of fire on ecosystem carbon storage depends on the rates of recovery for soil microbial communities and vegetation, as well as carbon loss via gaseous emissions versus addition in the form of pyrogenic carbon. Although both soil and living vegetation can contain substantial C stocks, the combined effects of drought and increased fire are reducing forest carbon uptake in some regions, creating a scenario in which severe, repeated disturbances may exceed the ability of some forests to recover (Brando et al., 2019).

Recently burned ash has substantially higher dissolved organic carbon (DOC) leaching potential than older ash (Revchuk & Suffet, 2014), but reported wildfire effects on DOC are generally minor (Smith et al., 2011; Robinne et al., 2020). Severe fires, however, can cause elevated DOC concentrations (Santos et al., 2019; Uzun et al., 2020).

Ecosystem C pools in forests may require decades to reestablish after fire, depending on the rate at which trees regrow (Johnson et al., 2007). A synthesis of boreal forest C stocks following fire found an average decrease of 60% (range <10% to >80%) as C in living trees and the soil organic layer declined (Palviainen et al., 2020). Total C stocks increased with time since fire, with increases slowing after about a century, although the rate of C stock recovery also depends on fire severity and the trajectory of vegetation succession following fire (Cahoon et al., 2022).

The details of boreal peatland C stocks following wildfire illustrate some of the complexities of predicting fire effects on these stocks. Peatlands cover more than 75% of the landscape in the southern portion of Canada's Northwest Territories. Within this region, peatlands on the Taiga shield are more susceptible to wildfire because they are smaller and hydrologically isolated within a rocky landscape (Bourgeau-Chavez et al., 2022). Larger areas of these peatlands burn at higher severity, releasing C, relative to peatlands on the Taiga plains that are hydrologically well-connected to groundwater. Peatlands on the plains also cover larger areas and are more likely to include unburned and lightly burned patches than peatlands on the shield, limiting C losses after fire. However, it becomes important to understand potential tipping points associated with drought that may reduce hydrologic connectivity and facilitate deeper burning of the organic soil layers in peatlands on the Taiga plains (Bourgeau-Chavez et al., 2022).

Trends in carbon stock with time after fire reflect interactions among vegetation growth rates, fire severity, and climate warming. Positive trends in gross primary productivity in Arctic boreal biomes are driven by warming, but higher temperatures also increase the risk of wildfires and associated loss of carbon over a period of 1–3 years (Madani et al., 2021). Mortality during severe wildfires reduces ecosystem productivity, but post-fire productivity can increase in moderately burned areas. In addition, carbon uptake during the rapid recovery of productivity after fire can be offset by the release of ancient carbon stored in the soil. The net effect is projected to be a decrease in primary productivity that amplifies climate warming as severity and frequency of wildfires increase in the boreal biome (Madani et al., 2021).

In ecosystems underlain by permafrost, fire has long-term consequences for the C balance (Holloway et al., 2020). Increased soil temperatures and active layer thickness make previously frozen soil organic matter available for decomposition by soil microbes that release gaseous forms of C into the atmosphere. Decomposition can be reduced as the active layer returns to pre-fire thickness and organic matter once again freezes, but a warming climate means that thawed permafrost will be vulnerable to microbial decomposition for longer periods. High fire severity results in a net loss of carbon, so increasing frequency and magnitude of fire, and associated changes to permafrost, will likely result in increased carbon loss from permafrost catchments (Holloway et al., 2020).

A synthesis of 50 post-fire studies found that the magnitude of organic carbon released to downstream waters during fires ranged from slightly negative values to +1400%, with a median just slightly above zero (Paul et al., 2022). In other words, with some spectacular exceptions, wildfire does not necessarily substantially change the export of organic carbon. Paul et al. (2022) also noted that at least some of the increases in aquatic systems (including lakes) below burned catchments are likely attributable to increases in algal biomass in response to dissolved nutrient export from burned areas. Changes in organic carbon can persist from 0 to 14 years, with a median value of less than 5 years (Paul et al., 2022).

Dissolved organic matter (*DOM*) includes C, N, P, and sulfur (S) compounds. A single DOM sample, however, can include tens of thousands of distinct molecules, each with its own chemical properties and reactivity in the environment (Xenopoulos et al., 2021). The production, transport, and degradation of terrestrial DOM influence C and nutrient cycling in upland soils and aquatic ecosystems receiving terrestrial DOM (Olefeldt et al., 2013).

Wildfire can alter the turnover of DOM in soils and the linkages between terrestrial and aquatic C cycling by altering terrestrial DOM production and composition. Soil C solubility can decrease with fire, for example, whereas soil N becomes more soluble following fire (Olefeldt et al., 2013). Fire also alters flow paths and sources of organic matter within the landscape, causing shifts in DOM composition (Xenopoulos et al., 2021). The dynamics of DOM following fire vary over time and spatially in relation to burn severity, such that the N content and stability of DOM can increase in rivers draining catchments burned at greater severity (Roebuck et al., 2022). The reduced stability of DOM pools in a landscape slowly recovering from fire reflects the reduction in new organic matter inputs to soil and the dominance of microbially derived, more bioavailable soil organic matter (Xenopoulos et al., 2021).

The effect of fire on DOM in uplands partly depends on soil water interaction times. Low-severity wildfires in northern peatlands create ash and char that enter the peat matrix and reduce pore diameters, promoting subsurface soil moisture retention and longer pore water residence times (Ackley et al., 2021). This leads to elevated dissolved solutes and DOM in pore water.

DOM can also be influenced by atmospheric transport during and after wildfire. Atmospheric inputs of pyrogenic carbon, for example, can influence marine DOM in coastal areas with large wildfires (Coward et al., 2022). These details of fire-altered characteristics of DOM in surface and subsurface upland waters and inputs to rivers are important because of their influence on nutrient availability and water quality.

As noted earlier for water, human activities that alter land cover and topography in upland environments have the potential to alter post-fire fluxes of solutes. Cropping and grazing, fire suppression, and other alterations of land cover influence the chemical, physical, and biological properties of soils (e.g., Wills et al., 2016), for example, and thus the behavior of soils after fire with respect to infiltration and nutrient release. Analogously, changes in vegetation community composition influence soil characteristics and nutrients available to be either consumed by wildfire or released downslope following fire (e.g., Mustard et al., 2012). Human alterations may be the dominant influence on nutrient dynamics following wildfire. A comparison of multiple burned catchments in semiarid regions of Utah, USA, for example, indicated that nutrient concentrations were higher in catchments with greater urban and agricultural land use, regardless of burn status (Crandall et al., 2021).

In summary, wildfire can alter (i) the amounts of solutes stored in different components of a catchment, such as living and dead vegetation and soil, (ii) the surface and subsurface flow paths of water and hence the reaction times of water with the matrix or substrate and the solute chemistry of the water, (iii) biogeochemical processes influenced by soil microorganisms and living vegetation, and (iv) the fluxes of particulate material (ash, mineral sediment, organic matter) that carry chemical compounds downslope. The magnitude of these alterations depends on factors including the severity of the fire, the return interval between fires, and the topography and pre-fire vegetation community of the burned site. Generally, the more severe the fire, the greater the alterations of solute fluxes. In reviewing modeling studies of water quality in burned areas, Basso et al. (2022) identified the transport of ash, which can be a source of nutrients and contaminants, as a major gap in existing work.

3.2.4. Sediment Yields

Sediment yields from uplands to river corridors in high-relief terrains commonly increase substantially following severe fire, as sediment is more readily mobilized through individual grain movements, as well as mass movements (Text Box 3.2). These increases typically last less than a decade (e.g., Reneau et al., 2007; Cole et al., 2020) but can result in annual sediment yields more than an order of magnitude greater than pre-fire levels (Lamb et al., 2011). These post-fire changes are most pronounced in high-relief

Text Box 3.2 How Do We Know? Rates of Post-fire Sedimentation

Measures of post-fire sedimentation (Moody & Martin, 2009; Moody et al., 2013; East et al., 2021) can be derived from:

i. hillslope point and plot measurements, typically conducted over relatively short timespans of days to a few years, point measurements employ erosion pins or repeat survey transects (e.g., Morris et al., 2012); hillslope plots commonly use sediment fences to retain sediment from upslope (e.g., Wilson et al., 2021);
ii. channel measurements of suspended sediment transport;
iii. comparisons of topographic change over time reveal volumes of sediment eroded or deposited in diverse locations within a burned catchment, typically utilizing terrestrial and/or airborne lidar (Pelletier & Orem, 2014; Brogan et al., 2019; Rengers et al., 2021); and
iv. sedimentation in depositional areas such as artificial reservoirs, lakes, or alluvial fans (e.g., Meyer & Wells, 1997; Blake et al., 2004).

environments. Wildfires of low severity and those occurring in low-relief environments may create negligible changes in sediment yields.

Post-fire sediment yields have been equated to thousands of years of background sediment yields in the absence of fire (e.g., Meyer et al., 2001) and up to half of temporally averaged sediment production on steep hillslopes (Roering & Gerber, 2005), but the spatial scale over which measurements are made is critically important. Shakesby and Doerr (2006) noted an approximately inverse relationship between the amount of erosion and the scale of measurement. Part of this discrepancy may reflect spatial and temporal variations in connectivity. Rills (small, linear erosional features on hillslopes, sometimes distinguished from gullies as being ≤0.3 m deep; Thwaites et al., 2022) along a recently burned slope after low-intensity rainfall exhibit longitudinally discontinuous sediment transport, for example, because of locally transport-limited segments of the rills (Wester et al., 2014). Higher-magnitude rainstorms might overcome these limitations and create enhanced downslope sediment connectivity, but relatively small, sub-meter to meter-scale discontinuities of this type might not be reflected in plot-scale sediment yields.

The commonly observed trend of post-fire increases in sediment yield that peak during the first rainstorms and then decline over time is largely based on research conducted in the western United States (Swanson, 1981). The period of elevated post-fire yields, termed the *window of disturbance* by Prosser and Williams (1998), can be as brief as 3 months but is more commonly 3–10 years long (Shakesby, 2011; Rengers et al., 2020). As in the case of solute yields to rivers, lower relief, low-intensity rainfall, or snowmelt runoff, less severe fire and limited hydrologic connectivity between uplands and rivers can all restrict post-fire sediment yields to rivers (e.g., Jordan, 2012; Mast et al., 2016).

Severe wildfire increases sediment availability by (i) promoting rock weathering; (ii) consuming the canopy, understory, litter, and duff layers, thus exposing large areas of bare soil; (iii) creating heat that penetrates the soil and changes aggregate stability in a manner that reduces the critical shear stress needed to initiate sediment motion (e.g., soil water repellency and soil sealing by ash or sediment particles that substantially reduce hydraulic conductivity and infiltration); and (iv) generating heat that reduces root cohesion, the secretion of cohesive compounds by soil microbes, and stabilizing hyphae of soil fungi (Neary et al., 1999; DeBano, 2000; Shakesby & Doerr, 2006; Larsen et al., 2009; Moody et al., 2013).

As erosion depletes the sediment supply and vegetation regrows, sediment availability decreases over time, resulting in an initial post-fire phase in which sediment yields are transport-limited, transitioning to a supply-limited phase (Shakesby & Doerr, 2006) (Figure 3.7). This effect may be most pronounced in semiarid landscapes, which typically exhibit

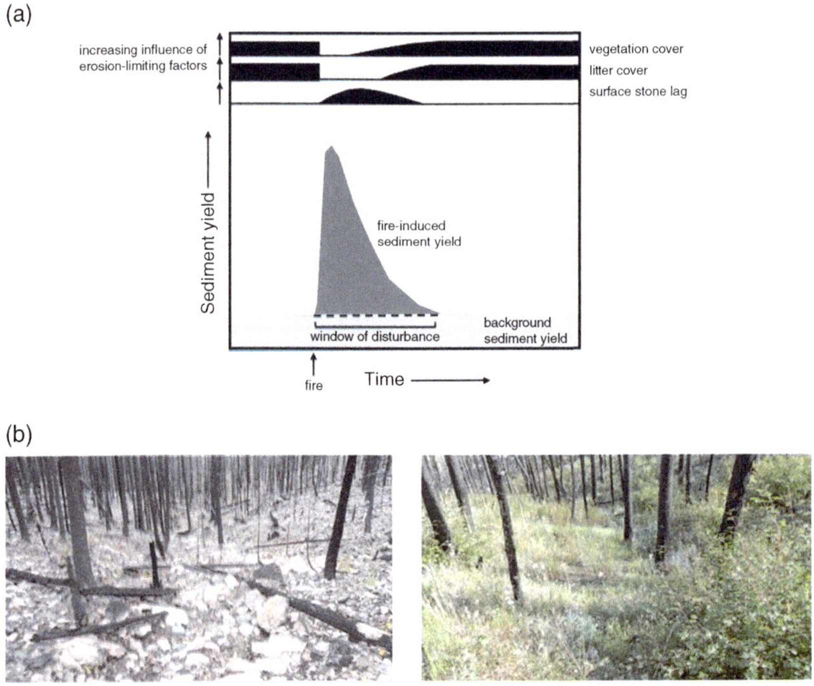

Figure 3.7 (a) Hypothetical decline in sediment yield following a wildfire, illustrating the role of vegetation cover, litter cover, and stone lag in reducing erosion rates. (Shakesby & Doerr, 2006 / with permission of ELSEVIER.) (b) Downslope views of the same hillslope shortly after the fire in 2012 (left) and in 2015 (right), showing regrowth of ground cover. Skin Gulch catchment, Colorado, USA.

high sediment yields, sufficient vegetation to sustain large fires, and relatively slow regrowth of vegetation following fire (Goode et al., 2012). Correlations such as that between loss of surface cover and increased sediment yields have been documented over spatial scales ranging from 1 m^2 to 100 km^2, but are nonlinear (Larsen et al., 2009).

Deposition of ash on hillslopes can also change runoff generation and erodibility in disparate ways (Smith et al., 2011). Ash can increase infiltration capacity (e.g., Shakesby, 2011), or ash clogging of soil macropores and the formation of surface crusts can reduce soil infiltration rates (e.g., Gabet & Sternberg, 2008). Saturation of the ash layer during rainfall may lead to incorporation of ash into surface runoff, causing increased density and viscosity of the fluid and potentially helping to generate debris flows (Gabet & Sternberg, 2008; Smith et al., 2011).

As with runoff and infiltration, bioturbation on burned slopes can increase or decrease upland erosion. Disruption of the ground surface by mound-building ants and termites, by birds that scratch the soil, and by burrowing mammals can increase the supply of mobile sediment. Mounds can also increase surface roughness, and tunnels can provide sinks on water-repellent soil, thereby limiting erosion (Shakesby & Doerr, 2006).

Fallen wood in upland environments can also influence surface roughness and connectivity. Sediment can be trapped upslope of fire-produced or pre-fire downed logs (Adams et al., 2023). Burned roots and snapped stems of burned trees can reduce the effectiveness of tree-throw in displacing sediment when a fallen tree pulls sediment from the surface via its exposed roots (Adams et al., 2023).

Human-induced alterations of topography and surface characteristics associated with containing a wildfire, including digging fire breaks and containment lines with hand tools or heavy machinery (Backer et al., 2004), and limiting post-fire erosion using techniques such as contour felling (Robichaud et al., 2008), can also increase or decrease upland erosion. Among the most ubiquitous human alterations of topography in forested regions may be unpaved roads that can substantially alter downslope movement of water, solutes, and sediment in burned (Sosa-Pérez & MacDonald, 2017) or unburned (e.g., Larsen & Parks, 1997) areas.

Processes of downslope sediment movement include individual grain movements driven by gravity, wind, and water, as well as water-driven mass movements. *Dry ravel*, which has been reported mainly for burn sites in the western United States (Shakesby & Doerr, 2006), involves rapid, dry, particle-to-particle rolling, sliding, and bouncing of sediment and some organic matter under the force of gravity (Florsheim et al., 1991).

Dry ravel is sensitive to the angle of repose and sediment supply on the hillslope (Moody et al., 2013), as well as surface roughness (Roth et al., 2020). As might be expected, the process tends to occur on steep, relatively smooth slopes that receive the greatest insolation (south-facing in the Northern Hemisphere; Wondzell & King, 2003). Dry ravel also occurs in unburned terrain, and mobilized sediment can be stored upslope of vegetation (Lamb et al., 2013) until the consumption of the vegetation by wildfire (DiBiase & Lamb, 2013). Ravel-mobilized sediment can form depositional cones when it reaches slopes less than the angle of repose (Lamb et al., 2011). Sediment mobilized by dry ravel in burned or unburned terrain also tends to accumulate in hillslope concavities and ephemeral channels. Sediment can be mobilized from these locations by runoff post-fire (Florsheim et al., 1991), sometimes helping to initiate debris flows (DiBiase & Lamb, 2020) (Figure 3.8).

Wind erosion of burned organic matter and fine sediment (Shakesby & Doerr, 2006; Shakesby, 2011) can be substantial in drylands until vegetation regrows (e.g., Sankey et al., 2009). Wildfire in Chihuahuan desert shrubland facilitated wind erosion of sediment that was greater by a

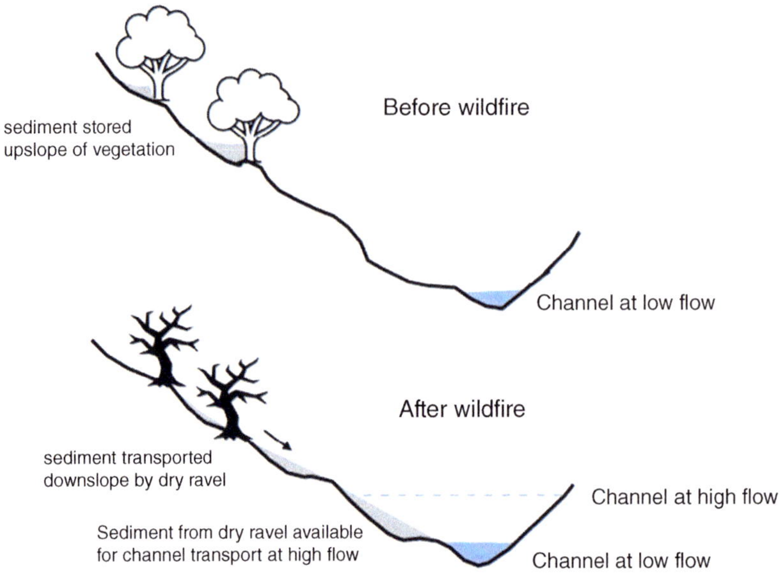

Figure 3.8 Hillslope profile showing the effect of wildfire on vegetation and downslope transfer of sediment by dry ravel. (After Shakesby & Doerr, 2006, Figure 12 & Florsheim et al., 1991.)

factor of three, averaged over several months after the fire, relative to comparable unburned areas, and rising to 70 times as great during periods of high-velocity winds immediately following the fire (Whicker et al., 2002). Burned areas can also create regional sources of dust (Miller et al., 2012; Hahnenberger & Nicoll, 2014).

Water-driven grain movements include the processes of rainsplash, sheetwash, and rill transport. *Rainsplash*, or raindrop impact, occurs when the force of falling raindrops dislodges individual particles up to sand size. This detachment makes particles more susceptible to wind and water erosion (Gabet & Dunne, 2003). Rainsplash tends to act uniformly across a slope, but its effectiveness is reduced as fine sediment is detached and mobilized, and coarser sediment (e.g., cobbles) is revealed (Shakesby & Doerr, 2006), or as soil is compacted and sealed as soil pores clog (Swanson, 1981). Rainsplash erosion can be particularly effective on poorly aggregated, water-repellent soils (Shakesby & Doerr, 2006) and, in some scenarios, can constitute the majority of sediment mobilized on burned slopes (McGuire et al., 2016).

Once particles are detached by rainsplash or overland flow, they can be transported downslope by *sheetflow*, which is distributed relatively uniformly across the ground surface (e.g., Meyer et al., 2001), or by flow that is concentrated into specific paths that develop into rills (Wagenbrenner et al., 2010).

Figure 3.9 Examples of rills and gullies formed in recently burned areas in Colorado, USA. Arrows indicate flow direction. The rill in the upper left photo is about 50 cm across and could be considered a very small, shallow debris flow based on the presence of levees; the rill in the right photo is about 10 cm across.

Bioturbation can alter sheetflow erosion both by supplying subsurface sediment and by providing sinks for erosive overland flow (Blake et al., 2009).

Rill formation has been described for both water-repellent burned soil (DeBano, 2000) and wettable post-fire soils (Shakesby & Doerr, 2006) (Figure 3.9). The spatial extent and temporal duration of rills can be constrained by bioturbation, the permeability of underlying bedrock, rapid vegetation regrowth, exhaustion of sediment supply, and increased infiltration capacity over time since the fire (Wondzell & King, 2003; Shakesby & Doerr, 2006), but rill formation can be the predominant form of hillslope erosion following wildfire (e.g., Robichaud et al., 2010; Alessio et al., 2021).

Increased sediment yield in many burned areas originates primarily from uplands rather than from increased erosion of river beds and banks (e.g., Blake et al., 2009; Wilkinson et al., 2009; Smith et al., 2011; DeLong et al., 2018; Rengers et al., 2021) (Text Box 3.3). The efficiency of hillslope-to-river delivery ratios can increase with the proportion of the catchment that burns at high severity (Wilson et al., 2021), and channels—especially headwater tributaries—can contribute to increased sediment yields (e.g., Santi et al., 2008; Moody & Martin, 2009; Nyman et al., 2020) (Figure 3.10). The rate of increase in sediment yields can also correlate

Text Box 3.3 How Do We Know? Sediment Tracing

Sediment tracing can be used to understand where in the catchment the fine sediment (typically, <63 microns) entering a river following a wildfire originates. Sediment tracing methodologies may incorporate fallout radionuclides (^{137}Cs, excess ^{210}Pb, 7Be, $^{239,240}Pu$), soil geochemical properties, mineral magnetics, color (Kieta et al., 2023), and organic compounds. After selecting a set of tracer properties that effectively discriminate target sources, multivariate mixing models based on a series of linear equations can be employed to determine the proportional contributions from each source (Wilkinson et al., 2009; Smith et al., 2013). Alternatively, repeat terrestrial and airborne lidar datasets can be analyzed using Digital Elevation Model (DEM) differencing to identify areas of erosion and deposition, either pre- and post-fire, provided pre-fire data are available, or over time following the fire (DeLong et al., 2018; Rengers et al., 2021). Additionally, landscape evolution models can be used to estimate the magnitude, spatial distribution, and duration of post-fire increases in sediment yield (e.g., Follmi et al., 2022).

Figure 3.10 An example of a small tributary channel contributing substantial sediment to a larger channel during post-fire erosion. This channel in the Colorado Rocky Mountains, USA, stored substantial sediment before the fire. After the fire, the channel incised more than 1 m vertically in places (to weathered bedrock) and widened more than 1 m on each side via bank erosion. The mainstem is downslope (behind the photographer) in this view.

with drainage area and average terrain slope (Pelletier & Orem, 2014). Upland sediment yields that change over time post-fire are correlated with maximum rainfall intensity thresholds for generating runoff or sediment yield, which also vary with time post-fire (Wilson et al., 2018). However, areas that burn at mild to moderate severity and/or that do not experience intense rainfall may not exhibit post-fire increases in either runoff or sediment yield (e.g., Prosser & Williams, 1998).

Post-fire sediment transport can also occur in episodic mass movements, including debris flows, landslides, and soil creep. *Soil creep* is a slow, seasonal downslope movement of sediment (Pawlik and Šamonil, 2018) that can be enhanced post-fire by increased soil wetness following removal of litter and vegetation (Helvey et al., 1985; Shakesby & Doerr, 2006).

Debris flows occur when masses of poorly sorted, saturated sediment surge downslope, with both solids and fluids transferring momentum to sustain motion (Costa, 1984; Iverson et al., 1997). Debris flows typically have 50–77% sediment concentration by volume, but may be as low as 40% (McGuire et al., 2017), creating a viscous slurry capable of transporting huge boulders (Cannon & DeGraff, 2009; Moody et al., 2013) (Figure 3.11).

Landslides are mass movements in which motion is more rigid, with deformation localized along persistent slip surfaces or shear zones, although the distinction between debris flows and landslides is gradational (Iverson et al., 1997). Post-fire reduction in root cohesion can particularly contribute to landslides (Goode et al., 2012). Root strength can reach a minimum 8–12 years after tree death in fire, as the old roots decay and new roots start to grow; this can lead to an increase in mass movements (Jackson & Roering, 2009). In southern California, USA, debris flows occurred during the first year following wildfire, but shallow landsliding replaced debris flows in subsequent years (Rengers et al., 2020; Thomas et al., 2021). Multiple landslides occurring three years after the fire mobilized a much smaller total volume of sediment, however, than the debris flows during the first year after the fire (Rengers et al., 2020).

Post-fire landslides have also been attributed to increases in soil moisture (e.g., Cannon et al., 1998), as well as to increased peak flows following wildfires that facilitate accelerated bank erosion and increased bank-side failures (Wondzell & King, 2003). Santi and Rengers (2020) noted that few well-controlled data exist to evaluate the relative importance of these potential post-fire landslide triggers.

Both debris flows and landslides occur in steep terrain that has not burned, but wildfire can greatly lower the threshold conditions for mass movements. In some burned areas, even brief rainfall of small magnitude and high return interval can trigger a landslide or debris flow (Cannon et al., 2008), but local peaks in short-duration (≤30 minute) rainfall

Figure 3.11 (a) A debris flow sourced from a small channel tributary to Little Beaver Creek, Colorado, USA. The view is across the debris-flow fan toward the tributary. The orange field notebook is 19 × 12 cm in size. (b) Small debris flow triggered by a January 2021 atmospheric river storm in the burn zone from the 2020 CZU Lightning Fire Complex, Santa Cruz Mountains, California, USA. Photograph courtesy of Amy E. East. (c) Damage caused by the 9 January 2018 Montecito debris flow, which was triggered by rainfall 3 weeks after the Thomas Fire, both in California, USA. Photo ID626, courtesy of Kean et al., 2019.

intensity are most commonly associated with debris-flow occurrence (Kean et al., 2011). Historical distributions of wildfire and precipitation frequency across southern California, USA, reveal that at least small debris flows occur almost every year in burned areas, with major debris flows occurring every 10–13 years, largely in response to storms that exceed a threshold of precipitation intensity (Kean & Staley, 2021). Sensitivity analysis of possible future changes suggests that debris-flow occurrence in the region is more sensitive to increases in precipitation intensity than to increases in fire frequency and severity (Kean & Staley, 2021).

Debris flows in burned landscapes can initiate through runoff-driven flow concentration and progressive sediment bulking, wherein clear water flow transitions to hyperconcentrated flow and subsequently to a debris flow (Meyer et al., 1995; Cannon, 2001; Cannon et al., 2010), typically as it flows down a channel (e.g., Meyer et al., 2001; Santi et al., 2008).

Alternatively, debris flows may occur primarily within the headwater portion of a river network, transitioning to hyperconcentrated flows downstream as channel gradient decreases and discontinuities, such as channel confluences, create depositional areas (Nyman et al., 2020). Mass movements can only occur where sufficient sediment is available and dry ravel following fire can be important in concentrating sediment within rills and steep, ephemeral channels that can subsequently give rise to debris flows during rainfall (DiBiase & Lamb, 2020). Maintaining a sediment concentration greater than 30% can be difficult without an additional perturbation, however, so processes like bulking and channel-bed failure may be necessary to create and maintain the high sediment concentrations that sustain debris flows (McGuire et al., 2017).

Thin debris flows (Gabet, 2003) can form when saturated soil above the water-repellent layer flows downslope. Entrainment of ash can be important during this process as a means of providing sufficient fine-grained material (Gabet & Sternberg, 2008). Alternatively, debris flows can result from saturation-initiated failure of a landslide that transitions downslope into a debris flow (Cannon & Gartner, 2005). Debris flows in burned areas can also initiate where runoff cascading over exposed bedrock mobilizes colluvium or channel beds downslope, a process that has been called the "firehose effect" (Johnson, 1984; Larsen, 2003).

The primary source of debris-flow sediment can also change with time after a fire. Following a 2018 wildfire in the Santa Ana Mountains of California, USA, evacuation of dry ravel and pre-fire colluvium produced half of the sediment transported during the initial runoff-generated debris flows, which also removed much of the sediment stored in headwater river corridors (Guilinger et al., 2020). Subsequent storms of equal or greater rainfall intensity mobilized more sediment via rill and interrill erosion on upland hillslopes, although the volume of sediment mobilized was much less than during the initial storms. Guilinger et al. (2020) interpreted these patterns as indicative of a more rapid depletion of sediment sources in river corridors than on hillslopes, resulting in decreasing sediment fluxes over time from burned headwater catchments subjected to repeated runoff.

Surface vegetation recovery appears to be a primary influence on the duration over which debris flows occur following a fire. Utilizing data from 25 fires in the western United States, Graber et al. (2023) found that more than 98% of post-fire debris flows occur when satellite-derived leaf area index in the burn area is less than two-thirds of pre-fire values. The time required to reach this level of vegetation recovery varies among sites, from less than 1 year to 4 years after the fire.

Much of the literature on post-fire debris flows comes from field investigations in the western United States, although debris flows have also been documented in other fire-prone regions, including Australia (Nyman et al., 2020), China (Jin et al., 2022), Greece (Diakakis et al., 2023), and Italy (Esposito et al., 2017, 2023). Field investigations of post-fire debris flows in the western United States are commonly undertaken as part of hazard assessment and mitigation efforts. Early warning systems for post-fire debris flows during intense rainfall have traditionally relied on regionally specific, empirical rainfall intensity-duration thresholds that require years to develop (e.g., Gartner et al., 2014). This approach is feasible in part because regional differences in the rainfall intensity threshold for triggering hillslope instability are well constrained within the western United States. Since 2017, however, the U.S. Geological Survey has employed a single method for the entire western region of the country (Staley et al., 2017).

Recent research on post-fire debris flows has focused on logistic regression models that can predict precipitation thresholds for areas lacking data (e.g., Staley et al., 2017); analysis and comparison of debris-flow runout models (e.g., Barnhart et al., 2021); reduced-complexity models of debris-flow inundation (Gorr et al., 2022); and simulations examining how rainfall intensity-duration thresholds for triggering debris flows increase over time since burning as a result of ongoing changes in soil infiltration capacity, canopy interception, hydraulic roughness, and the grain size of sediment entrained in an incipient debris flow (Rengers et al., 2019; Tang et al., 2019; Hoch et al., 2021).

In summary, post-wildfire sediment fluxes reflect precipitation, infiltration, runoff, and sediment availability on upland slopes and in small, ephemeral channels, as well as sediment erosion and transport. Each of these factors is linked by states (e.g., soil moisture, surface roughness) and by feedback mechanisms (Moody et al., 2013). These states and feedbacks, which are strongly influenced by fire severity, change through time after a fire, primarily as a function of vegetation regrowth and depletion of available sediment. Precipitation and gravity provide the forces that mobilize upland sediment, but vegetation, and vegetation regrowth after a fire, control sediment mobility by mediating the interaction between these forces and the supply of sediment available for movement.

3.2.5. Particulate Organic Matter

Particulate organic matter (POM) entering river corridors from uplands after fire can increase, but the proportion of burned (ash, charcoal, charred) versus unburned material depends on the spatial extent and severity of the fire within the uplands contributing to a specific river reach or to the entire catchment (e.g., Wagner et al., 2015).

Figure 3.12 Example of particulate organic matter mobilized from burned area 1 year after fire in the Little Beaver Creek catchment, Colorado, USA. These lines of POM outline the extent of weakly channelized overland flow occurring after an intense summer convective storm. The arrow indicates flow direction.

Very few studies have examined POM export from uplands. Instead, the focus has been on the effects of changing POM abundance in rivers. However, Pinto et al. (2008) noted that many partially burned leaves and conifer needles will fall to the forest floor or remain weakly attached to tree branches following fires of lower severity. This unburned or lightly charred POM can be rapidly transported downslope during periods of overland runoff (Figure 3.12). This creates POM fluxes with different nutrient compositions, as most or all of the volatile components were consumed by fire. Post-fire POM transport also reflects differences in input timing, as inputs shift from the normal seasonal leaf-fall to the period immediately post-fire (Pinto et al., 2008). In fires of higher severity, POM fluxes from uplands decrease until the vegetation regrows.

3.2.6. Coarse Woody Debris

Wood is at the center of multiple fluxes following wildfire: the flux of living trees killed or injured by the fire that become snags (standing dead wood) or coarse woody debris (downed, dead wood outside of river corridors) (Peterson et al., 2023); the flux of coarse woody debris downslope

(Rosso et al., 2007); the ability of coarse woody debris to trap and store sediment (Adams et al., 2023); the incorporation of woody debris into sediment debris flows and the influence of wood on the debris-flow properties (Lancaster et al., 2003; Rengers et al., 2023); and the flux of carbon and nutrients from coarse woody debris into the soil as the wood decays (Marañón-Jiménez & Castro, 2013) and into living organisms such as microbes, fungi, and invertebrates that promote wood decomposition (Li et al., 2007). This section reviews what is known about variations in coarse woody debris, or CWD, abundance over time following fire and the ways in which CWD influences burned ecosystems.

CWD is commonly defined as including dead, woody material in all stages of decay with a diameter >7.5 cm. Some investigators include both snags and fallen wood (e.g., Zhou et al., 2007; Acker et al., 2013), whereas others restrict CWD to wood resting on the ground surface or partly buried (e.g., Hyde et al., 2011) or refer to fallen dead wood as downed CWD (e.g., Tinker & Knight, 2000). Here, I use CWD to refer to fallen dead wood, as distinct from snags (Figure 3.13).

CWD is always present in natural (as opposed to managed) forests. Very decayed CWD is more likely to be consumed during a fire than sound

Figure 3.13 Example of coarse woody debris (most or all of these pieces were on the ground prior to the fire) about 2 weeks after the fire, Little Beaver Creek catchment, Colorado, USA.

CWD if other conditions, such as fire severity, are similar (Hyde et al., 2012), but studies in lodgepole pine (*Pinus contorta*) forests in Wyoming, USA, indicate that even crown fires and severe surface fires may consume or convert to charcoal < 20% of the CWD present before the fire (Tinker & Knight, 2000). Studies in other forest types in the western United States indicate much greater consumption of CWD in even low-severity fires (e.g., Keyser et al., 2008; Acker et al., 2017). The proportion of CWD consumed during wildfire reflects site-specific characteristics of CWD such as decay state and moisture content, as well as fire severity and local topography.

Trees newly killed or damaged by wildfire can add to the supply of CWD, although there may be a substantial delay of 8–10 years or more (Passovoy & Fulé, 2006; Acker et al., 2013) following fire before a snag topples. Snag persistence typically increases with tree size. Some tree species are also more likely to remain standing as snags (Brown et al., 2013).

As snags fall or damaged trees lose branches during the first 20 years or so following the fire, CWD mass in diverse types of conifer forests in western North America increases relative to unburned stands of similar age (Tinker & Knight, 2000; Roccaforte et al., 2012; Stevens-Rumann et al., 2015). The greatest increases occur where pre-fire stand density was high (Schoennagel et al., 2008). Fires that occur in rapid succession, however, whether stand-replacing fires (Spies et al., 1988; Donato et al., 2016) or low-intensity prescribed fires (Aponte et al., 2014), can decrease the abundance of CWD because after the first fire, new trees are not killed, but CWD continues to be consumed by fire.

From a maximum value within a few years after wildfire (e.g., Roccaforte et al., 2012), CWD declines to a minimum at approximately 100–200 years after a stand-replacing fire, during the period when trees are regrowing. CWD abundance then starts to increase again as large trees die (Spies & Cline, 1988; Spies et al., 1988) (Figure 3.14). Fire in some Australian *Eucalyptus* forests, however, appears to create no significant changes in CWD abundance (Fairman et al., 2017). Topography can also influence fire severity, tree mortality, and CWD abundance, creating substantial variations in CWD abundance post-fire over relatively small distances (Bassett et al., 2015).

Although combustion can release carbon stored within CWD in gaseous form, CWD that is not completely consumed influences post-fire carbon and nutrient cycling. CWD can be up to 60% of a forest carbon budget but is more likely to compose 5–10% across diverse forested environments (Hyde et al., 2011). Where post-fire nutrient levels in soil limit vegetation regrowth, the abundance of relatively nutrient-rich CWD—which Marañón-Jiménez et al. (2013) found could contain

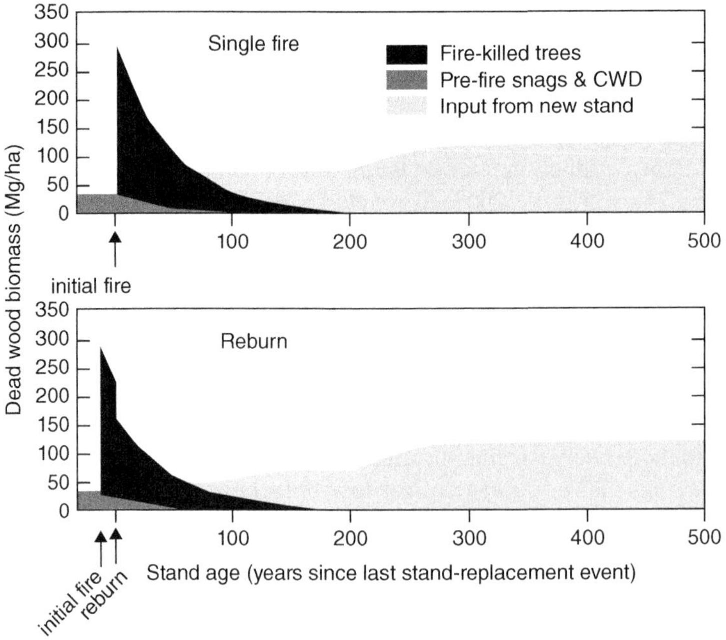

Figure 3.14 Projections of dead wood biomass with time for a temperate conifer forest in Oregon, USA. (Donato et al., 2016 / John Wiley & Sons / CC BY 4.0.)

2–9 times more nutrients than the soil in a pine forest in southeastern Spain—is an important source of N, as well as the micronutrients Na, Mn, Fe, Zn, and Cu as the wood decomposes. Four years after a fire in Spain, nutrients released by CWD accounted for approximately 40% of the soil N content, 65% of P, 18% greater soil organic matter, 42% greater total C, and 36% greater microbial biomass (Marañón-Jiménez & Castro, 2013). CWD is also important to the ectomycorrhizal fungi that assist trees in obtaining nutrients from soil (Graham et al., 1994). Decayed CWD can retain water that is important to fungi and plants during dry periods (Harmon et al., 1986).

CWD decomposition, during which the complex organic structure of wood is reduced to its mineral form, occurs via biological respiration, leaching, and fragmentation (Zhou et al., 2007). Biological respiration represents the invasion of CWD by diverse species of microbes and invertebrates and the respiration by these organisms, which can release up to half of the stored carbon in gaseous form. Leaching occurs as water transports nutrients and solutes from CWD into the soil. Fragmentation describes the reduction of CWD volume during decay via physical forces such as breakage or freezing and thawing, as well as biological processes such as wood-boring by insects, pecking by birds, or demolition by bears seeking insects within the decaying wood (Harmon et al., 1986). CWD exposed at the surface experiences fluctuations in temperature and moisture, causing the wood to repeatedly shrink and expand in a manner that detaches wood fragments via cracks. Organisms such as fungi and termites can then colonize the newly exposed surfaces along the cracks.

CWD also increases the structural diversity of the forest (Brown et al., 2013), benefiting both plants and animals. CWD provides vital habitat for a wide range of microbes, fungi, and invertebrates (Harmon et al., 1986). The decaying wood provides seed germination sites. Early successional tree species such as *Salix* and *Populus* spp. are favored by browsing animals such as elk (*Cervus canadensis*), which can limit the regeneration of these species. Accumulations of CWD can protect seedlings from browsers (Ripple & Larsen, 2001; Chantal & Granström, 2007). Voles and other small mammals use CWD for cover from predators while traveling, as well as for den and forage sites (Maser et al., 1979; Bull, 2002; Ucitel et al., 2003). CWD can also create a physical barrier that limits soil erosion post-fire (Hyde et al., 2011; Adams et al., 2023).

Schoennagel et al. (2008) emphasized the interactions between CWD and the heterogeneity present after the large, severe wildfires of 1988 in the northern Rocky Mountains of Wyoming, USA. Post-fire heterogeneities in stand age, CWD abundance, and microbial and understory communities reflect interactions between pre-fire conditions and patterns of fire severity. These heterogeneities may enhance resilience to large, severe disturbances by providing variability in species, forest structure, and ecological functions that can increase redundancy. (In ecology, functional redundancy implies that species loss is compensated by other species contributing similarly to ecosystem functioning (Fetzer et al., 2015).)

In summary, the abundance of CWD can be increased or reduced by wildfire and typically varies over time after a fire. Where it is present, CWD influences sediment movement downslope, soil chemistry and moisture, and habitat for diverse organisms, including microbes, fungi, plants, and various types of animals. Interactions between CWD and post-fire processes can also enhance resilience to future fires by creating spatial heterogeneity and functional redundancy in forested ecosystems.

3.2.7. Summary and Synthesis

Increased post-fire fluxes from uplands into river networks tend to be relatively transient (months to years), episodic, variable in space and time, and nonlinear (Moody et al., 2013). The diverse processes involved in these fluxes are measured using different methods that cover various spatial scales, timespans, and spatial and temporal resolutions. Consequently, generalizing responses across study areas or predicting responses within a study area remains challenging.

Moody et al. (2013) summarized factors that limit continued progress in understanding and predicting post-fire fluxes: (i) responses are transient and influenced by variables such as burn severity, speed of vegetation recovery, post-wildfire precipitation patterns, sediment availability, and catchment morphology; (ii) high-magnitude responses tend to be episodic, brief, and destructive, making it difficult to collect field data; (iii) responses are nonlinear and controlled by multiple thresholds, as well as being highly dependent on the sequence and characteristics of precipitation inputs; (iv) completely burned catchments are commonly limited to less than a few tens of square kilometers in size and nested within larger catchments with spatially heterogeneous burn severity; and (v) a lack of consistent methods and scales of measurement across studies limits quantitative comparisons. The one generalization that can be made is that severe wildfires, especially in steep terrain, increase the fluxes of all the materials considered here.

3.3. Terrestrial Forms

Fluxes of material moving downslope following wildfires may be sufficient to change upland topography over various timespans. When averaged over centuries to millennia, for example, the majority of hillslope lowering may occur immediately following wildfires (Santi et al., 2013). Three examples of post-fire topographic changes include colluvial hollows, talus cones, and channel heads. (Alluvial fans are commonly strongly influenced by post-fire processes and provide a stratigraphic record of past fires, but they will be addressed in the next chapter.)

Colluvial hollows are concave or convergent areas within uplands (Figure 3.15). These hollows tend to accumulate subsurface water, particularly when associated with more densely fractured bedrock or sediment accumulations (Gabrielli et al., 2012). Sediment gradually accumulates in hollows through processes such as dry ravel (Jackson & Roering, 2009), creep (Hales et al., 2012), and slopewash (Field & Banning, 1998). During post-fire precipitation, hollows can be the initiation points for landslides or debris flows that evacuate stored sediment, thereby accentuating the concavity of the slope at that location (Hack & Goodlett, 1960; Dietrich & Dunne, 1978; Benda & Dunne, 1997). Sediment mobilized from hollows may be redeposited downslope in regions of lower hillslope gradient, forming alluvial fans, or may move into and through river networks.

(a) (b)

(c) (d)

Figure 3.15 Examples of colluvial hollows. (a) Showing more rapid vegetation growth in spring the year after a wildfire than the steeper slopes at the rear of the view. The hillslope steepens abruptly and drops to the active channel just beyond the foreground (out of view) in this photograph. Little Beaver Creek catchment, Colorado, USA. (b) Sitkum River catchment, Washington, USA. (c) Sitkum River catchment. (d) Mt. Washington area, Washington, USA. Because the Washington images are in unburned and fully forested hillslopes, it is difficult to see the colluvial hollows, but the steep angles of these slopes represent settings in which colluvial hollows can give rise to debris flows. Courtesy of Daniel N. Scott.

Talus cones are accumulations of colluvial material resulting from rockfall onto low-gradient surfaces, such as valley floors. Wildfires can expedite the development of talus cones through heat-induced spalling and diminished hillslope stability (Santi et al., 2013).

A *channel head* is defined as the upstream boundary where concentrated water flow and sediment transport occur between definable banks (Montgomery & Dietrich, 1988). This point marks the transition from diffuse, downslope movement of water and sediment to channelized surface flow, delineating the start of river networks (Figure 3.16).

Figure 3.16 Examples of post-fire channel heads. These are small, subtle features. Dashed outlines indicate upper and lower edges of banks. Orange hand tape is about 3 cm × 3 cm in size. Arrows indicate flow direction.

Channel heads reflect the minimum drainage area necessary to concentrate downslope flow sufficiently to erode a channel (Montgomery & Dietrich, 1992). This minimum area is influenced by characteristics of precipitation, infiltration, land cover, rainfall-runoff, and subsurface flow. Wildfire can at least temporarily alter many of these characteristics and can cause upslope migration of channel heads during post-fire runoff. Following a wildfire in Colorado, USA, the mean drainage area above channel heads decreased by more than two orders of magnitude as compared to unburned sites (Wohl, 2013a) but then gradually migrated downslope as vegetation regrew. Within four years, the drainage area at channel heads had returned to values comparable to those found at unburned sites in the region (Wohl & Scott, 2017).

3.4. Terrestrial Biota

For living organisms, wildfire can create the best of times or the worst of times (Figure 3.17). Some forms of life are displaced by fire or killed during the fire, and the post-fire environment is so hostile to their survival that members of the species may not be able to inhabit the burned area for decades. Other life forms thrive on the conditions created by wildfire or even require wildfire for reproduction. Pyrodiversity—fire-derived habitat heterogeneity (He et al., 2019)—creates temporal and spatial niche diversity for a wide array of organisms, and both plant and animal biodiversity, as well as soil microbial biodiversity, commonly increase with pyrodiversity (e.g., Dawe et al., 2022; Jones & Tingley, 2022; Porter et al., 2023). However, the specialized strategies that organisms have evolved to resist, promote, or

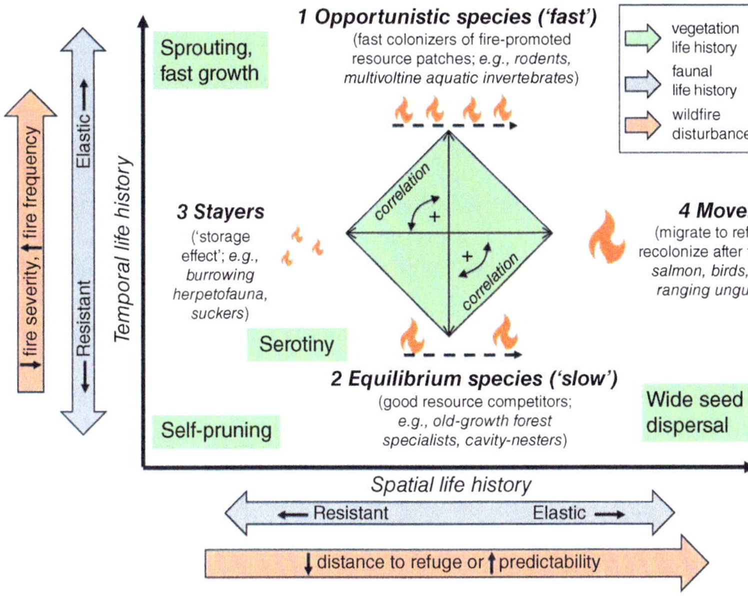

Figure 3.17 Species life history traits in time (y-axis) and space (x-axis) reflect the wildfire regimes to which they are adapted. Opportunistic species show rapid responses to habitat disturbance and dominate habitats with relatively frequent and unpredictable disturbances. Equilibrium-type species thrive in stable habitats experiencing less exposure to abiotic stress (e.g., fire refugia). *Stayers* persist through disturbance using resistant strategies. *Movers* are populations that recover through high mobility. *Serotiny* refers to the release of seeds after exposure to fire. (Jager et al., 2021 / John Wiley & Sons / CC BY 4.0.)

recover from fire may prove detrimental if the fire regime changes (Bowman et al., 2020).

Overall, wildfire is a major disturbance that influences the structure of most deciduous, coniferous, and mixed forest types at temperate and boreal latitudes (Wright & Bailey, 1982; Attiwill, 1994; Angelstam, 1998). Fire is also one of the key factors in maintaining grasslands and savannas at temperate and tropical latitudes, as well as Mediterranean-type shrublands (Moreno & Oechel, 1994; McPherson, 1997; Ruan et al., 2020). Forests may exhibit greater spatial heterogeneity of soil moisture and burn severity, whereas grasslands can display highly uniform patterns of soil moisture, vegetation, and burn severity (Hamilton & Burton, 2023).

Reviewing studies primarily from western North America, Swanson et al. (2011) emphasized the diversity of species, processes, and structure of early successional forest ecosystems that develop after stand-replacing fire.

Early succession is the only period when tree canopies do not dominate the forest site, which facilitates high productivity of herbs and shrubs, complex food webs, large nutrient fluxes, and high structural and spatial complexity. Species diversity is enhanced by the presence of survivors (organisms that survive severe disturbances), opportunists (species that can quickly exploit new resources as they arise), and habitat specialists (species that directly utilize the legacy of recently killed trees) that require the distinctive conditions present after fire. The presence of living trees, snags, and coarse woody debris creates diverse habitats for surviving and colonizing organisms (Swanson et al., 2011).

In the context of fire ecology, ecosystems can be classified as fire-dependent, fire-sensitive, or fire-independent (Shlisky et al., 2007). Fire-dependent ecosystems are those in which the majority of species have evolved in the presence of fire. Fire is essential for conserving biodiversity, and alterations in the fire regime can substantially affect these ecosystems, which include savannas and some temperate coniferous forests (Shlisky et al., 2009). Fire-sensitive ecosystems are those in which most of the species have not evolved in the presence of fire. Although wildfires may play a lesser role in maintaining ecosystem structure and function, an increase in fire activity may have negative implications for biodiversity in these ecosystems, which include tropical moist broadleaf forests. Fire-independent ecosystems, such as deserts and tundras, naturally inherently lack sufficient fuel or ignition sources to sustain fire as an evolutionary driver (Shlisky et al., 2009). Similar to fire-sensitive ecosystems, an increase in fire can impact these ecosystems by altering community composition (e.g., Steers & Allen, 2011).

In any ecosystem, smoke generated during a fire creates hazardous conditions for numerous animal species. In a review of 41 published studies, Sanderfoot et al. (2021) found that smoke inhalation can lead to both acute and chronic health problems, including carbon monoxide poisoning, respiratory distress, neurological impairment, respiratory and cardiovascular diseases, oxidative stress, and immunosuppression in aquatic and terrestrial species. Some species utilize smoke as a cue to avoid fire or to conserve energy, but Sanderfoot et al. (2021) noted adverse health effects for birds, reptiles, mammals (including dolphins in coastal areas near wildfires), and certain insects. In this context, it is important to understand that the chemical composition of fine particulate matter ($PM_{2.5}$, defined as particles smaller than 2.5 μm in aerodynamic diameter) in wildfire smoke is more toxic than that of urban ambient $PM_{2.5}$ (Sanderfoot et al., 2021).

Whelan et al. (2002) and Engstrom (2010) distinguished three levels at which fire affects animals: (i) death, injury, or encouraged migration into or out of the fire area, typically occurring over a period of days or weeks;

(ii) post-fire processes such as starvation, predation, or immigration, which can determine population viability over timespans based on fire characteristics and return interval; and (iii) the ways in which fire regimes can modify a species, fostering adaptations to fire over evolutionary timespans.

Most wildfires have unburned or minimally burned patches that provide refugia from the effects of fire for various organisms. The size, distribution, and connectivity between these patches can exert an important influence on biotic responses after fire. Examining burned areas in Newfoundland, Canada, for example, Mallik (2022) described how unburned patches of black spruce (*Picea mariana*) forest provide sites for spruce regeneration, whereas Kalmia (*Kalmia angustifolia*) heath that develops in non-severe burn areas inhibits spruce regeneration and fosters a shrub savanna community as an alternate vegetation state.

The following sections review existing knowledge regarding the effects of fire on different groups of organisms that live in terrestrial environments affected by wildfire, starting with soil organisms and proceeding through fungi, plants, insects and other pollinators, reptiles and amphibians, birds, and mammals. Lake ecosystems are addressed under a separate subheading due to their distinctive biotic community, which differs from that of the river network within a catchment.

3.4.1. Soil Organisms

Soils constitute the heart of Earth's critical zone, which extends from the vegetation canopy to groundwater, and is critical to living organisms. Soils support crucial processes, including water and biogeochemical cycling, as well as hosting reservoirs of water and nutrients that nourish both belowground and aboveground organisms. Soil ecosystems encompass a range of organisms, from the invisible—bacteria, nematodes, fungi, cyanobacteria, and others that are critical to nutrient cycling—to the visible—earthworms, ants, termites, rodents, and larger burrowing animals, all of which influence the decomposition of plant litter and soil turnover and structure.

Fire duration is the critical parameter for soil communities. Due to the insulating properties of soil, intense but short-duration fires do not transfer as much heat to the soil as do slow-moving fires (Neary et al., 1999). Comparisons of pre- and post-fire soil bacterial assemblages in native California sage scrub and grassland habitats, for example, suggest minimal effects from fast-moving fires (Cox et al., 2022), although a different study in the same environment reports reduced soil bacterial and fungal abundance, along with altered bacterial and fungal composition following fire (Glassman et al., 2023). Research conducted in the boreal forest and taiga plains ecoregions of northern Canada, five years post-fire, also indicates resilience within soil bacterial communities across a spectrum of fire severity (Whitman et al., 2022).

Burning fuels transfer heat to soil organic and mineral layers via radiation, convection, conduction, mass transport, and vaporization and condensation (Chandler et al., 1983; Campbell et al., 1995). Vaporization and condensation, which involve the phase changes of water and organic compounds distilled by combustion, are particularly important in rapidly transferring heat into soils (Neary et al., 1999). When water vaporizes due to the heat from a fire, the vapor moves rapidly through soil pores, releasing heat as it condenses and thereby raising soil temperature. Combustion and soil conditions vary enormously, but surface organic layers and dry soil conditions generally retard soil heating, whereas bare mineral soil surfaces that are moist can heat rapidly, but only up to 100 °C until all water has evaporated (Neary et al., 1999; Santín & Doerr, 2016). Figure 3.18 illustrates the ranges of surface and subsurface temperatures created

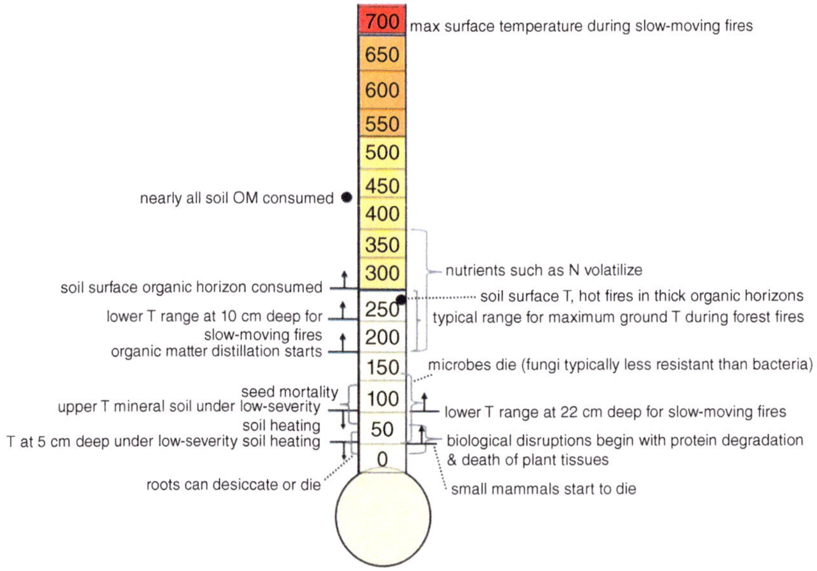

Figure 3.18 Schematic illustration of changes in soil communities in relation to temperatures (°C) reached in the soil during a wildfire, although temperatures are rarely elevated above 100 °C beyond depths of 10 cm. (Massman et al., 2002; Rengers et al., 2017; Neary et al., 1999 / with permission of ELSEVIER.)

during fire and the associated effects on soil organisms and physico-chemical processes.

Soil temperatures can remain elevated for an hour or less following low-intensity fires, but elevated temperatures can persist for more than five days after severe fires (Neary et al., 1999). Reviewing multiple studies across diverse environments, Neary et al. (1999) concluded that temperatures in surface mineral soils can remain elevated above normal levels for years where fire consumes the entire organic horizon and shading from the overstory, allowing direct solar radiation to heat exposed soil. These persistently elevated soil temperatures and associated lower water contents can reduce soil microbial activity for years following the fire in a hot, dry climate such as central Texas, USA (Cooperdock et al., 2020).

Post-fire changes in soil temperature, moisture, structure, and nutrient availability can kill microbes or alter their reproductive capabilities. These effects are greatest in soil organic horizons and the upper 1–2 cm of soil, where microbes are most abundant and where fire-induced soil heating is greatest (Neary et al., 1999). Changes in the microbial community then alter soil nutrient cycling. Heterotrophic microbes that utilize organic carbon as food usually decline in abundance, causing a loss of C, whereas autotrophic microbes that use sunlight for photosynthesis increase, accelerating the loss of N (Neary et al., 1999). Mycorrhizal soil fungi that form a symbiotic association with a plant can either increase in abundance if the fire stimulates subsequent plant growth or decrease if there is widespread plant death (Neary et al., 1999).

Soil invertebrates that live in litter or soil surface horizons and are not highly mobile are most susceptible to the direct effects of fire (DeBano et al., 1998). Invertebrates in fire-adapted ecosystems are more likely to be highly mobile and have adaptations for conserving water and resisting heat (Neary et al., 1999). Soil vertebrates may fare better during fire because they are sufficiently mobile to escape by burrowing deeper or fleeing on the surface, although post-fire loss of habitat, exposure of burrow openings, and increased predation can continue for several years (Ream, 1981).

Biocrusts form a distinctive subset of soil organisms. Biocrusts consist of communities of microorganisms, lichens, and bryophytes bound to the soil surface. This means they are directly exposed to fires. Studies of the biocrust community in coastal California, USA grasslands found that biocrusts can require 6 years to return to pre-fire characteristics of community composition and function (Palmer et al., 2023).

In summary, fire can affect soil organisms in diverse ways. Most of the effects of fire on soil communities are greatest in the organic horizon and upper few centimeters of soil, and where fire-induced soil heating is greatest and/or where fire-induced changes in vegetation continue to affect solar radiation reaching the soil after the fire. Fire-induced alterations of soil communities can have secondary effects ranging from the physical characteristics of soil (infiltration capacity, cohesion) and its ability to resist erosion, to biogeochemical reactions and nutrient availability for other organisms, which in turn affects processes such as rate of plant regrowth or community composition of vegetation.

3.4.2. Fungi

Biologists classify fungi as their own kingdom, separate from plants, animals, and other organisms. Fungi are eukaryotic (their cells have a nucleus) organisms that range from yeasts and molds to mushrooms. Fungi, like animals, are heterotrophs that acquire food by absorbing dissolved molecules, usually by secreting digestive enzymes into the environment. They are critical both for decomposing dead plants and for making nutrients more accessible to living plants and other organisms. If plants are the "gatekeepers" between precipitation and downslope movement of sediment, fungi are the gatekeepers between plants and any nutrients the plants cannot obtain through photosynthesis. In this role, the ability of fungi to withstand or recover after wildfire exerts a key control on post-fire vegetation regrowth and thus on most of the processes occurring in terrestrial environments.

The effects of fire on fungi vary as a function of: (i) fire characteristics such as timing and intensity; (ii) modifications of the substrate (soil, living plants, coarse woody debris), habitat, and nutrients for fungi (Schwilk et al., 2009; Wang et al., 2012); and (iii) life history traits of the fungi. Thermotolerant fungi (Suryanayanan et al., 2011) include fire-adapted plant-fungal associations (Baynes et al., 2011). These fungi can produce heat-resistant compounds (Sharkey et al., 2001) and can fruit frequently and in great abundance after fire.

Pyrophilous or fire-loving fungi appear after fire. These can be (i) fire-resistant species that are heat resistant for some aspects of their life cycle; (ii) fire-responsive species that are triggered to grow and/or fruit by heat or by chemical changes caused by heat; or (iii) fire-adapted fungi with specific traits related to fire, which may require fire to complete their life cycle, although these three classifications overlap (Fox et al., 2022).

Some fungal communities are maintained by fire, whereas infrequent fires can significantly alter other fungal communities for decades (Dooley & Treseder, 2012). Biotrophic fungi, which feed on their plant host without killing the host, are affected directly by the fire and the responses of their host. Saprotrophic fungi that feed on dead organic matter can benefit from newly available, fire-generated substrates. Pathogens can benefit from hosts stressed by fire (Fox et al., 2022).

Fire can directly kill fungi, particularly saprotrophic fungi within burning plant litter. Although some fungi can survive temperatures greater than 50 °C and even up to 145 °C (Kipfer et al., 2010; Suryanayanan et al., 2011), 60 °C is typically considered lethal for most soil organisms (Neary et al., 1999). Mineral soil and coarse woody debris can provide a fire refugium for fungi (Meddens et al., 2018). Fire consumption of plants and CWD can reduce fungal habitat and substrate (Bowman et al., 2009). Plant death can also kill plant-associated fungi such as mycorrhizal species, even if the fungi are insulated from the heat of the fire (Dove & Hart, 2017). Frequent fires can drive fungal adaptations such as post-fire fruiting due to loss of substrate or host plants (Kuo et al., 2014), evolution of heat-resistant spores, and fire-induced germination and growth.

Indirect effects of fire include modification of soil properties such as infiltration, soil moisture, soil albedo and temperature, chemistry, and nutrient availability (Certini, 2005; Glassman et al., 2017; Tahovská et al., 2020), all of which can affect fungal physiology, growth, and community composition (Fox et al., 2022). Fire can also indirectly affect fungi by modifying fungal food web connections via alterations of microbial biomass and the composition of soil-dwelling invertebrates (Certini et al., 2021) that can reduce competition for surviving fungi or destroy key symbionts for mycorrhizal or saprotrophic fungi (Zak & Wicklow, 1980; Semenova-Nelsen et al., 2019; Yang et al., 2020).

Niches created by pyrodiversity can increase landscape-level fungal richness, even if fire reduces richness within smaller landscape patches (Fox et al., 2022). Rapid heat attenuation with depth during fire, for example, can promote fungal community heterogeneity with depth (Massman, 2012), especially within the so-called "Goldilocks zone" in which the effects of fire temperature and duration stimulate rather than kill fungi (Kipfer et al., 2010; Bruns et al., 2020).

In summary, fungi provide an example of the idea that the largely invisible (to humans) members of an ecosystem can be among the most important. Non-ecologists may be most likely to think of mushrooms when hearing the word "fungi," but the ongoing discoveries of the critical importance of the rhizosphere—the zone of chemical, biological, and physical influence generated by root growth and activity—and the "wood-wide web" (e.g., Rhodes, 2017; Angrish, 2022) to the existence of more visible, aboveground vegetation reflect the crucial role of fungi in ecosystems. Consequently, the response of fungal communities to wildfire critically influences vegetation regrowth and thus downslope fluxes of diverse materials.

3.4.3. Plants

Like some soil-dwelling organisms, plants have evolved various adaptations to survive fire. Resprouters have belowground energy storage organs that provide resources for rapid regrowth following fire (Clarke et al., 2013). Fire-stimulated flowering species produce seeds abundantly or, in some cases, solely in response to fire (Lamont & Downes, 2011). Some species produce seeds that are relatively tolerant of high temperatures (Tangney et al., 2019). Other species create persistent seed banks in which seeds can survive for longer than a year in soil or in closed woody cones or fruit in the canopy before resprouting. Some species produce seeds that require heat to break dormancy or smoke to promote germination (Pausas & Keeley, 2014). These adaptations allow the plant to either survive fire or release offspring that can take advantage of the post-fire environment (Fox et al., 2022).

Plant adaptations to fire, however, can depend on the seasonal cycles that plants follow, and the adaptations can be less effective if fire occurs outside of the normal season (Miller et al., 2019) (Figure 3.19). As described in the first chapter, the fire season is lengthening in many parts of the world, and this is likely to alter post-fire plant community compositions in the future. This shift in community composition to more fire-resilient communities has already been observed in some regions, including dry forests (Nemens et al., 2022) and sagebrush steppe (Davies et al., 2012) in the western United States.

A meta-analysis of the effects of fire on soil seed banks suggests that species richness tends to decline, whereas abundance is less likely to be altered (Shi et al., 2022). Wildfires with high soil burn severity are more likely to cause declines in soil seed-bank richness and abundance, but details differ between plant types and ecosystems. Fire effects are more likely to be negative in forests (especially tropical forests) and positive in Mediterranean ecosystems, and the richness and abundance of tree species typically decrease after fire in all environments.

In addition to the more frequent and severe disturbances associated with changing fire regimes, plants in fire-prone environments may be under stress from warmer and drier conditions that reduce plant growth. Reduced recruitment, growth, and survival, especially for fire-intolerant woody species and systems in which fuel load accumulates rapidly, may narrow the fire-interval window compatible with population persistence. This scenario was described by Enright et al. (2015) as "interval squeeze." They predicted that interval squeeze will increase woody plant extinction risk and change ecosystem structure, composition, and carbon storage.

Understory plant communities. Reviewing the effects of fire and other disturbances on western North American mixed-conifer forests, Abella

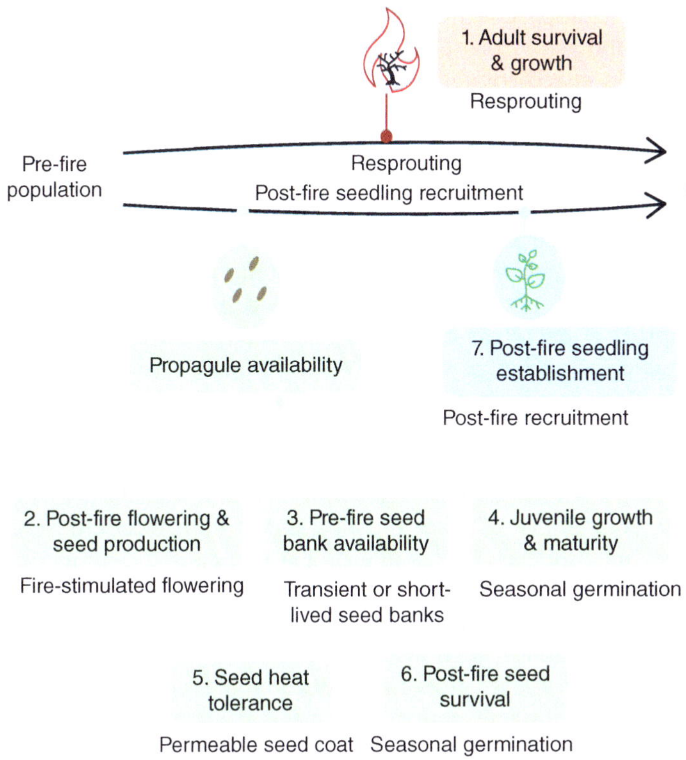

Figure 3.19 Conceptual diagram of seven key mechanisms driving fire seasonality effects on plant populations. Fire seasonality can alter population dynamics via impacts on adult survival, propagule availability, and post-fire seedling establishment. Resprouting and post-fire seedling recruitment are two major strategies for population persistence in fire-prone environments. Text under each of the seven mechanisms identifies plant or seed traits that make species vulnerable to decline. (Miller et al., 2019 / with permission of ELSEVIER.)

and Springer (2015) noted that understory plant abundance (cover, biomass, or density) commonly declines in the first four years after fire but then increases in subsequent years. The abundance of understory plants can be more sensitive to disturbance than overall species richness (Griffis et al., 2002), and a year after the fire, dominant plant cover type tends to be similar to the unburned vegetation—grasses in chaparral systems, forbs in mixed-conifer forests, and shrubs in boreal forests (Lentile et al., 2007). Fires benefit understory species that are absent or uncommon in undisturbed forest, and some groups of native species (e.g., *Epilobium* spp.) appear to depend on fire (Abella & Springer, 2015).

Removal of canopy cover also changes the microclimatic extremes experienced by understory plants and juvenile trees post-fire. Sites burned at high severity in mixed-conifer and subalpine forests in Montana, USA have significantly higher daily maximum temperatures and daily maximum vapor pressure deficit for at least 2 years post-fire (Wolf et al., 2021). In western North America, cool-wet forests are more vulnerable than warm-dry forests to fire-caused changes in microclimate (Wolf et al., 2021). Plant-water relations are particularly sensitive to extremes in aridity (Reyer et al., 2013), so fire-induced losses of forest buffering capacity increase the vulnerability of understory plants to climatic stressors and may impact patterns of plant regeneration by favoring species that are more drought-tolerant or able to regenerate through resprouting (Davis et al., 2018). Standing dead trees can help to moderate surface temperatures after a fire (Wolf et al., 2021).

Forest succession. Stand-killing or stand-replacing wildfires, as the names imply, kill all or most of the trees within a stand, which is defined as a contiguous community of trees of sufficiently uniform composition, structure, age, and size class to distinguish it from adjacent communities. Such fires are part of the natural range of variability in some regions, but their spatial extent and frequency may be increasing as a result of the effects of changing climate, high live tree densities created by decades of fire suppression, and/or large amounts of dead biomass caused by disturbances such as drought and insect infestations of trees (e.g., Stephens et al., 2022).

Forest cover returns in stages following stand-killing wildfire (Bartels et al., 2016). The first stage, stand initiation, lasts for several years as vegetation establishes and regenerates. The stem exclusion stage is characterized by intense competition for growing space and resources among regenerated species, as lack of space prevents further establishment of new stems. The canopy transition stage represents a mature forest with vertical canopy stratification and crown closure, as individual tree canopies overlap. Finally, the gap dynamics stage represents old growth with a mosaic canopy because of individual tree senescence (Bartels et al., 2016).

The rate of forest regrowth reflects local site conditions, regional climate, disturbance history, regional species pool, and species life histories. Using Canada's diverse forests as an example, canopy cover returns to approximately 40% as early as 5 years after disturbance and remains at this level for about 50 years, but then decreases with changing species composition (Bartels et al., 2016). Stand basal area increases until old-growth conditions are reached at about 200 years after disturbance and then decreases with time, as natural tree senescence creates more open forest stands (Bartels et al., 2016).

Resources such as light and soil nutrients may be abundant following severe fire, and the establishment and abundance of species with overlapping functional traits such as photophilia (thrives in sunlight) are high (Pacala & Rees, 1998). Competition intensifies and site resources become increasingly limited as communities develop (Glenn-Lewin et al., 1992). Communities transition from fast-growing, early succession colonizers to slower-growing, shade-tolerant species, with overlapping mixtures of early and late-succession species, stronger partitioning of niches, and greater diversity of contrasting functional traits (Reich, 2014). (Functional traits are any morphological, physiological, or phenological plant characteristics affecting overall plant fitness through their influence on survival, growth, and reproduction (Violle et al., 2007).) In the long-term absence of severe disturbance, climax communities of late-succession species with high functional redundancy, such as shade-tolerance, may dominate forests (Chen & Popadiouk, 2002). These observations lead to a hump-shaped model of general changes in forest functional diversity with time since disturbance in North American boreal forests, although tree species exhibit individualistic responses to succession-driven changes in species diversity (Taylor et al., 2020).

Shifting fire regimes have fueled concern that competition with rapidly reestablishing shrubs may shift tree composition toward shade-tolerant species after fire in western North America (Hessburg et al., 2019). Tubbesing et al. (2022) found no evidence of this in conifer forests in which ponderosa pine (*Pinus ponderosa*) emerged at a faster rate than shade-tolerant white fir (*Abies concolor*) for the first two decades after fire until fir emergence rates caught up over longer timespans. Investigations in other North American forest types and in other regions of the world may indicate different patterns, but this remains to be seen.

Stand-killing fires can be rare under conditions in relatively moist forests (as opposed to human ignitions of fire that alter environmental conditions in these forests). Fires in tropical moist broadleaf forests, for example, are mostly of low to intermediate frequency and low intensity (Ruan et al., 2020). Bowman et al. (2020) noted that fire activity has a broadly unimodal relationship with primary productivity. Tropical savanna and other areas of intermediate primary productivity, abundant fuel, and reliable dry seasons burn at very high frequency. Fire is rare in rainforests with high primary productivity but wet fuel. Arid regions have fire weather but low primary productivity and limited fuel.

Grassland communities. Grasslands are among the most fire-prone biomes (e.g., Zhai et al., 2023). Grasslands have distinctive characteristics relative to forests with respect to vegetation response to fire, although the details of plant response depend on factors such as the timing of fire relative to plant lifecycle phases such as annual seed development (Ruckman et al., 2012).

In general, net primary productivity of grasslands increases following fire because of increased availability of nutrients such as plant-assimilated nitrogen released by the burning of biomass (Bond, 2001). The primary productivity of some perennial herbaceous plants is positively affected by fire because of the decreased competition for water and light. Perennial herbaceous vegetation is generally more tolerant of fire than other plant forms (Stavi, 2019). The underground rhizome buds of perennial herbaceous plants remain unharmed during typical grassland fires (Bond, 2001), for example, allowing rapid regrowth after fire.

If fire severity and frequency increase, this can facilitate the growth of invasive plant species and can encourage the dominance of pyrophytes, or fire-resistant species (Reeves et al., 2016). Passive pyrophytes include species that resist fire with hard-coated seeds or underground storage organs (Perevolotsky et al., 1992). Fire accelerates the growth of active pyrophytes via mechanisms such as increased rates of seed dispersal (Perevolotsky et al., 1992).

In general, grasslands may be more resilient to wildfires than forests in the sense that grasslands are less likely to exhibit persistent declines in vegetation because of fire. A synthesis across North America's grassland biome indicated that all vegetation functional groups exhibit relatively rapid recovery to pre-fire ranges of variation, with only local persistent declines in vegetation that represent extreme cases within larger fires (Donovan et al., 2020).

In summary, individual plant species, as well as plant communities, exhibit diverse responses to wildfire that depend partly on the seasonality and intensity of the fire and partly on the adaptations (or lack thereof) of the plants to fire. Even in fire-prone ecosystems, some species are adapted to benefit from fire, whereas other species are at least temporarily harmed by fire. The details of how and how fast plants respond to fire strongly influence terrestrial fluxes of water, solutes, sediment, and POM after fire, as well as the resources available to other living organisms.

3.4.4. Arthropods, Insects, and Other Pollinators

Most flowering plants (85%) rely on animal pollinators (primarily insects and birds, but occasionally mammals such as bats) for sexual reproduction (Ollerton et al., 2011). Many pollinators are good flyers or have generalist feeding habits, allowing them to recolonize or forage in burned areas as soon as flowers are available (Carbone & Aguilar, 2017; Garcia et al., 2018). Ground-nesting pollinators that survived the fire can benefit from lower competition for flowers and lower predation levels in burned areas. These characteristics help to explain the results of a global synthesis indicating that wildfires tend to increase pollinator abundance and richness because of increased flowering after fire (Carbone et al., 2019). Hymenoptera (wasps, bees, ants, and others) particularly benefit from fires. The exceptions to these patterns are short fire intervals, which may reduce pollinators, and a general decrease in Lepidoptera (butterflies, moths) and bird pollinators after fires. The timespan over which these responses occur also varies with vegetation community. Grassland species may flower quickly after fire, whereas woody vegetation may take longer to recover.

In general, the effects of fire on arthropods (millipedes, beetles, springtails, flies, ticks and mites, hemiptera, and others) vary, depending on many interactive factors, including fire characteristics, removal of biomass, vegetative regrowth, and the type of arthropod under consideration. Fires reduce the abundance of some species of ticks and mites that are not on their host animals at the time of the fire, for example, but tick and mite species that eat plants and detritus can increase during the increased productivity of vegetation following grassland fire (Warren et al., 1987). Grassland spiders typically seek refuge in the soil or under nonflammable debris during the fire, but then depend on prey density and diversity to survive after the fire. Overall, grassland arthropod survival post-fire depends on the degree to which food and habitat are destroyed and subsequently recover, and this can vary among species and types of arthropods (Warren et al., 1987; Yang et al., 2022).

Forests have slightly more complicated interactions among fire, vegetation, and insects. Insect outbreaks that kill trees can create fuel accumulation for subsequent fires (Stuart et al., 1989), although the relationship between dead trees and fire occurrence is not straightforward, especially in stands of trees that have been dead for years (Hicke et al., 2012; Harvey et al., 2014; Hart et al., 2015). Fire may also predispose trees to subsequent insect attacks (Stocks, 1987). Alternatively, insect outbreaks followed by fire can redirect vegetation succession (Amman & Schmitz, 1988).

Mountain pine beetle (*Dendroctonus ponderosae*) and lodgepole pine (*Pinus contorta*) provide an example of insect and fire effects on vegetation succession (McCullough et al., 1998). Outbreaks of mountain pine beetles in senescing lodgepole pine stands can result in large fuel buildups (Stuart et al., 1989). Subsequent intense wildfire opens serotinous lodgepole pine cones, as well as creating conditions for post-fire lodgepole pine germination and growth, setting up future beetle food supplies (Amman & Schmitz, 1988). When fire does not occur, lodgepole pines can be replaced by other coniferous tree species (Mitchell, 1990).

Overall, the effects of fire on forest insects vary with insect type. Most insects are absent immediately after fire, but scorched and partially burned trees provide habitat for diverse bark beetle, wood-borer, and related beetle groups, so the populations and diversity of these beetles can be high for up to five years after fire before subsequently declining (McCullough et al., 1998). The species composition of ground beetles and other insects changes through time post-fire as plant species composition, vertical distribution of plants, and habitat complexity change (McCullough et al., 1998). Arthropods that can fly or balloon (some spiders) can rapidly recolonize burned areas as suitable habitat develops (Koponen, 2005).

3.4.5. Reptiles and Amphibians

The limited number of studies that have focused specifically on amphibians and reptiles suggests that these animals typically fare relatively well during fires by sheltering in wetlands, seeking refuge under less flammable surface objects, such as damp coarse woody debris or large rocks, and burrowing into the soil or seeking cover in existing burrows (e.g., Lips, 1991). More mobile species may also flee (Patterson, 1984).

Not all animals in a burn area survive the fire, of course, but the greater challenges to reptiles and amphibians may arise post-fire when they re-emerge into a world with altered habitat and food sources. The details of herpetofauna survival post-fire depend on fire severity and extent, and on whether the ecosystem is fire-dependent, in which case the animals have likely adapted to burned landscapes. Lizards sheltering under objects during fire, for example, selectively use burned and charred branches for perching after the fire. Other individuals invade burned areas from adjacent unburned areas (Lillywhite et al., 1977).

In fire-dependent ecosystems, fire maintains and restores vegetation species composition and structure, as well as the temporally shifting landscape mosaic to which many reptiles and amphibians are adapted. Reptile species diversity in these ecosystems correlates strongly with vegetation density, and fires can increase total reptile and amphibian diversity (Russell et al., 1999), although some species may decline in population density immediately after fire (e.g., Friend, 1993).

Fire can also be vital to maintaining non-alluvial isolated wetlands that are critical habitats for amphibian breeding and larval development, as well as for reptile (e.g., turtle) cover, foraging, and hibernation (Russell et al., 1999). Specifically, fire limits the accumulation of organic matter that can allow succession of wetlands to shrub thickets and closed-canopy forest stands and even permanent drying of wetlands with increased transpiration from trees (Russell et al., 1999).

3.4.6. Birds

Comparisons among individual studies of bird responses to wildfire are complicated by differences in the scale, severity, and frequency of fire; time elapsed between the fire and the study; and study design (Murphy et al., 2021). However, as with other types of organisms, the response of birds clearly depends on the environment (e.g., fire-dependent or fire-independent), the severity and spatial extent of the fire, and the habitat and food resource needs of individual bird species. Given the high mobility of birds, the composition of bird communities immediately after fire seems to depend on resource availability rather than capacity to survive the actual fire (e.g., Lyon, 1997), although observations of dead but unburned birds and other animals immediately after fire suggest that birds can be killed by smoke inhalation (A. East, U.S. Geological Survey, Jan 2024 pers. comm.). Although many avian communities can recover rapidly after severe fire, increases in fire frequency can limit recovery and appear to have driven some species to extinction because of changes in plant communities, vegetative structure, and food resources for birds (Woinarski & Recher, 1997).

Bird species richness can decline and community composition can change even in ecosystems with frequent fires, such as southeastern Australia, where bird communities had not recovered seven years after a fire (Murphy et al., 2021). In this example, 18 bird species were sensitive to fire, including those that ate insects, grains, and fruits; 9 species, including predators, responded positively; and another 11 species preferred burned areas close to the fire edge. Other studies suggest that bird communities in semiarid shrubland and woodland may require many decades to centuries to recover after fire (Watson et al., 2012; Davis et al., 2016).

In comparing bird populations for five years before fire and three years after fire in Montana, USA, Smucker et al. (2005) emphasized the influence of fire severity and time since fire on bird abundances. Some species decrease after high-severity forest fires but increase in abundance after low-severity fires. Among bird species that changed after fire, twice as many increased as decreased, and abundance changed with time since fire (Smucker et al., 2005).

Cavity-nesting birds flourish after wildfires kill trees and increase the abundance of snags, but the abundance of individual species varies through time after fire. Over a period of 12 years after fire in Idaho, USA, nest densities of open-space foragers increased with time, whereas nest densities of wood-foragers peaked 4–5 years after the fire (Saab et al., 2007). Species that obtain insects from wood rapidly colonize post-fire forests and then decline over time as bark and wood-boring beetles decline, whereas species that catch insects on the wing remain abundant for as long as 25 years after fire (Saab & Vierling, 2001), perhaps because shrub regrowth hosts flying insects whereas the slow regrowth of canopy cover improves conditions for aerial foraging (Hannon & Drapeau, 2005).

Some bird species, including the endangered red-cockaded woodpecker (*Picoides borealis*) in the southeastern United States, require habitats that are maintained by frequent, low-severity fires (Stephens et al., 2019). The abundances of many bird species adapted to disturbance-mediated habitats have declined and in North America these declines have been more severe and common than those of species associated with less frequently disturbed mature or closed-canopy forests (Brawn et al., 2001). Fire strongly influences vegetation structure and composition and bird communities in systems as diverse as the coniferous and boreal forests of the western United States and southern Canada (Schieck & Song, 2006); oak savannas of the midwestern United States; pine savannas of the southeastern United States; North American grasslands; diverse Australian ecosystems (Reilly, 1991); Mediterranean-type ecosystems (Moreno & Oechel, 1994); and Cerrado (tropical savanna) in Brazil (Cavalcanti, 1999).

3.4.7. Mammals

Small mammals represent the majority of mammalian species in North American forests, play important roles in the forest food web (Tallmon et al., 2003), and are viewed as indicators of forest-floor function (Carey & Harrington, 2001). Reviewing the effects of fire on small mammals with a range of body mass from 1 to 1000 g (rodents, squirrels, shrews, small marsupials, and others), Griffiths and Brook (2014) found that fire reduces abundance, particularly in species with body size in the range of 100–1000 g and those with habitat requirements sensitive to fire, such as dense ground cover.

Comparing the responses of deer mice (*Peromyscus maniculatus*) and red-backed voles (*Myodes gapperi*) in the same post-fire environment indicates some of the subtleties behind these generalizations. Deer mice, which prefer open microhabitats, increase in abundance after fire. Red-backed voles, which require high moisture levels and prefer structurally complex microhabitats, decline after fire (Zwolak, 2009).

Body mass patterns might reflect either the inability of medium-sized small mammals to find refuge during a fire, or to have lower reproductive rates or increased predation following fire (Griffiths & Brook, 2014). The smallest species might be able to find micro-refugia and have lower energy needs for reproduction, whereas the largest species might be able to move to avoid death by fire and to locate new food after the fire. Fast-moving wildfires, however, can kill both small and large mammals.

In grasslands, large ungulates such as elk (*Cervus canadensis*) and bison (*Bison bison*) in western North America prefer grazing in burned areas with rapid vegetation regrowth (Pearson et al., 1995). Grazing in burned grassland and mountain shrub communities results in a measurable increase in protein concentration in the winter diets of animals such as bighorn sheep (*Ovis canadensis*) and mule deer (*Odocoileus hemionus*) (Tracy & McNaughton, 1997).

Little is known about the effects of fire on the numbers and activity of predators (Fisher & Wilkinson, 2005), although they presumably respond to fluctuations in prey following fire. In the savanna environment of the Brazilian Cerrado, for example, herbivores return first to burned areas, followed by an increase in the relative abundance of carnivores during the second year after fire (Nunes et al., 2023).

3.4.8. Effects on Biotic Communities

The effects of wildfire on upland biotic communities are diverse. These effects depend on fire severity, spatial extent, and recurrence interval, as well as the species under consideration. As noted previously, some species require the habitat created by fire or are stimulated to reproduce by the triggers of heat or smoke. Some species have physiological or behavioral adaptations that allow them to avoid the direct effects of fire and then return to or recolonize burned areas. Other species are precluded from burned areas until pre-fire conditions of habitat and resources reestablish. In general, episodic fires help to maintain diverse habitat and resources that can sustain diverse biotic communities, but a change from natural fire regimes, either through increased frequency and greater severity or through lack of fires because of human fire suppression, may cause biotic communities to also change with respect to species present and age distribution of individuals within a species.

3.5. Summary

The primary effects of wildfire in upland portions of a catchment are to (i) reduce infiltration capacity and the stability of regolith, especially in areas with steep terrain that have experienced severe fire, and (ii) alter soil microbial communities and aboveground plant biomass. These effects combine to influence biogeochemical processes and, in steep, severely burned environments, typically greatly increase downslope fluxes of diverse materials for timespans that can last from months to many years or decades.

The details of how wildfire affects terrestrial environments vary with pre-existing conditions and with the characteristics of the fire. Pre-existing conditions include topography, thickness and characteristics of regolith, and plant and soil microbial communities. Characteristics of the fire that influence post-fire effects include intensity, severity, return interval, spatial extent, and the relation of a particular fire to the fire regime—did the fire occur outside of the typical fire season, for example, or at unusual intensity? In severely burned and steep terrain, fluxes from uplands to the river network and depositional environments increase substantially and persist for varying lengths of time, whereas the morphology and stability of upland landforms are only locally affected. The responses of individual species and biotic communities vary substantially as a function of their evolutionary adaptation to wildfire and the characteristic or atypical nature of the fire.

The details of biotic responses matter to individual species and communities, but also to the entire landscape and ecosystem. The degree to which pre-fire microbial communities are altered during and after a fire influences biogeochemical reactions in the uplands and fluxes to river networks. The rate and extent to which vegetation species and communities colonize uplands strongly influence the stability of the uplands and the associated fluxes of material downslope and into rivers, and both reflect and influence soil microbial communities. The return of animals, from insects to birds and mammals, after a fire influences vegetation regrowth by affecting propagule dispersal, pollination, soil stability and nutrient content, and herbivory and parasitism.

The presence of large predators in a post-fire landscape might not seem to matter much for river processes and forms, for example, yet the well-documented example of Yellowstone National Park in the western United States clearly reflects the connection between predators and rivers (Wolf et al., 2007; Ripple & Beschta, 2012). The presence of predators can alter the behavior of large herbivores. The behavior of large herbivores can alter riparian vegetation. Riparian vegetation can alter nitrate uptake in the hyporheic zone, riverbank stability, and dissipation of energy during peak flows. Dissipation of peak flow energy can alter the grain-size distribution and stability of riverbed substrate. Riverbed substrate can alter the periphyton, benthic insects, and fish communities present in the river, and so on.

The 19th-century American environmentalist John Muir famously wrote, "When we try to pick out anything by itself, we find it hitched to everything else in the Universe." I cannot think of a better description for understanding wildfires and post-fire landscapes and ecosystems.

References

Abbott, B. W., Rocha, A. V., Shogren, A., Zarnetske, J. P., Iannucci, F., Bowden, W. B., et al. (2021). Tundra wildfire triggers sustained lateral nutrient loss in Alaskan Arctic. *Global Change Biology*, *27*, 1408–1430. https://doi.org/10.1111/gcb.15507

Abella, S. R., & Springer, J. D. (2015). Effects of tree cutting and fire on understory vegetation in mixed conifer forests. *Forest Ecology and Management*, *335*, 281–299. https://doi.org/10.1016/j.foreco.2014.09.009

Aber, J. D., Nadelhoffer, K. J., Steudler, P., & Melillo, J. M. (1989). Nitrogen saturation in northern forest ecosystems. *Bioscience*, *39*(6), 378–386. http://www.jstor.org/stable/1311067?origin=JSTOR-pdf

Acker, S. A., Kertis, J., Bruner, H., O'Connell, K., & Sexton, J. (2013). Dynamics of coarse woody debris following wildfire in a mountain hemlock (*Tsuga mertensiana*) forest. *Forest Ecology and Management*, *302*, 231–239. https://doi.org/10.1016/j.foreco.2013.03.013

Acker, S. A., Kertis, J. A., & Pabst, R. J. (2017). Tree regeneration, understory development, and biomass dynamics following wildfire in a mountain hemlock (*Tsuga mertensiana*) forest. *Forest Ecology and Management*, *384*, 72–82. https://doi.org/10.1016/j.foreco.2016.09.047

Ackley, C., Tank, S. E., Haynes, K. M., Rezanezhad, M. C. C., & Quinton, W. L. (2021). Coupled hydrological and geochemical impacts of wildfire in peatland-dominated regions of discontinuous permafrost. *Science of the Total Environment*, *782*, 146841. https://doi.org/10.1016/j.scitotenv.2021.146841

Adams, K. V., Dixon, J. L., Wilcox, A. C., & McWethy, D. (2023). Fire-produced coarse woody debris and its role in sediment storage on hillslopes. *Earth Surface Processes and Landforms*, *48*, 1665–1678. https://doi.org/10.1002/esp.5573

Alessio, P., Dunne, T., & Morell, K. (2021). Post-wildfire generation of debris-flow slurry by rill erosion on colluvial hillslopes. *Journal of Geophysical Research: Earth Surface*, *126*, e2021JF006108. https://doi.org/10.1029/2021JF006108

Amman, G. D., & Schmitz, R. F. (1988). Mountain pine beetle-lodgepole pine interactions and strategies for reducing tree losses. *Ambio*, *17*, 62–68. https://www.jstor.org/stable/4313420

Angelstam, P. (1998). Maintaining and restoring biodiversity in European boreal forests by developing natural disturbance regimes. *Journal of Vegetation Science*, *9*, 593–602. https://doi.org/10.2307/3237275

Angrish, R. (2022). The wood wide web. *Resonance*, *27*, 1429–1441. https://doi.org/10.1007/s12045-022-1435-x

Aponte, C., Tolhurst, K. G., & Bennett, L. T. (2014). Repeat prescribed fires decrease stocks and change attributes of coarse woody debris in a temperate eucalypt forest. *Ecological Applications*, *24*, 976–989. https://doi.org/10.1890/13-1426.1

Attiwill, P. M. (1994). The disturbance of forest ecosystems: the ecological basis for conservative management. *Forest Ecology and Management*, *63*, 247–300. https://doi.org/10.1016/0378-1127(94)90114-7

Atwood, A., Hille, M., Clark, M. K., Rengers, F., Ntarlagiannis, D., Townsend, K., & West, A. J. (2023). Importance of subsurface water for hydrological response during storms in a post-wildfire bedrock landscape. *Nature Communications*, *14*, 3814. https://doi.org/10.1038/s41467-023-39095-z

Babur, E., Dindaroglu, T., Danish, S., Haggblom, M. M., Ozlu, E., Gozukara, G., & Uslu, O. S. (2022). Spatial responses of soil carbon stocks, total nitrogen, and microbial indices to post-wildfire in the Mediterranean red pine forest. *Journal of Environmental Management*, *320*, 115939. https://doi.org/10.1016/j.jenvman.2022.115939

Backer, D. M., Jensen, S. E., & McPherson, G. R. (2004). Impacts of fire-suppression activities on natural communities. *Conservation Biology*, *18*, 937–946. https://doi.org/10.1111/j.1523-1739.2004.494_1.x

Barnhart, K. R., Jones, R. P., George, D. L., McArdell, B. W., Rengers, F. K., Staley, D. M., & Kean, J. W. (2021). Multi-model comparison of computed debris flow runout for the 9 January 2018 Montecito, California post-wildfire event. *Journal of Geophysical Research Earth Surface*, *126*, e2021JF006245. https://doi.org/10.1029/2021JF006245

Bartels, S. F., Chen, H. Y. H., Wulder, M. A., & White, J. C. (2016). Trends in post-disturbance recovery rates of Canada's forests following wildfire and harvest. *Forest Ecology and Management*, *361*, 194–207. https://doi.org/10.1016/j.foreco.2015.11.015

Bassett, M., Chia, E. K., Leonard, S. W. J., Nimmo, D. G., Holland, G. J., Ritchie, E. G., et al. (2015). The effects of topographic variation and the fire regime on coarse woody debris: insights from a large wildfire. *Forest Ecology and Management*, *340*, 126–134. https://doi.org/10.1016/j.foreco.2014.12.028

Basso, M., Mateus, M., Ramos, T. B., & Vieira, D. C. S. (2021). Potential post-fire impacts on a water supply reservoir: an integrated watershed-reservoir approach. *Frontiers in Environmental Science*, *9*, 684703. https://doi.org/10.3389/fenvs.2021.684703

Basso, M., Serpa, D., Mateus, M., Keizer, J. J., Catarina, D., & Vieira, S. (2022). Advances on water quality modeling in burned areas: a review. *PLOS Water*, *1*, e0000025. https://doi.org/10.1371/journal.pwat.0000025

Basso, M., Vieira, D. C. S., Ramos, T. B., & Mateus, M. (2020). Assessing the adequacy of SWAT model to simulate postfire effects on watershed hydrological regime and water quality. *Land Degradation & Development*, *31*, 619–631. https://doi.org/10.1002/ldr.3476

Baynes, M., Newcombe, G., Dixon, L., Castlebury, L., & O'Donnell, K. (2011). A novel plant-fungal mutualism associated with fire. *Fungal Biology*, *116*, 133–144. https://doi.org/10.1016/j.funbio.2011.10.008

Benda, L., & Dunne, T. (1997). Stochastic forcing of sediment supply to channel networks form landsliding and debris flow. *Water Resources Research*, *33*, 2849–2863. https://doi.org/10.1029/97WR02388

Benedict, K. B., Prenni, A. J., Carrico, C. M., Sullivan, A. P., Schichtel, B. A., & Collett, J. L. (2017). Enhanced concentrations of reactive nitrogen species in wildfire smoke. *Atmospheric Environment*, *148*, 8–15. https://doi.org/10.1016/j.atmosenv.2016.10.030

Birkeland, P. W. (1999). *Soils and geomorphology* (3rd ed.). New York: Oxford University Press.

Bitner, K., Gallaher, B., & Mullen, K. (2001). *Review of wildfire effects on chemical water quality*. Los Alamos, NM: Los Alamos National Laboratory. LA-13826-MS

Blackwelder, E. (1927). Fire as an agent in rock weathering. *Journal of Geology*, *35*, 134–140.

Blake, W. H., Theocharopoulos, S. P., Skoulikidis, N., Clark, P., Tountas, P., Hartley, R., & Amaxidis, Y. (2010). Wildfire impacts on hillslope sediment and phosphorus yields. *Journal of Soils and Sediments*, *10*, 671–682. https://doi.org/10.1007/s11368-010-0201-y

Blake, W. H., Wallbrink, P. J., Doerr, S. H., Shakesby, R. A., & Humphreys, G. S. (2004). Sediment redistribution following wildfire in the Sydney region, Australia: a mineral magnetic tracing approach. In V. Golosov, V. Belyaev, & D. E. Walling (Eds.), *Sediment transfer through the fluvial system* (Vol. 288, pp. 52–59). Wallingford: IAHS Publ.

Blake, W. H., Wallbrink, P. J., Wilkinson, S. N., Humphreys, G. S., Doerr, S. H., Shakesby, R. A., & Tomkins, K. M. (2009). Deriving hillslope sediment budgets in wildfire-affected forests using fallout radionuclide tracers. *Geomorphology*, *104*, 105–116. https://doi.org/10.1016/j.geomorph.2008.08.004

Blank, R. R., & Zamudio, D. C. (1998). The influence of wildfire on aqueous-extractable soil solutes in forested and wet meadow ecosystems along the eastern front of the Sierra-Nevada Range, California. *International Journal of Wildland Fire*, *8*, 79–85. https://doi.org/10.1071/WF9980079

Bodí, M. B., Martin, D. A., Balfour, V. N., Santin, C., Doerr, S. H., Pereira, P., et al. (2014). Wildland fire ash: production, composition and eco-hydro-geomorphic effects. *Earth-Science Reviews*, *130*, 103–127. https://doi.org/10.1016/j.earscirev.2013.12.007

Bond, W. (2001). Ecological effects of fires. In S. A. Levin (Ed.), *Encyclopedia of biodiversity* (Vol. Volume 2, pp. 745–753). Cambridge, MA: Academic Press.

Bonell, M., & Williams, J. (1987). Infiltration and redistribution of overland flow and sediment on a low relief landscape of semi-arid, tropical Queensland. In R. H. Swanson, P. Y. Bernier, & P. D. Woodward (Eds.), *Forest hydrology and watershed management* (pp. 199–211). Wallingford, UK: IAHS-AISH Publication no. 167.

Bourgeau-Chavez, L. L., Graham, J. A., Vander Bilt, D. J. L., & Battaglia, M. J. (2022). Assessing the broadscale effects of wildfire under extreme drought conditions to boreal peatlands. *Frontiers in Forests and Global Change*, *5*, 965605. https://doi.org/10.3389/ffgc.2022.965605

Bowman, D. M. J. S., Balch, J. K., Artaxo, P., Bond, W. J., et al. (2009). Fire in the Earth system. *Science*, *324*, 481–484. https://doi.org/10.1126/science.1163886

Bowman, D. M. J. S., Kolden, C. A., Abatzoglou, J. T., Johnston, F. H., van der Werf, G. R., & Flannigan, M. (2020). Vegetation fires in the Anthropocene. *Nature Reviews Earth and Environment*, *1*, 501–515. https://doi.org/10.1111/pce.13916

Bowman, D. M. J. S., Williamson, G. J., Price OF, Ndalila, M. N., & Bradstock, R. A. (2021). Australian forests, megafires and the risk of dwindling carbon stocks. *Plant, Cell and Environment*, *44*, 347–355. https://doi.org/10.1111/pce.13916

Bracken, L. J., Turnbull, L., Wainwright, J., & Bogaart, P. (2015). Sediment connectivity: a framework for understanding sediment transfer at multiple scales. *Earth Surface Processes and Landforms*, *40*, 177–188. https://doi.org/10.1002/esp.3635

Bracken, L. J., Wainwright, J., Ali, G. A., Tetzlaff, D., Smith, M. W., Reaney, S. M., & Roy, A. G. (2013). Concepts of hydrological connectivity: research approaches, pathways and future agendas. *Earth-Science Reviews*, *119*, 17–34. https://doi.org/10.1016/j.earscirev.2013.02.001

Brando, P. M., Paolucci, L., Ummenhofer, C. C., Em, O., Hartmann, H., Cattau, M. E., et al. (2019). Droughts, wildfires, and forest carbon cycling: a pantropical synthesis. *Annual Reviews of Earth and Planetary Science*, *47*, 555–581. https://doi.org/10.1146/annurev-earth-082517-010235

Brantley, S. L., Eissenstat, D. M., Marshall, J. A., Godsey, S. E., Balogh-Brunstad, Z., Karwan, K. L., et al. (2017). Reviews and syntheses: on the role trees play in building and plumbing the critical zone. *Biogeosciences*, *14*, 5115–5142. https://doi.org/10.5194/bg-14-5115-2017

Brawn, J. D., Robinson, S. K., & Thompson, F. R. (2001). The role of disturbance in the ecology and conservation of birds. *Annual Review of Ecology and Systematics*, *32*, 251–276. https://doi.org/10.1146/annurev.ecolsys.32.081501.114031

Brogan, D. J., Nelson, P. A., & MacDonald, L. H. (2019). Spatial and temporal patterns of sediment storage and erosion following a wildfire and extreme flood. *Earth Surface Dynamics*, *7*, 563–590. https://doi.org/10.5194/esurf-7-563-2019

Brown MJ, Kertis J, Huff MH. 2013. *Natural Tree Regeneration and Coarse Woody Debris Dynamics After a Forest Fire in the* Western Cascade Range. USDA Forest Service Research Paper PNW-RP-592, Portland, OR.

Bruns, T. D., Chung, J. A., Carver, A. A., & Glassman, S. I. (2020). A simple pyrocosm for studying soil microbial response to fire reveals a rapid, massive response by *Pyronema* species. *PloS One*, *15*, e0222691. https://doi.org/10.1371/journal.pone.0222691

Buckman, S., Morris, R. H., & Bourman, R. P. (2021). Fire-induced rock spalling as a mechanism of weathering responsible for flared slope and inselberg development. *Nature Communications*, *12*, 2150. https://doi.org/10.1038/s41467-021-22451-2

Bull, E. L. (2002). The value of coarse woody debris to vertebrates in the Pacific Northwest. In W. F. Laudenslayer, P. J. Shea, B. E. Valentine, C. P. Weatherspoon, & T. E. Lisle (Eds.), *Proceedings of the Symposium on the Ecology and Management of Dead Wood in Western Forests*. USDA Forest Service General Technical Report PSW-GTR-181, Albany, CA (pp. 171–178).

Burles, K., & Boon, S. (2011). Snowmelt energy balance in a burned forest plot, Crowsnest Pass, Alberta, Canada. *Hydrological Processes*, *25*, 3012–3029. https://doi.org/10.1002/hyp.8067

Cahoon, S. M. P., Sullivan, P. F., & Gray, A. N. (2022). Interactions among wildfire, forest type and landscape position are key determinants of boreal forest carbon stocks. *Journal of Ecology*, *110*, 2475–2492. https://doi.org/10.1111/1365-2745.13963

Campbell, G. S., Jungbauer, J. D., Bristow, K. L., & Hungerford, R. D. (1995). Soil temperature and water content beneath a surface fire. *Soil Science*, *159*, 363–374.

Cannon, S. H. (2001). Debris-flow generation from recently burned watersheds. *Environmental and Engineering Geoscience*, *7*, 321–341. https://doi.org/10.2113/gseegeosci.7.4.321

Cannon, S. H., & DeGraff, J. (2009). The increasing wildfire and post-fire debris-flow threat in western USA, and implications for consequences of climate change. In K. Sassa & P. Canuti (Eds.), *Landslides – disaster risk reduction* (pp. 177–190). Berlin: Springer-Verlag.

Cannon S. H., Ellis W. L., Godt J. W. 1998. *Evaluation of the landside potential in Capulin Canyon following the Dome Fire, Bandelier National Monument, New Mexico*. U.S. Geological Survey Open-File Report 98-42, 17 pp.

Cannon, S. H., & Gartner, J. E. (2005). Wildfire-related debris flow from a hazards perspective. In O. Hungr & M. Jacob (Eds.), *Debris flow hazards and related phenomena* (pp. 363–385). Berlin: Praxis, Springer-Verlag.

Cannon, S. H., Gartner, J. E., Rupert, M. G., Michael, J. A., Rea, A. H., & Parrett, C. (2010). Predicting the probability and volume of postwildfire debris flows in the intermountain western United States. *Geological Society of America Bulletin*, *122*, 127–144. https://doi.org/10.1130/B26459.1

Cannon, S. H., Gartner, J. E., Wilson, R. C., Bowers, J. C., & Laber, J. L. (2008). Storm rainfall conditions for floods and debris flows for recently burned areas in south-western Colorado and southern California. *Geomorphology*, *96*, 250–269. https://doi.org/10.1016/j.geomorph.2007.03.019

Cao, L., Elliott, W., & Long, J. W. (2021). Spatial simulation of forest road effects on hydrology and soil erosion after a wildfire. *Hydrological Processes*, *35*, E14139. https://doi.org/10.1002/hyp.14139

Caon, L., Vallejo, V. R., Ritsema, C. J., & Geissen, V. (2014). Effects of wildfire on soil nutrients in Mediterranean ecosystems. *Earth-Science Reviews*, *139*, 47–58. https://doi.org/10.1016/j.earscirev.2014.09.001

Carbone, L. M., & Aguilar, R. (2017). Fire frequency effects on soil and pollinators: what shapes sexual plant reproduction? *Plant Ecology*, *218*, 1283–1297. https://doi.org/10.1007/s11258-017-0768-0

Carbone, L. M., Tavella, J., Pausas, J. G., & Aguilar, R. (2019). A global synthesis of fire effects on pollinators. *Global Ecology and Biogeography*, *28*, 1487–1498. https://doi.org/10.1111/geb.12939

Carey, A. B., & Harrington, C. A. (2001). Small mammals in young forests: implications for management for sustainability. *Forest Ecology and Management*, *154*, 289–309. https://doi.org/10.1016/S0378-1127(00)00638-1

Cavalcanti, R. B. (1999). Bird species richness and conservation in the Cerrado region of central Brazil. *Studies in Avian Biology*, *19*, 244–249.

Certini, G. (2005). Effects of fire on properties of forest soils: a review. *Oecologia*, *143*, 1–10. https://doi.org/10.1007/s00442-004-1788-8

Certini, G., Moya, D., Lucas-Borja, M. E., & Mastrolonardo, G. (2021). The impact of fire on soil-dwelling biota: a review. *Forest Ecology and Management*, *488*, 118989. https://doi.org/10.1016/j.foreco.2021.118989

Chandler, C., Cheney, P., Thomas, P., Trabaud, L., & Williams, D. (1983). *Fire in forestry*. Forest Fire Behavior and Effects (Vol. 1). New York: John Wiley & Sons.

Chantal, M., & Granström, A. (2007). Aggregations of dead wood after wildfire act as browsing refugia for seedlings of *Populus tremula* and *Salix caprea*. *Forest Ecology and Management*, *250*, 3–8. https://doi.org/10.1016/j.foreco.2007.03.035

Chapin, F. S., McFarland, J., McGuire, A. D., Euskirchen, E. S., Ruess, R. W., & Kielland, K. (2009). The changing global carbon cycle: linking plant-soil carbon dynamics to global consequences. *Journal of Ecology*, *97*, 840–850. https://doi.org/10.1111/j.1365-2745.2009.01529.x

Chapman, D. S., Gunn, I. D. M., Pringle, H. E. K., Siriwardena, G. M., Taylor, P., Thackeray, S. J., et al. (2020). Invasion of freshwater ecosystems is promoted by network connectivity to hotspots of human activity. *Global Ecology and Biogeography*, *29*, 645–655. https://doi.org/10.1111/geb.13051

Chen, H. Y. H., & Popadiouk, R. V. (2002). Dynamics of North American boreal mixed woods. *Environmental Research*, *10*, 137–166. https://doi.org/10.1139/a02-007

Chen, L., Liu, L., Mao, C., Qin, S., Wang, J., Liu, F., et al. (2018). Nitrogen availability regulates topsoil carbon dynamics after permafrost thaw by altering microbial metabolic efficiency. *Nature Communications*, *9*, 3951. https://doi.org/10.1038/s41467-018-06232-y

Clarke, P. J., Lawes, M. J., Midgely, J. J., Lamont, B. B., Ojeda, F., Burrows, G. E., et al. (2013). Resprouting as a key functional trait: how buds, protection and resources drive persistence after fire. *New Phytologist*, *197*, 19–35. https://doi.org/10.1111/nph.12001

Cole, R. P., Bladon, K. D., Wagenbrenner, J. W., & Coe, D. B. R. (2020). Hillslope sediment production after wildfire and post-fire forest management in northern California. *Hydrological Processes*, *34*, 5242–5259. https://doi.org/10.1002/hyp.13932

Cook, E. M., Sponseller, R., Grimm, N. B., & Hall, S. J. (2018). Mixed method approach to assess atmospheric nitrogen deposition in arid and semi-arid ecosystems. *Environmental Pollution*, *239*, 617–630. https://doi.org/10.1016/j.envpol.2018.04.013

Cooperdock, S. C., Hawkes, C. V., Xu, D. R., & Breecker, D. O. (2020). Soil water content and soil respiration rates are reduced for years following wildfire in a hot and dry climate. *Global Biogeochemical Cycles*, *34*, e2020BG006699. https://doi.org/10.1029/2020GB006699

Costa, J. E. (1984). Physical geomorphology of debris flows. In J. E. Costa & P. J. Fleisher (Eds.), *Developments and applications of geomorphology* (pp. 268–317). Berlin: Springer-Verlag.

Costa, M. R., Calvao, A. R., & Aranha, J. (2014). Linking wildfire effects on soil and water chemistry of the Marão River watershed, Portugal, and biomass changes detected from Landsat imagery. *Applied Geochemistry*, *44*, 93–102. https://doi.org/10.1016/j.apgeochem.2013.09.009

Cotrufo, M. F., Boot, C. M., Kampf, S., Nelson, P. A., Brogan, D. J., Covino, T., et al. (2016). Redistribution of pyrogenic carbon from hillslopes to stream corridors following a large montane wildfire. *Global Biogeochemical Cycles*, *30*, 1348–1355. https://doi.org/10.1002/2016GB005467

Coward, E. K., Seech, K., Carter, M. L., Flick, R. E., & Grassian, V. H. (2022). Of sea and smoke: evidence of marine dissolved organic matter deposition from 2020 western United States wildfires. *Environmental Science and Technology Letters*, *9*, 869–876. https://doi.org/10.1021/acs.estlett.2c00383

Cox, E., Cavalcanti, A. R. O., Crane, E. J., & Meyer, W. M. (2022). Soil bacterial assemblage responses to wildfire in low elevation southern California habitats. *PloS ONE*, *17*, e0266256. https://doi.org/10.1371/journal.pone.0266256

Crandall, T., Jones, E., Greenhalgh, M., Frei, R. J., Griffin, N., Severe, E., et al. (2021). Megafire affects stream sediment flux and dissolved organic matter reactivity, but land use dominates nutrient dynamics in semiarid watersheds. *PLoS ONE*, *16*, e0257733. https://doi.org/10.1371/journal.pone.0257733

Davies, G. M., Bakker, J. D., Dettweiler-Robinson, E., Dunwiddie, P. W., Hall, S. A., Downs, J., & Evans, J. (2012). Trajectories of change in sagebrush steppe vegetation communities in relation to multiple wildfires. *Ecological Applications*, *22*, 1562–1577. https://doi.org/10.1890/10-2089.1

Davis, K. T., Higuera, P. E., & Sala, A. (2018). Anticipating fire-mediated impacts of climate change using a demographic framework. *Functional Ecology*, *32*, 1729–1745. https://doi.org/10.1111/1365-2435.13132

Davis, R. A., Doherty, T. S., van Etten, E. J. B., et al. (2016). Conserving long unburnt vegetation is important for bird species, guilds and diversity. *Biodiversity and Conservation*, *25*, 2709–2722. https://doi.org/10.1007/s10531-016-1196-5

Dawe, D. A., Parisien, M. A., Van Dongen, A., & Whitman, E. (2022). Initial succession after wildfire in dry boreal forests of northwestern North America. *Plant Ecology*, *223*, 789–809. https://doi.org/10.1007/s11258-022-01237-6

DeBano, L. F. (2000). The role of fire and soil heating on water repellency in wildland environments: a review. *Journal of Hydrology*, *231–232*, 195–206. https://doi.org/10.1016/S0022-1694(00)00194-3

DeBano, L. F., Neary, D. G., & Ffolliott, P. F. (1998). *Fire effects on ecosystems*. NY: John Wiley & Sons.

DeBano L. F., Rice R. M., Eugene C. C. 1979. *Soil heating in chaparral fires: effects on soil properties, plant nutrients, erosion, and runoff*. USDA Forest Service Research Paper PSW-RP-145, Berkeley, CA. https://www.fs.usda.gov/research/treesearch/29933

DeLong, S. B., Youberg, A. M., DeLong, W. M., & Murphy, B. P. (2018). Post-wildfire landscape change and erosional processes from repeat terrestrial lidar in a steep headwater catchment, Chiricahua Mountains, Arizona, USA. *Geomorphology*, *300*, 13–30. https://doi.org/10.1016/j.geomorph.2017.09.028

Diakakis, M., Mavroulis, S., Vassilakis, E., & Chalvatzi, V. (2023). Exploring the application of a debris flow likelihood regression model in Mediterranean post-fire environments, using field observations-based validation. *Land*, *12*, 555. https://doi.org/10.3390/land12030555

DiBiase, R. A., & Lamb, M. P. (2013). Vegetation and wildfire controls on sediment yield in bedrock landscapes. *Geophysical Research Letters*, *40*, 1093–1097. https://doi.org/10.1002/grl.50277

DiBiase, R. A., & Lamb, M. P. (2020). Dry sediment loading of headwater channels fuels post-wildfire debris flows in bedrock landscapes. *Geology*, *48*, 189–193. https://doi.org/10.1130/G46847.1

Dietrich, W. E., & Dunne, T. (1978). Sediment budget for a small catchment in mountainous terrain. *Zeitschrift fur Geomorphologie Supplement*, *29*, 191–206.

Dijkstra, F. A., Jenkins, M., de Courcelles, V. R., Keitel, C., Barbour, M. M., Kayler, Z. E., & Adams, M. A. (2017). Enhanced decomposition and nitrogen mineralization sustain rapid growth of *Eucalyptus regnans* after wildfire. *Journal of Ecology*, *105*, 229–236. https://doi.org/10.1111/1365-2745.12663

Doerr, S. H., Shakesby, R. A., & Walsh, R. P. D. (2000). Soil water repellency: its causes, characteristics and hydro-geomorphological significance. *Earth-Science Reviews*, *51*, 33–65. https://doi.org/10.1016/S0012-8252(00)00011-8

Doerr, S. H., Woods, S. W., Martin, D. A., & Casimiro, M. (2009). 'Natural background' soil water repellency in conifer forests of the north-western USA: its prediction and relationship to wildfire occurrence. *Journal of Hydrology*, *371*, 12–21. https://doi.org/10.1016/j.jhydrol.2009.03.011

Donato, D. C., Fontaine, J. B., & Campbell, J. L. (2016). Burning the legacy? Influence of wildfire reburn on dead wood dynamics in a temperate conifer forest. *Ecosphere*, *7*, e01341. https://doi.org/10.1002/ecs2.1341

Donovan, V. M., Twidwell, D., Uden, D. R., Tadesse, T., Wardlow, B. D., Bielski, C. H., et al. (2020). Resilience to large, "catastrophic" wildfires in North America's grassland biome. *Earth's Future*, *8*, e2020EF001487. https://doi.org/10.1029/2020EF001487

Dooley, S. R., & Treseder, K. K. (2012). The effect of fire on microbial biomass: a meta-analysis of field studies. *Biogeochemistry*, *109*, 49–61. https://doi.org/10.1007/s10533-011-9633-8

Dove, N. C., & Hart, S. C. (2017). Fire reduces fungal species richness and in situ mycorrhizal colonization: a meta-analysis. *Fire Ecology*, *13*, 37–65. https://doi.org/10.4996/fireecology.130237746

East, A. E., Logan, J. B., Dartnell, P., Lieber-Kotz, O., Cavagnaro, D. B., McCoy, S. W., & Lindsay, D. N. (2021). Watershed sediment yield following the 2018 Carr Fire, Whiskeytown National Recreation Area, northern California. *Earth and Space Science*, *8*, e2021EA001828. https://doi.org/10.1029/2021EA001828

Ebel, B. A., Hinckley, E. S., & Martin, D. A. (2012). Soil-water dynamics and unsaturated storage during snowmelt following wildfire. *Hydrology and Earth System Sciences*, *16*, 1401–1417. https://doi.org/10.5194/hess-16-1401-2012

Ebel, B. A., & Moody, J. A. (2017). Synthesis of soil-hydraulic properties and infiltration timescales in wildfire-affected soils. *Hydrological Processes*, *31*, 324–340. https://doi.org/10.1002/hyp.10998

Eckdahl, J. A., Kristensen, J. A., & Metcalfe, D. B. (2023). Climate and forest properties explain wildfire impact on microbial community and nutrient mobilization in boreal soil. *Frontiers in Forests and Global Change*, *6*, 1136354. https://doi.org/10.3389/ffgc.2023.1136354

Emelko, M. B., Stone, M., Silins, U., Allin, D., Collins, A. L., Williams, C. H. S., et al. (2016). Sediment-phosphorus dynamics can shift aquatic ecology and cause downstream legacy effects after wildfire in large river systems. *Global Change Biology*, *22*, 1168–1184. https://doi.org/10.1111/gcb.13073

Engstrom, R. T. (2010). First-order fire effects on animals: review and recommendations. *Fire Ecology*, *6*, 115–130. https://doi.org/10.4996/fireecology.0601115

Enright, N. J., Fontaine, J. B., Bowman, D. M. J. S., Bradstock, R. A., & Williams, R. J. (2015). Interval squeeze: altered fire regimes and demographic responses interact to threaten woody species persistence as climate changes. *Frontiers in Ecology and Environment*, *13*, 265–272. https://doi.org/10.1890/140231

Eppes, M. C., & Keanini, R. (2017). Mechanical weathering and rock erosion by climate-dependent subcritical cracking. *Reviews of Geophysics*, *55*, 470–508. https://doi.org/10.1002/2017RG000557

Esposito, G., Gariano, S. L., Masi, R., Alfano, S., & Giannatiempo, G. (2023). Rainfall conditions leading to runoff-initiated post-fire debris flows in Campania, Southern Italy. *Geomorphology*, *423*, 108557. https://doi.org/10.1016/j.geomorph.2022.108557

Esposito, G., Matano, F., Molisso, F., Ruoppolo, G., Di Benedetto, A., & Sacchi, M. (2017). Post-fire erosion response in a watershed mantled by volcaniclastic deposits, Sarno Mountains, southern Italy. *Catena*, *152*, 227–241. https://doi.org/10.1016/j.catena.2017.01.009

Estrany, J., Ruiz, M., Calsamiglia, A., Carriqui, M., Garcia-Comendador, J., Nadal, M., et al. (2019). Sediment connectivity linked to vegetation using UAVs: High-resolution imagery for ecosystem management. *Science of the Total Environment*, *671*, 1192–1205. https://doi.org/10.1016/j.scitotenv.2019.03.399

Fairman, T. A., Bennett, L. T., Tupper, S., & Nitschke, C. R. (2017). Frequent wildfires erode tree persistence and alter stand structure and initial composition of a fire-tolerant sub-alpine forest. *Journal of Vegetation Science*, *28*, 1151–1165. https://doi.org/10.1111/jvs.12575

Fenn, M. E., Jovan, S., Yuan, F., Geiser, L., Meixner, T., & Gimeno, B. S. (2008). Empirical and simulated critical loads for nitrogen deposition in California mixed conifer forests. *Environmental Pollution*, *155*, 492–511. https://doi.org/10.1016/j.envpol.2008.03.019

Fernelius, K. J., Madsen, M. D., Hopkins, B. G., Bansal, S., Anderson, V. J., Eggett, D. L., & Roundy, B. A. (2017). Post-fire interactions between soil water repellency, soil fertility and plant growth in soil collected from a burned piñon-juniper woodland. *Journal of Arid Environments*, *144*, 98–109. https://doi.org/10.1016/j.jaridenv.2017.04.005

Ferreira, A. J. D., Coelho, C. O. A., Boulet, A. K., & Lopes, F. P. (2005). Temporal patterns of solute loss following wildfires in central Portugal. *International Journal of Wildland Fire*, *14*, 401–412. https://doi.org/10.1071/WF05043

Fetzer, I., Johst, K., Schäwe, R., & Chatzinotas, A. (2015). The extent of functional redundancy changes as species' roles shift in different environments. *Proceedings National Academy of Sciences*, *112*, 14888–14893. https://doi.org/10.1073/pnas.1505587112

Field, J., & Banning, E. B. (1998). Hillslope processes and archaeology in Wadi Ziqlab, Jordan. *Geoarchaeology: An International Journal*, *13*, 595–616. https://doi.

org/10.1002/(SICI)1520-6548(199808)13:6%3C595::AID-GEA4%3E3.0.CO;2-U

Fisher, J. T., & Wilkinson, L. (2005). The response of mammals to forest fire and timber harvest in the North American boreal forest. *Mammal Review*, *35*, 51–81.

Fisk, M. C., Zak, D. R., & Crow, T. R. (2002). Nitrogen storage and cycling in old- and second-growth northern hardwood forests. *Ecology*, *83*, 73–87. https://doi.org/10.1890/0012-9658(2002)083%5b0073:NSACIO%5d2.0.CO;2

Florsheim, J. L., Keller, E. A., & Best, D. W. (1991). Fluvial sediment transport in response to moderate storm flows following chaparral wildfire, Ventura County, southern California. *Geological Society of America Bulletin*, *103*, 504–511. https://doi.org/10.1130/0016-7606(1991)103%3C0504:FSTIRT%3E2.3.CO;2

Follmi, D., Baartman, J., Benali, A., & Nunes, J. P. (2022). How do large wildfires impact sediment redistribution over multiple decades? *Earth Surface Processes and Landforms*, *47*, 3033–3050. https://doi.org/10.1002/esp.5441

Fonseca, F., Silva, D., Bueno, P., Hernandez, Z., Royer, A. C., & de Figueiredo, T. (2022). Temporal dynamics of carbon storage in a Mediterranean mountain scrubland managed by prescribed fire. *Catena*, *202*, 106107. https://doi.org/10.1016/j.catena.2022.106107

Fox, S., Sikes, B. A., Brown, S. P., Cripps, C. L., Glassman, S. I., Hughes, K., et al. (2022). Fire as a driver of fungal diversity – A synthesis of current knowledge. *Mycologia*, *114*, 215–241. https://doi.org/10.1080/00275514.2021.2024422

Francos, M., Pereira, P., Alcaniz, M., Mataix-Solera, J., & Ubeda, X. (2014). Impact of an intense rainfall event on soil properties following a wildfire in a Mediterranean environment (north-east Spain). *Science of the Total Environment*, *572*, 1353–1362. https://doi.org/10.1016/j.scitotenv.2016.01.145

Friend, G. R. (1993). Impact of fire on small vertebrates in mallee woodlands and heathlands of temperate Australia: a review. *Biological Conservation*, *65*, 99–114. https://doi.org/10.1016/0006-3207(93)90439-8

Fryirs, K. (2013). (Dis)Connectivity in catchment sediment cascades: a fresh look at the sediment delivery problem. *Earth Surface Processes and Landforms*, *38*, 30–46. https://doi.org/10.1002/esp.3242

Gabet, E. J. (2003). Sediment transport by dry ravel. *Journal of Geophysical Research: Solid Earth*, *108*, 2049. https://doi.org/10.1029/2001JB001686

Gabet, E. J., & Dunne, T. (2003). Sediment detachment by rain power. *Water Resources Research*, *39*, ESG 1-1-ESG 1-12. https://doi.org/10.1029/2001WR000656

Gabet, E. J., & Sternberg, P. (2008). The effects of vegetative ash on infiltration capacity, sediment transport, and the generation of progressively bulked debris flows. *Geomorphology*, *101*, 666–673. https://doi.org/10.1016/j.geomorph.2008.03.005

Gabrielli, C. P., McDonnell, J. J., & Jarvis, W. T. (2012). The role of bedrock groundwater in rainfall-runoff response at hillslope and catchment scales. *Journal of Hydrology*, *450-451*, 117–133. https://doi.org/10.1016/j.jhydrol.2012.05.023

Gallaway, J. M., Martin, Y. E., & Johnson, E. A. (2009). Sediment transport due to tree root throw: integrating tree population dynamics, wildfire and geomorphic response. *Earth Surface Processes and Landforms*, *34*, 1255–1269. https://doi.org/10.1002/esp.1813

Garcia, Y., Castellanos, M. C., & Pausas, J. G. (2018). Differential pollinator response underlies plant reproductive resilience after fires. *Annals of Botany*, *122*, 961–971. https://doi.org/10.1093/aob/mcy122

Gartner, J. E., Cannon, S. H., & Santi, P. M. (2014). Empirical models for predicting volumes of sediment deposited by debris flows and sediment-laden floods in the Transverse Range of southern California. *Engineering Geology*, *176*, 45–56. https://doi.org/10.1016/j.enggeo.2014.04.008

Gersh, M., Gleason, K. E., & Surunis, A. (2022). Forest fire effects on landscape snow albedo recovery and decay. *Remote Sensing*, *14*, 4079. https://doi.org/10.3390/rs14164079

Giovando, J., & Niemann, J. D. (2022). Wildfire impacts on snowpack phenology in a changing climate within the western US. *Water Resources Research*, *58*, e2021WR031569. https://doi.org/10.1029/2021WR031569

Glassman, S. I., Randolph, J. W. J., Saroa, S. S., Capocchi, J. K., Walters, K. E., & Pulido-Chavez, L. L. (2023). Prescribed versus wildfire impacts on exotic plants and soil microbes in California grasslands. *Applied Soil Ecology*, *185*, 104795. https://doi.org/10.1016/j.apsoil.2022.104795

Glassman, S. I., Wang, I. J., & Bruns, T. D. (2017). Environmental filtering by pH and soil nutrients drives community assembly in fungi at fine spatial scales. *Molecular Ecology*, *26*, 6960–6973. https://doi.org/10.1111/mec.14414

Gleason, K. E., Nolin, A. W., & Roth, T. R. (2013). Charred forests increase snowmelt: Effects of burned woody debris and incoming solar radiation on snow ablation. *Geophysical Research Letters*, *40*, 4654–4661. https://doi.org/10.1002/grl.50896

Glenn-Lewin, D. C., Peet, R. K., & Veblen, T. T. (1992). *Plant succession: theory and prediction*. NY: Chapman & Hall.

Goode, J. R., Luce, C. H., & Buffington, J. M. (2012). Enhanced sediment delivery in a changing climate in semi-arid mountain basins: Implications for water resource management and aquatic habitat in the northern Rocky Mountains. *Geomorphology*, *139-140*, 1–15. https://doi.org/10.1016/j.geomorph.2011.06.021

Gorr, A. N., McGuire, L. A., Youberg, A. M., & Rengers, F. K. (2022). A progressive flow-routing model for rapid assessment of debris-flow inundation. *Landslides*, *19*, 2055–2073. https://doi.org/10.1007/s10346-022-01890-y

Graber, A. P., Thomas, M. A., & Kean, J. W. (2023). How long do runoff-generated debris-flow hazards persist after wildfire? *Geophysical Research Letters*, *50*, e2023GL105101. https://doi.org/10.1029/2023GL105101

Graham R. T., Harvey A. E., Jurgensen M. F., Jain T. B, Tonn J. R, Page-Dumroese D. S. 1994. *Managing coarse woody debris in forests of the Rocky Mountains. USDA Forest Service*, Intermountain Research Station, Research Paper INT-RP-477, Ogden, UT.

Granath, G., Evans, C. D., Strengbom, J., Folster, J., Grelle, A., Stromqvist, J., & Kohler, S. J. (2021). The impact of wildfire on biogeochemical fluxes and

water quality in boreal catchments. *Biogeosciences*, *18*, 3243–3261. https://doi.org/10.5194/bg-18-3243-2021
Griffis, K. L., Crawford, J. A., Wagner, M. R., & Moir, W. H. (2002). Understory response to management treatments in northern Arizona ponderosa pine forests. *Forest Ecology and Management*, *146*, 239–245. https://doi.org/10.1016/S0378-1127(00)00461-8
Griffiths, A. D., & Brook, B. W. (2014). Effect of fire on small mammals: a systematic review. *International Journal of Wildland Fire*, *23*, 1034–1043. https://doi.org/10.1071/WF14026
Grulke, N. E., Minnich, R. A., Paine, T. D., Seybold, S. J., Chavez, D. J., Fenn, M. E., et al. (2008). Air pollution increases forest susceptibility to wildfires: a case study in the San Bernardino Mountains in southern California. *Developments in Environmental Science*, *8*, 365–403. https://doi.org/10.1016/S1474-8177(08)00017-X
Guilinger, J. J., Gray, A. B., Barth, N. C., & Fong, B. T. (2020). The evolution of sediment sources over a sequence of postfire sediment-laden flows revealed through repeat high-resolution change detection. *Journal of Geophysical Research Earth Surface*, *125*, e2020JF005527. https://doi.org/10.1029/2020JF005527
Gustine, R. N., Hanan, E. J., Robichaud, P. R., & Elliot, W. J. (2022). From burned slopes to streams: how wildfire affects nitrogen cycling and retention in forests and fire-prone watersheds. *Biogeochemistry*, *157*, 51–68. https://doi.org/10.1007/s10533-021-00861-0
Hack, J. T., & Goodlett, J. C. (1960). Geomorphology and forest ecology of a mountain region in the Central Appalachians. *U.S. Geological Survey Professional Paper*, *347*. 66 p
Hahnenberger, M., & Nicoll, K. (2014). Geomorphic and land cover identification of dust sources in the eastern Great Basin of Utah, USA. *Geomorphology*, *204*, 657–672. https://doi.org/10.1016/j.geomorph.2013.09.013
Hales, T. C., Scharer, K. M., & Wooten, R. M. (2012). Southern Appalachian hillslope erosion rates measured by soil and detrital radiocarbon in hollows. *Geomorphology*, *138*, 121–129. https://doi.org/10.1016/j.geomorph.2011.08.030
Hamilton, N. P., & Burton, P. J. (2023). Wildfire disturbance reveals evidence of ecosystem resilience and precariousness in a forest-grassland mosaic. *Ecosphere*, *14*, e4460. https://doi.org/10.1002/ecs2.4460
Hanan, E. J., D'Antonio, C. M., Roberts, D. A., & Schimel, J. P. (2016). Factors regulating nitrogen retention during the early stages of recovery from fire in coastal chaparral ecosystems. *Ecosystems*, *19*(5), 910–926. https://doi.org/10.1007/s10021-016-9975-0
Hanan, E. J., Tague, C. N., & Schimel, J. P. (2017). Nitrogen cycling and export in California chaparral: the role of climate in shaping ecosystem responses to fire. *Ecological Monographs*, *87*, 76–90. https://doi.org/10.1002/ecm.1234
Hannon, S. J., & Drapeau, P. (2005). Bird responses to burning and logging in the boreal forest of Canada. *Studies in Avian Biology*, *30*, 97–115.

Harmon, M. E., Franklin, J. F., Swanson, F. J., Sollins, P., Gregory, S. V., Lattin, J. D., et al. (1986). Ecology of coarse woody debris in temperate ecosystems. *Advances in Ecological Research*, *15*, 133–302. https://doi.org/10.1016/S0065-2504(08)60121-X

Hart, S. J., Schoennagel, T., Veblen, T. T., & Chapman, T. B. (2015). Area burned in the western United States is unaffected by recent mountain pine beetle outbreaks. *Proceedings of the National Academy of Sciences*, *112*, 4375–4380. https://doi.org/10.1073/pnas.1424037112

Harvey, B. J., Donato, D. C., & Turner, M. G. (2014). Recent mountain pine beetle outbreaks, wildfire severity, and postfire tree regeneration in the US Northern Rockies. *Proceedings of the National Academy of Sciences*, *111*, 15120–15125. www.pnas.org/cgi/doi/10.1073/pnas.1411346111

Hatchett, B. J., Koshkin, A. L., Guirguis, K., Rittger, K., Nolin, A. W., Heggli, A., et al. (2023). Midwinter dry spells amplify post-fire snowpack decline. *Geophysical Research Letters*, *50*, e2022GL101235. https://doi.org/10.1029/2022GL101235

Hauer, F. R., & Spencer, C. N. (1998). Phosphorus and nitrogen dynamics in streams associated with wildfire: a study of intermediate and long-term effects. *International Journal of Wildland Fire*, *8*, 183–198. https://doi.org/10.1071/WF9980183

Haynes, R. J. (1986). *Mineral nitrogen in the plant-soil system*. Orlando, FL: Academic Press, Inc.

He, T., Lamont, B. B., & Pausas, J. G. (2019). Fire as a key driver of Earth's biodiversity. *Biological Reviews*, *94*, 1983–2010. https://doi.org/10.1111/brv.12544

Heath, J. T., Chafer, C. J., Bishop, T. F. A., & Van Ogtrop, F. F. (2015). Wildfire effects on soil carbon and water repellency under eucalyptus forest in East Australia. *Soil Research*, *53*, 12–23. https://doi.org/10.1071/SR13170

Heathwaite, A. L., & Dils, R. M. (2000). Characterising phosphorus loss in surface and subsurface hydrological pathways. *Science of the Total Environment*, *251*(252), 523–538. https://doi.org/10.1016/S0048-9697(00)00393-4

Heimsath, A. M., Dietrich, W. E., Nishiizumi, K., & Finkel, R. C. (1997). The soil production function and landscape equilibrium. *Nature*, *388*, 358–361. https://doi.org/10.1038/41056

Heindel, R. C., Murphy, S. F., Repert, D. A., Wetherbee, G. A., Liethen, A. E., Clow, D. W., & Halamka, T. A. (2022). Elevated nitrogen deposition to fire-prone forests adjacent to urban and agricultural areas, Colorado Front Range, USA. *Earth's Future*, *10*, e2021EF002373. https://doi.org/10.1029/2021EF002373

Helvey, J. D., Tiedemann, A. R., & Anderson, T. D. (1985). Plant nutrient losses by soil erosion and mass movement after wildfire. *Journal of Soil and Water Conservation*, *40*, 168–173. https://www.jswconline.org/content/40/1/168.short

Hessburg, P. F., Miller, C. L., Parks, S. A., Povak, N. A., Taylor, A. H., Higuera, P. E., et al. (2019). Climate, environment, and disturbance history govern resilience of western North American forests. *Frontiers in Ecology and Evolution*, *7*, 239. https://doi.org/10.3389/fevo.2019.00239

Hewitt, R. E., Day, N. J., DeVans, M. R., & Taylor, D. L. (2022). Wildfire impacts on root-associated fungi and predicted plant-soil feedbacks in the boreal forest: research progress and recommendations. *Functional Ecology*, *37*, 2110–2125. https://doi.org/10.1111/1365-2435.14205

Hicke, J. A., Johnson, M. C., Hayes, J. L., & Preisler, H. K. (2012). Effects of bark beetle-caused tree mortality on wildfire. *Forest Ecology and Management*, *271*, 81–90. https://doi.org/10.1016/j.foreco.2012.02.005

Hoch, O. J., McGuire, L. A., Youberg, A. M., & Rengers, F. K. (2021). Hydrogeomorphic recovery and temporal changes in rainfall thresholds for debris flows following wildfire. *Journal of Geophysical Research: Earth Surface*, *126*, e2021JF006374. https://doi.org/10.1029/2021JF006374

Holden, J., Palmer, S. M., Johnston, K., Wearing, C., Irvine, B., & Brown, L. E. (2015). Impact of prescribed burning on blanket peat hydrology. *Water Resources Research*, *51*, 6472–6484. https://doi.org/10.1002/2014WR016782

Holloway, J. E., Lewkowicz, A. G., Douglas, T. A., Li, X., Turetsky, M. R., Baltzer, J. L., & Jin, H. (2020). Impact of wildfire on permafrost landscapes: a review of recent advances and future prospects. *Permafrost and Periglacial Processes*, *31*, 371–382. https://doi.org/10.1002/ppp.2048

Homyak, P. M., Sickman, J. O., Miller, A. E., Melack, J. M., Meixner, T., & Schimel, J. P. (2014). Assessing nitrogen-saturation in a seasonally dry chaparral watershed: limitations of traditional indicators of N-saturation. *Ecosystems*, *17*, 1286–1305. https://doi.org/10.1007/s10021-014-9792-2

Hou, Y., Wei, X., Zhang, M., Creed, I. F., McNulty, S. G., & Ferraz, S. F. B. (2023). A global synthesis of hydrological sensitivities to deforestation and forestation. *Forest Ecology and Management*, *529*, 120718. https://doi.org/10.1016/j.foreco.2022.120718

Hyde, J. C., Smith, A. M. S., & Ottmar, R. D. (2012). Properties affecting the consumption of sound and rotten coarse woody debris in northern Idaho: a preliminary investigation using laboratory fires. *International Journal of Wildland Fire*, *21*, 596–608. https://doi.org/10.1071/WF11016

Hyde, J. C., Smith, A. M. S., Ottmar, R. D., Alvarado, E. C., & Morgan, P. (2011). The combustion of sound and rotten coarse woody debris: a review. *International Journal of Wildland Fire*, *20*, 163–174. https://doi.org/10.1071/WF09113

Ibáñez, T. S., Wardle, D. A., Gundale, M. J., & Nilsson, M. C. (2022). Effects of soil abiotic and biotic factors on tree seedling regeneration following a boreal forest wildfire. *Ecosystems*, *25*, 471–487. https://doi.org/10.1007/s10021-021-00666-0

Iverson, R. M., Reid, M. E., & LaHusen, R. G. (1997). Debris-flow mobilization from landslides. *Annual Reviews of Earth and Planetary Sciences*, *25*, 85–138. https://doi.org/10.1146/annurev.earth.25.1.85

Jackson, M., & Roering, J. J. (2009). Post-fire geomorphic response in steep, forested landscapes: Oregon Coast Range, USA. *Quaternary Science Reviews*, *28*, 1131–1146. https://doi.org/10.1016/j.quascirev.2008.05.003

Jager, H. I., Long, J. W., Malison, R. L., Murphy, B. P., Rust, A., Silva, L. G. M., et al. (2021). Resilience of terrestrial and aquatic fauna to historical and future wildfire regimes in western North America. *Ecology and Evolution*, *11*, 12259–12284. https://doi.org/10.1002/ece3.8026

Jin, T., Hu, X., Liu, B., Xi, C., He, K., Cao, X., et al. (2022). Susceptibility prediction of post-fire debris flows in Xichang, China, using a logistic regression model from a spatiotemporal perspective. *Remote Sensing*, *14*, 1306. https://doi.org/10.3390/rs14061306

Johnson, A. M. (1984). Debris flows. In D. Brunsden & D. B. Prior (Eds.), *Slope instability* (pp. 257–361). NY: Wiley.

Johnson, D., Murphy, J. D., Walker, R. F., Glass, D. W., & Miller, W. W. (2007). Wildfire effects on forest carbon and nutrient budgets. *Ecological Engineering*, *31*, 183–192. https://doi.org/10.1016/j.ecoleng.2007.03.003

Johnson, D., & Turner, J. (2014). Nitrogen budgets of forest ecosystems: a review. *Forest Ecology and Management*, *318*, 370–379. https://doi.org/10.1016/j.foreco.2013.08.028

Jones, G. M., & Tingley, M. W. (2022). Pyrodiversity and biodiversity: a history, synthesis, and outlook. *Diversity and Distributions*, *28*, 386–403. https://doi.org/10.1111/ddi.13280

Jones, W. M. K., Pollard, W. H., & Jones, B. M. (2019). Rapid initialization of retrogressive thaw slumps in the Canadian High Arctic and their response to climate and terrain factors. *Environmental Research Letters*, *14*, 055006. https://doi.org/10.1088/1748-9326/ab12fd

Jordan, P. (2012). Sediment yields and water quality effects of severe wildfires in southern British Columbia. In, Wildfire and Water Quality: Processes. *Impacts and Challenges. IAHS Publication*, *354*, 25–35.

Jung, H. Y., Hogue, T. S., Rademacher, L. K., & Meixner, T. (2009). Impact of wildfire on source water contributions in Devil Creek, CA: evidence from end-member mixing analysis. *Hydrological Processes*, *23*, 183–200. https://doi.org/10.1002/hyp.7132

Kean, J. W., & Staley, D. M. (2021). Forecasting the frequency and magnitude of postfire debris flows across southern California. *Earth's Future*, *9*, e202EF001735. https://doi.org/10.1029/2020EF001735

Kean, J. W., Staley, D. M., & Cannon, S. H. (2011). In situ measurements of post-fire debris flows in southern California: comparisons of the timing and magnitude of 24 debris-flow events with rainfall and soil moisture conditions. *Journal of Geophysical Research*, *116*, F04019. https://doi.org/10.1029/2011JF002005

Kean J. W., Staley D. M., Lancaster J. T., Rengers F. K., Swanson B. J., Coe J. A, Hernandez J. L., Sigman A. J., Allstadt K., Lindsay D. N. 2019. *Debris-flow inundation and damage data from the 9 January 2018 Montecito debris-flow event*. U.S. Geological Survey data release. https://doi.org/10.5066/P9JQJU0E

Kelley, C. J., Keller, C. K., Brooks, E. S., Smith, J. L., Orr, C. H., & Evans, R. D. (2017). Water and nitrogen movement through a semiarid dryland agricultural catchment: seasonal and decadal trends. *Hydrological Processes*, *31*, 1889–1899. https://doi.org/10.1002/hyp.11152

Kemter, M., Fischer, M., Luna, L. V., Schöenfeld, E., Vogel, J., Banerjee, A., et al. (2021). Cascading hazards in the aftermath of Australia's 2019/2020 Black

Summer wildfires. *Earth's Future*, *9*, e2020EF001884. https://doi.org/10.1029/2020EF001884

Keyser, T. L., Lentile, L. B., Smith, F. W., & Shepperd, W. D. (2008). Changes in forest structure after a large, mixed-severity wildfire in ponderosa pine forests of the Black Hills, South Dakota, USA. *Forest Science*, *54*, 328–338. https://doi.org/10.1093/forestscience/54.3.328

Kieta, K. A., Owens, P. N., & Petticrew, E. L. (2023). Determination of sediment sources following a major wildfire and evaluation of the use of color properties and polycyclic aromatic hydrocarbons (PAHs) as tracers. *Journal of Soils and Sediments*, *23*, 4187–4207. https://doi.org/10.1007/s11368-023-03565-0

Kipfer, T., Egli, S., Ghazoul, J., Moser, B., & Wohlgemuth, T. (2010). Susceptibility of ectomycorrhizal fungi to soil heating. *Fungal Biology*, *114*, 467–472. https://doi.org/10.1016/j.funbio.2010.03.008

Komonen, A., & Müller, J. (2018). Dispersal ecology of deadwood organisms and connectivity conservation. *Conservation Biology*, *32*, 535–545. https://doi.org/10.1111/cobi.13087

Koponen, S. (2005). Early succession of a boreal spider community after forest fire. *The Journal of Arachnology*, *33*, 230–235. https://doi.org/10.1636/CT04-112.1

Köster, K., Aaltonen, H., Berninger, F., Heinonsalo, J., Köster, E., Ribeiro-Kumara, C., et al. (2021). Impacts of wildfire on soil microbiome in boreal environments. *Current Opinion in Environmental Soil and Health*, *22*, 100258. https://doi.org/10.1016/j.coesh.2021.100258

Kuo, H. C., Hui, S., Choi, J., Asiegbu, F. O., Valkonen, J. P., & Lee, Y. H. (2014). Secret lifestyles of *Neurospora crassa*. *Scientific Reports*, *4*, 5135. https://doi.org/10.1038/srep05135

Kurth, V. J., Hart, S. C., Ross, C. S., Kaye, J. P., & Fulé, P. Z. (2014). Stand-replacing wildfires increase nitrification for decades in southwestern ponderosa pine forests. *Oecologia*, *175*, 395–407. https://doi.org/10.1007/s00442-014-2906-x

Lamb, M. P., Levina, M., DiBiase, R. A., & Fuller, B. M. (2013). Sediment storage by vegetation in steep bedrock landscapes: theory, experiments, and implications for postfire sediment yield. *Journal of Geophysical Research Earth Surface*, *118*, 1147–1160. https://doi.org/10.1002/jgrf.20058

Lamb, M. P., Scheingross, J. S., Amidon, W. H., Swanson, E., & Limaye, A. (2011). A model for fire-induced sediment yield by dry ravel in steep landscapes. *Journal of Geophysical Research Earth Surface*, *116*, F03006. https://doi.org/10.1029/2010JF001878

Lamont, B. B., & Downes, K. S. (2011). Fire-stimulated flowering among resprouters and geophytes in Australia and South Africa. *Plant Ecology*, *212*, 2111–2125. https://doi.org/10.1007/s11258-011-9987-y

Lancaster, S. T., Hayes, S. K., & Grant, G. E. (2003). Effects of wood on debris flow runout in small mountain watersheds. *Water Resources Research*, *39*, 1168. https://doi.org/10.1029/2001WR001227

Lane, P. N. J., Sheridan, G. J., Noske, P. J., & Sherwin, C. B. (2008). Phosphorus and nitrogen exports from SE Australian forests following wildfire. *Journal of Hydrology*, *361*, 186–198. https://doi.org/10.1016/j.jhydrol.2008.07.041

Larsen I. J., 2003. *From the rim to the river – the geomorphology of debris flows in the Green River Canyons of Dinosaur National Monument*, Colorado and Utah. MS thesis, Utah State University, Logan, UT, 196 pp.

Larsen, I. J., MacDonald, L. H., Brown, E., Rough, D., Welsh, M. J., Pietraszek, J. H., et al. (2009). Causes of post-fire runoff and erosion: water repellency, cover, or soil sealing? *Soil Science Society of America Journal*, *73*, 1393–1407. https://doi.org/10.2136/sssaj2007.0432

Larsen, M. C., & Parks, J. E. (1997). How wide is a road? The association of roads and mass-wasting in a forested montane environment. *Earth Surface Processes and Landforms*, *22*, 835–848. https://doi.org/10.1002/(SICI)1096-9837(199709)22:9%3C835::AID-ESP782%3E3.0.CO;2-C

Lentile, L. B., Morgan, P., Hudak, A. T., Bobbitt, M. J., Lewis, S. A., Smith, A. M. S., & Robichaud, P. R. (2007). Post-fire burn severity and vegetation response following eight large wildfires across the western United States. *Fire Ecology*, *3*, 91–108. https://doi.org/10.4996/fireecology.0301091

Lewkowicz, A. G., & Harris, C. (2005). Frequency and magnitude of active-layer detachment failures in discontinuous and continuous permafrost, northern Canada. *Permafrost and Periglacial Processes*, *16*, 115–130. https://doi.org/10.1002/ppp.522

Li, X. Y., Jin, H. J., Wang, H. W., Marchenko, S. S., Shan, W., Luo, D. L., et al. (2021). Influences of forest fires on the permafrost environment: a review. *Advances in Climate Change Research*, *12*, 48–65. https://doi.org/10.1016/j.accre.2021.01.001

Li, Z., Li-min, D., Hui-yan, G., & Lei, Z. (2007). Review on the decomposition and influence factors of coarse woody debris in forest ecosystem. *Journal of Forestry Research*, *18*, 48–54. https://doi.org/10.1007/s11676-007-0009-9

Lillywhite, H. B., Friedman, G., & Ford, N. (1977). Color matching and perch selection by lizards in recently burned chaparral. *Copeia*, *1977*, 115–121. https://doi.org/10.2307/1443512

Lin, J., Compton, J. E., Leibowitz, S. G., Mueller-Warrant, G., Matthews, W., Schoenholtz, S. H., et al. (2019). Seasonality of nitrogen balances in a Mediterranean climate watershed, Oregon, US. *Biogeochemistry*, *142*, 247–264. https://doi.org/10.1007/s10533-018-0532-0

Lindaas, J., Pollack, I. B., Garofalo, L. A., Pothier, M. A., Farmer, D. K., Kreidenweis, S. M., et al. (2021). Emissions of reactive nitrogen from Western U.S. wildfires during summer 2018. *Journal of Geophysical Research*, *126*, e2020JD032657. https://doi.org/10.1029/2020JD032657

Lips, K. R. (1991). Vertebrates associated with tortoise (*Gopherus polyphemus*) burrows in four habitats in south-central Florida. *Journal of Herpetology*, *25*, 477–481. https://doi.org/10.2307/1564772

Liu, J., Qiu, L., Wang, X., Wei, X., Gao, H., Zhang, Y., & Cheng, J. (2018). Effects of wildfire and topography on soil nutrients in a semiarid restored grassland. *Plant and Soil*, *428*, 123–136. https://doi.org/10.1007/s11104-018-3659-9

López-Cruz, S. C., Aryal, D. R., Velazquez-Sanabria, C. A., Guevara-Hernandez, F., Venegas-Sandoval, A., Casanova-Lugo, F., et al. (2022). Effect of prescribed burning on tree diversity, biomass stocks and soil organic carbon storage in tropical highland forests. *Forests*, *13*, 2164. https://doi.org/10.3390/f13122164

Lu, X., Li, Y., Washington-Allen, R. A., & Li, Y. (2019). Structural and sedimentological connectivity on a rilled hillslope. *Science of the Total Environment*, *655*, 1479–1494. https://doi.org/10.1016/j.scitotenv.2018.11.137

Lyon, R. H. (1997). Effects of an extensive wildfire on birds in far eastern Victoria. *Pacific Conservation Biology*, *3*, 221–234. https://doi.org/10.1071/PC970221

MacNeille, R. B., Lohse, K. A., Godsey, S. E., Perdrial, J. N., & Baxter, C. V. (2020). Influence of drying and wildfire on longitudinal chemistry patterns and processes of intermittent streams. *Frontiers in Water*, *2*, 563841. https://doi.org/10.3389/frwa.2020.563841

Madani, N., Parazoo, N. C., Kimball, J. S., Reichle, R. H., Chatterjee, A., Watts, J. D., et al. (2021). The impacts of climate and wildfire on ecosystem gross primary productivity in Alaska. *Journal of Geophysical Research: Biogeosciences*, *126*, e2020JG006078. https://doi.org/10.1029/2020JG006078

Mallik, A. U. (2022). Post-fire habitat heterogeneity leads to black spruce-Kalmia L. shrub savannah alternate state. *Forests*, *13*, 570. https://doi.org/10.3390/f13040570

Mansilha, C., Melo, A., Martins, Z. E., Ferreira, I. M. P. L. V. O., Pereira, A. M., & Marques, J. E. (2020). Wildfire effects on groundwater quality from springs connected to small public supply systems in a peri-urban forest area (Braga Region, NW Portugal). *Water*, *12*, 1146. https://doi.org/10.3390/w12041146

Marañón-Jiménez, S., & Castro, J. (2013). Effect of decomposing post-fire coarse woody debris on soil fertility and nutrient availability in a Mediterranean ecosystem. *Biogeochemistry*, *112*, 519–535. https://doi.org/10.1007/s10533-012-9744-x

Marañón-Jiménez, S., Castro, J., Fernández-Ondoño, E., & Zamora, R. (2013). Charred wood remaining after a wildfire as a reservoir of macro- and micronutrients in a Mediterranean pine forest. *International Journal of Wildland Fire*, *22*, 681–695. https://doi.org/10.1071/WF12030

Martí-Roura, M., Casals, P., & Romanyà. (2011). Temporal changes in soil organic C under Mediterranean shrublands and grasslands: impact of fire and drought. *Plant and Soil*, *338*, 280–300. https://doi.org/10.1007/s11104-010-0485-0

Maser, S., Anderson, R. G., Cromack, K., Williams, J. T., & Cummins, K. W. (1979). Dead and down woody material. In J. Thomas (Ed.), *Wildlife habitats in managed forests, the Blue Mountains of Oregon and Washington*. Washington, DC: USDA Forest Service Agricultural Handbook 553.

Massman, W. (2012). Modeling soil heating and moisture transport under extreme conditions: forest fires and slash pile burns. *Water Resources Research*, *48*, W10548. https://doi.org/10.1029/2011WR011710

Massman, W. J., Frank, J. M., Shepperd, W. D., & Platten, M. J. (2002). In situ soil temperature and heat flux measurements during controlled surface burns at a southern Colorado forest site. In P. N. Omi & L. A. Joyce (Eds.), *Fire, Fuel*

Treatments, and Ecological Restoration: Conference Proceedings. USDA Forest Service Proceedings RMRS-P-29, Fort Collins, CO (pp. 69–90).

Mast, M. A., Murphy, S. F., Clow, D. W., Penn, C. A., & Sexstone, G. A. (2016). Water-quality response to a high-elevation wildfire in the Colorado Front Range. *Hydrological Processes*, *30*, 1811–1823. https://doi.org/10.1002/hyp.10755

McCullough, D. G., Werner, R. A., & Neumann, D. (1998). Fire and insects in northern and boreal forest ecosystems of North America. *Annual Reviews in Entomology*, *43*, 107–127. https://doi.org/10.1146/annurev.ento.43.1.107

McDonnell, J. J., Spence, C., Karran, D. J., van Meerveld, H. J., & Harman, C. J. (2021). Fill-and-spill: A process definition of runoff generation at the scale of the beholder. *Water Resources Research*, *57*, e2020WR027514. https://doi.org/10.1029/2020WR027514

McGuire, L. A., Kean, J. W., Staley, D. M., Rengers, F. K., & Wasklewicz, T. A. (2016). Constraining the relative importance of raindrop- and flow-driven sediment transport mechanisms in postwildfire environments and implications for recovery time scales. *Journal of Geophysical Research: Earth Surface*, *121*, 2211–2237. https://doi.org/10.1002/2016JF003867

McGuire, L. A., Rengers, F. K., Kean, J. W., & Staley, D. M. (2017). Debris flow initiation by runoff in a recently burned basin: Is grain-by-grain sediment bulking or en masse failure to blame? *Geophysical Research Letters*, *44*, 7310–7319. https://doi.org/10.1002/2017GL074243

McGuire, L. A., Rengers, F. K., Youberg, A. M., Gorr, A. N., Hoch, O. J., Beers, R., & Porter, R. (2024). Characteristics of debris flow prone watersheds and triggering rainstorms following the Tadpole Fire, New Mexico, USA. *EGUsphere*. preprint. https://doi.org/10.5194/egusphere-2023-1672

McGuire, L. A., & Youberg, A. M. (2020). What drives spatial variability in rainfall intensity-duration thresholds for post-wildfire debris flows? Insights from the 2018 Buzzard Fire, NM, USA. *Landslides*, *17*, 2385–2399. https://doi.org/10.1007/s10346-020-01470-y

McPherson, G. R. (1997). *Ecology and management of North American Savannas*. Tucson: University of Arizona Press.

Meddens, A. J., Kolden, C. A., Lutz, J. A., Smith, A. M., Cansler, C. A., Abatzoglou, J. T., et al. (2018). Fire refugia: what are they, and why do they matter for global change? *BioScience*, *68*, 944–954. https://doi.org/10.1093/biosci/biy103

Meyer, G. A., Pierce, J. L., Wood, S. H., & Jull, A. J. T. (2001). Fire, storms, and erosional events in the Idaho batholith. *Hydrological Processes*, *15*, 3025–3038. https://doi.org/10.1002/hyp.389

Meyer, G. A., & Wells, S. G. (1997). Fire-related sedimentation events on alluvial fans, Yellowstone National Park, USA. *Journal of Sedimentary Research*, *67*, 776–791. https://doi.org/10.1306/D426863A-2B26-11D7-8648000102C1865D

Meyer, G. A., Wells, S. G., & Jull, A. J. T. (1995). Fire and alluvial chronology in Yellowstone National Park: climatic and intrinsic controls on Holocene geomorphic processes. *Geological Society of America Bulletin*, *107*, 1211–1230. https://doi.org/10.1130/0016-7606(1995)107%3C1211:FAACIY%3E2.3.CO;2

Miller, M. E., Bowker, M. A., Reynolds, R. L., & Goldstein, H. L. (2012). Post-fire land treatments and wind erosion – Lessons from the Milford Flat Fire, UT, USA. *Aeolian Research*, *7*, 29–44. https://doi.org/10.1016/j.aeolia.2012.04.001

Miller, R. G., Tangney, R., Enright, N. J., Fontaine, J. B., Merritt, D. J., Ooi, M. K. J., et al. (2019). Mechanisms of fire seasonality effects on plant populations. *Trends in Ecology and Evolution*, *34*, 1104–1117. https://doi.org/10.1016/j.tree.2019.07.009

Mishra, A., Alnahit, A., & Campbell, B. (2021). Impact of land uses, drought, flood, wildfire, and cascading events on water quality and microbial communities: A review. *Journal of Hydrology*, *596*, 125707. https://doi.org/10.1016/j.jhydrol.2020.125707

Mitchell, R. G. (1990). Effects of prescribed fire on insect pests. In J. D. Walstad, S. R. Radosevich, & D. V. Sandberg (Eds.), *Natural and prescribed fire in Pacific Northwest forests* (pp. 111–116). Corvallis, OR: Oregon State University Press.

Moghimian, N., Jalali, S. G., Kooch, Y., & Rey, A. (2020). Downed logs improve soil properties in old-growth temperate forests of northern Iran. *Pedosphere*, *30*(3), 378–389. https://doi.org/10.1016/S1002-0160(17)60424-7

Montgomery, D. R., & Dietrich, W. E. (1988). Where do channels begin? *Nature*, *336*, 232–234. https://doi.org/10.1038/336232a0

Montgomery, D. R., & Dietrich, W. E. (1992). Channel initiation and the problem of landscape scale. *Science*, *255*, 826–830. https://doi.org/10.1126/science.255.5046.826

Moody, J. A., & Martin, D. A. (2009). Synthesis of sediment yields after wildland fire in different rainfall regimes in the western United States. *International Journal of Wildland Fire*, *18*, 96–115. https://doi.org/10.1071/WF07162

Moody, J. A., Shakesby, R. A., Robichaud, P. R., Cannon, S. H., & Martin, D. A. (2013). Current research issues related to post-wildfire runoff and erosion processes. *Earth-Science Reviews*, *122*, 10–37. https://doi.org/10.1016/j.earscirev.2013.03.004

Moreno, J. M., & Oechel, W. C. (Eds.) (1994). *The role of fire in Mediterranean-type ecosystems*. Ecology Study No. 107. NY: Springer-Verlag.

Morris, R., Dragovich, D., & Ostendorf, B. (2012). Hillslope erosion and post-fire sediment trapping at Mount Bold, South Australia. In M. Stone, A. Collins, & M. Thoms (Eds.), *Wildfire and water quality: processes, impacts and challenges* (Vol. 354, pp. 42–50). Wallingford: IAHS Pub.

Murphy, B. P., Czuba, J. A., & Belmont, P. (2019). Post-wildfire sediment cascades: a modeling framework linking debris flow generation and network-scale sediment routing. *Earth Surface Processes and Landforms*, *44*, 2126–2140. https://doi.org/10.1002/esp.4635

Murphy, M. J., Jones, H. A., & Koen, T. (2021). A seven-year study of the response of woodland birds to a large-scale wildfire and the role of proximity to unburnt habitat. *Austral Ecology*, *46*, 1138–1155. https://doi.org/10.1111/aec.13049

Mustard, J. F., Defries, R. S., Fisher, T., & Moran, E. (2012). Land-use and land-cover change pathways and impacts. In G. Gutman, et al. (Eds.), *Land change science*. Remote Sensing and Digital Image Processing (Vol. 6). Dordrecht: Springer. https://doi.org/10.1007/978-1-4020-2562-4_24

Neary, D. G., Klopatek, C. C., DeBano, L. F., & Ffolliott, P. F. (1999). Fire effects on belowground sustainability: a review and synthesis. *Forest Ecology and Management*, *122*, 51–71. https://doi.org/10.1016/S0378-1127(99)00032-8

Neary, D. G., Ryan, K. C., & DeBano, L. F. (2005). *Wildland fire in ecosystems: effects of fire on soils and water (RMRSGTR-42-V4)*. Ft. Collins, CO: U.S. Department of Agriculture, Forest Service, Rocky Mountain Research Station.

Neely, A. B., DiBiase, R. A., Corbett, L. B., Bierman, P. R., & Caffee, M. W. (2019). Bedrock fracture density controls on hillslope erodibility in steep, rocky landscapes with patchy soil cover, southern California, USA. *Earth and Planetary Science Letters*, *522*, 186–197. https://doi.org/10.1016/j.epsl.2019.06.011

Neff, J. C., Harden, J. W., & Gleixner, G. (2005). Fire effects on soil organic matter content, composition, and nutrients in boreal interior Alaska. *Canadian Journal of Forest Research*, *35*, 2178–2187. https://doi.org/10.1139/x05-154

Nelson, A. R., Narrowe, A. B., Rhoades, C. C., Fegel, T. S., Daly, R. A., Roth, H. K., et al. (2022). Wildfire-dependent changes in soil microbiome diversity and function. *Nature Microbiology*, *7*, 1419–1430. https://doi.org/10.1038/s41564-022-01203-y

Nemens, D. G., Kidd, K. R., Varner, J. M., & Wing, B. (2022). Recurring wildfires provoke type conversion in dry western forests. *Ecosphere*, *13*, e4184. https://doi.org/10.1002/ecs2.4184

Neris, J., Santin, C., Lew, R., Robichaud, P. R., Elliot, W. J., Lewis, S. A., et al. (2021). Designing tools to predict and mitigate impacts on water quality following the Australian 2019/2020 wildfires: insights from Sydney's largest water supply catchment. *Integrated Environmental Assessment and Management*, *17*, 1151–1161. https://doi.org/10.1002/ieam.4406

Noske, P. J., Lane, P. N. J., & Sheridan, G. J. (2010). Stream exports of coarse matter and phosphorus following wildfire in NE Victoria, Australia. *Hydrological Processes*, *24*, 1514–1529. https://doi.org/10.1002/hyp.7616

Novara, A., Gristina, L., Ruhl, J., Pasta, S., D'Angelo, G., La Mantia, T., & Pereira, P. (2013). Grassland fire effect on soil organic carbon reservoirs in a semiarid environment. *Solid Earth*, *4*, 381–385. https://doi.org/10.5194/se-4-381-2013

Nunes, B. A., Cunha, A., & Vieira, E. M. (2023). Effects of a large wildfire on the community composition of medium and large mammals in a neotropical savannah. *International Journal of Wildland Fire*, *32*, 1109–1118. https://doi.org/10.1071/WF22023

Nyman, P., Box, W. A. C., Stout, J. C., Sheridan, G. J., Keesstra, S. D., Lane, P. N. J., & Langhans, C. (2020). Debris-flow-dominated sediment transport through a channel network after wildfire. *Earth Surface Processes and Landforms*, *45*, 1155–1167. https://doi.org/10.1002/esp.4785

Olefeldt, D., Turetsky, M. R., & Blodau, C. (2013). Altered composition and microbial versus UV-mediated degradation of dissolved organic matter in boreal soils following wildfire. *Ecosystems*, *16*, 1396–1412. https://doi.org/10.1007/s10021-013-9691-y

Ollerton, J., Winfree, R., & Tarrant, S. (2011). How many flowering plants are pollinated by animals? *Oikos*, *120*, 321–326. https://doi.org/10.1111/j.1600-0706.2010.18644.x

Ollier, C. D., & Ash, J. E. (1983). Fire and rock breakdown. *Zeitschrift fur Geomorphologie*, *27*, 363–374.

Ortega-Becerril, J. A., Suarez, C., Vazquez-Tarrio, D., Garrote, J., & Gomez-Heras, M. (2024). Sediment response after wildfires in mountain streams and their effects on cultural heritage: The case of the 2021 Navalcruz Wildfire (Avila, Spain). *Fire*, *7*, 52. https://doi.org/10.3390/fire7020052

Ortíz-Rodríguez, A., Muñoz-Robles, C., & Borselli, L. (2019). Changes in connectivity and hydrological efficiency following wildland fires in Sierra Madre Oriental, Mexico. *Science of the Total Environment*, *655*, 112–128. https://doi.org/10.1016/j.scitotenv.2018.11.236

Pacala, S. W., & Rees, M. (1998). Models suggesting field experiments to test two hypotheses explaining successional diversity. *American Naturalist*, *152*, 729–737. https://doi.org/10.1086/286203

Palmer, B., Lawson, D., & Lipson, D. A. (2023). Years after a fire, biocrust microbial communities are similar to unburned communities in a coastal grassland. *Microbial Ecology*, *85*, 1028–1044. https://doi.org/10.1007/s00248-022-02137-y

Palviainen, M., Lauren, A., Pumpanen, J., Bergeron, Y., Bond-Lamberty, B., Larjavaara, M., et al. (2020). Decadal-scale recovery of carbon stocks after wildfires throughout the boreal forests. *Global Biogeochemical Cycles*, *34*, e2020GB006612. https://doi.org/10.1029/2020GB006612

Parton, W. J., Holland, E. A., Grosso, S. J. D., Hartman, M. D., Martin, R. E., Mosier, A. R., et al. (2001). Generalized model for No_x and N_2O emissions from soils. *Journal of Geophysical Research*, *106*, 17403–17419. https://doi.org/10.1029/2001JD900101

Passovoy, M. D., & Fulé, P. Z. (2006). Snag and woody debris dynamics following severe wildfires in northern Arizona ponderosa pine forests. *Forest Ecology and Management*, *223*, 237–246. https://doi.org/10.1016/j.foreco.2005.11.016

Patterson, G. B. (1984). The effect of burning-off tussock grassland on the population density of common skinks. *New Zealand Journal of Zoology*, *11*, 189–194. https://doi.org/10.1080/03014223.1984.10423757

Paul, M. J., LeDuc, S. D., Lassiter, M. G., Moorhead, L. C., Noyes, P. D., & Leibowitz, S. G. (2022). Wildfire induces changes in receiving waters: a review with considerations for water quality management. *Water Resources Research*, *58*, e2021WR030699. https://doi.org/10.1029/2021WR030699

Pausas, J. G., & Keeley, J. E. (2014). Evolutionary ecology of resprouting and seeding in fire-prone ecosystems. *New Phytologist*, *204*, 55–65. https://doi.org/10.1111/nph.12921

Pawlik, L., & Šamonil, P. (2018). Soil creep: the driving factors, evidence and significance for biogeomorphic and pedogenic domains and systems – A critical literature review. *Earth-Science Reviews*, *178*, 257–278. https://doi.org/10.1016/j.earscirev.2018.01.008

Pearson, S. M., Turner, M. G., Wallace, L. L., & Romme, W. H. (1995). Winter habitat use by large ungulates following fire in northern Yellowstone National Park. *Ecological Applications*, *5*, 744–755. https://doi.org/10.2307/1941982

Pelletier, J. D., & Orem, C. A. (2014). How do sediment yields from post-wildfire debris-laden flows depend on terrain slope, soil burn severity class, and drainage basin area? Insights from airborne-LiDAR change detection. *Earth Surface Processes and Landforms*, *39*, 1822–1832. https://doi.org/10.1002/esp.3570

Perevolotsky A, Lahman A, Polak G. 1992. *Mediterranean Woodlands—Wildfire Impacts*. The Society for the Protection of Nature in Israel: Tel Aviv, Israel, 1992 (in Stavi, 2019)

Peterson, D. W., Dodson, E. K., & Harrod, R. J. (2023). Snag decomposition following stand-replacing wildfires alters wildlife habitat use and surface woody fuels through time. *Ecosphere*, *14*, e4635. https://doi.org/10.1002/ecs2.4635

Pingree, M. R. A., & DeLuca, T. H. (2017). Function of wildfire-deposited pyrogenic carbon in terrestrial ecosystems. *Frontiers in Environmental Science*, *5*, 53. https://doi.org/10.3389/fenvs.2017.00053

Pinto, P., Vaz, P., Robinson, C., & Morais, M. (2008). Wildfire impacts on aquatic ecosystems. In L. M. G. Duarte & P. Pinto (Eds.), *Sustainable development: energy, environment and natural disasters* (pp. 25–34). Sao Paolo, Brazil: Fundação Luis de Molina.

Porter, T. M., Smenderovac, E., Morris, D., & Venier, L. (2023). All boreal forest successional stages needed to maintain the full suite of soil biodiversity, community composition, and function following wildfire. *Scientific Reports*, *13*, 7978. https://doi.org/10.1038/s41598-023-30732-7

Prosser, I. P., & Williams, L. (1998). The effect of wildfire on runoff and erosion in native *Eucalyptus* forest. *Hydrological Processes*, *12*, 251–265. https://doi.org/10.1002/(SICI)1099-1085(199802)12:2%3C251::AID-HYP574%3E3.0.CO;2-4

Ream C. H. 1981. *The effects of fire and other disturbances on small mammals and their predators: An annotated bibliography*. USDA Forest Service General Technical Report INT-106.

Records, R. M., Wohl, E., & Arabi, M. (2016). Phosphorus in the river corridor. *Earth-Science Reviews*, *158*, 65–88. https://doi.org/10.1016/j.earscirev.2016.04.010

Reddy, K. R., Kadlec, R. H., Flaigh, E., & Gale, P. M. (1999). Phosphorus retention in streams and wetlands: a review. *Critical Reviews in Environmental Science and Technology*, *29*, 83–146. https://doi.org/10.1080/10643389991259182

Reeves, M. C., Manning, M. E., DiBenedetto, J. P., Palmquist, K. A., Lauenroth, W. K., Bradford, J. B., & Schlaepfer, D. R. (2016). Effects of climate change on rangeland vegetation in the Northern Rockies. In J. E. Halofsky & D. L. Peterson (Eds.), *Climate change and Rocky Mountain ecosystems: advances in global change research* (pp. 97–114). Berlin: Springer.

Reich, P. B. (2014). The world-wide 'fast-slow' plant economics spectrum: a traits manifesto. *Journal of Ecology*, *102*, 275–301. https://doi.org/10.1111/1365-2745.12211

Reilly, P. (1991). The effect of wildfire on bird populations in a Victorian coastal habitat. *Emu*, *91*, 100–106. https://doi.org/10.1071/MU9910100

Reisser, M., Purves, R. S., Schmidt, M. W. I., & Abiven, S. (2016). Pyrogenic carbon in soils: a literature-based inventory and a global estimation of its content in soil organic carbon and stocks. *Frontiers in Earth Science*, *4*, 80. https://doi.org/10.3389/feart.2016.00080

Reneau, S. L., Katzman, D., Kuyumjian, G. A., Lavine, A., & Malmon, D. V. (2007). Sediment delivery after a wildfire. *Geology*, *35*, 151–154. https://doi.org/10.1130/G23288A.1

Rengers, F. K., McGuire, L. A., Barnhart, K. R., Youberg, A. M., Cadol, D., Gorr, A. N., et al. (2023). The influence of large woody debris on post-wildfire debris flow sediment storage. *Natural Hazards and Earth System Sciences*, *23*, 2075–2088. https://doi.org/10.5194/nhess-23-2075-2023

Rengers, F. K., McGuire, L. A., Kean, J. W., Staley, D. M., Dobre, M., Robichaud, P. R., & Swetnam, T. (2021). Movement of sediment through a burned landscape: Sediment volume observations and model comparisons in the San Gabriel Mountains, California, USA. *Journal of Geophysical Research: Earth Surface*, *126*, e2020JF006053. https://doi.org/10.1029/2020JF006053

Rengers, F. K., McGuire, L. A., Kean, J. W., Staley, D. M., & Youberg, A. M. (2019). Progress in simplifying hydrologic model parameterization for broad applications to post-wildfire flooding and debris-flow hazards. *Earth Surface Processes and Landforms*, *44*, 3078–3092. https://doi.org/10.1002/esp.4697

Rengers, F. K., McGuire, L. A., Oakley, N. S., Kean, J. W., Staley, D. M., & Tang, H. (2020). Landslides after wildfire: initiation, magnitude, and mobility. *Landslides*, *17*, 2631–2641. https://doi.org/10.1007/s10346-020-01506-3

Rengers, F. K., Pagonis, V., & Mahan, S. A. (2017). Can thermoluminescence be used to determine soil heating from a wildfire? *Radiation Measurements*, *107*, 119–127. https://doi.org/10.1016/j.radmeas.2017.09.002

Restaino, J. C., & Peterson, D. L. (2013). Wildfire and fuel treatment effects on forest carbon dynamics in the western United States. *Forest Ecology and Management*, *303*, 46–60. https://doi.org/10.1016/j.foreco.2013.03.043

Revchuk, A. D., & Suffet, I. H. (2014). Effect of wildfires on physicochemical changes of watershed dissolved organic matter. *Water Environment Research*, *86*, 372. https://doi.org/10.2175/106143013X13736496909671

Reyer, C. P. O., Leuzinger, S., Rammig, A., Wolf, A., Bartholomeus, R. P., Bonfante, A., et al. (2013). A plant's perspective of extremes: terrestrial plant responses to changing climatic variability. *Global Change Biology*, *19*, 75–89. https://doi.org/10.1111/gcb.12023

Rhoades, C. C., Chow, A. T., Covino, T. P., Fegel, T. S., Pierson, D. N., & Rhea, A. E. (2019). The legacy of a severe wildfire on stream nitrogen and carbon in headwater catchments. *Ecosystems*, *22*, 643–657. https://doi.org/10.1007/s10021-018-0293-6

Rhoades, C. C., Entwistle, D., & Butler, D. (2011). The influence of wildfire extent and severity on streamwater chemistry, sediment and temperature following the Hayman Fire, Colorado. *International Journal of Wildland Fire*, *320*, 430–442. https://doi.org/10.1071/WF09086

Rhodes, C. J. (2017). The whispering world of plants: 'The Wood Wide Web'. *Science Progress, 100*, 331–337. https://doi.org/10.3184/003685017X14968299580423

Ripple, W. J., & Beschta, R. L. (2012). Trophic cascades in Yellowstone: the first 15 years after wolf reintroduction. *Biological Conservation, 145*, 205–213. https://doi.org/10.1016/j.biocon.2011.11.005

Ripple, W. J., & Larsen, E. J. (2001). The role of postfire coarse woody debris in aspen regeneration. *Western Journal of Applied Forestry, 16*, 61–64. https://doi.org/10.1093/wjaf/16.2.61

Robichaud, P. R., Wagenbrenner, J. W., & Brown, R. E. (2010). Rill erosion in natural and disturbed forests: 1. Measurements. *Water Resources Research, 46*, W10506. https://doi.org/10.1029/2009WR008314

Robichaud, P. R., Wagenbrenner, J. W., Brown, R. E., Wohlgemuth, P. M., & Beyers, J. L. (2008). Evaluating the effectiveness of contour-felled log erosion barriers as a post-fire runoff and erosion mitigation treatment in the western United States. *International Journal of Wildland Fire, 17*, 255–273. https://doi.org/10.1071/WF07032

Robichaud, P. R., Wagenbrenner, J. W., Pierson, F. B., Spaeth, K. E., Ashmun, L. E., & Moffet, C. A. (2016). Infiltration and interrill erosion rates after a wildfire in western Montana, USA. *Catena, 142*, 77–88. https://doi.org/10.1016/j.catena.2016.01.027

Robinne, F. N., Hallema, D. W., Bladon, K. D., & Buttle, J. M. (2020). Wildfire impacts on hydrologic ecosystem services in North American high-latitude forests: a scoping review. *Journal of Hydrology, 581*, 124360. https://doi.org/10.1016/j.jhydrol.2019.124360

Roccaforte, J. P., Fule, P. Z., Chancellor, W. W., & Laughlin, D. C. (2012). Woody debris and tree regeneration dynamics following severe wildfires in Arizona ponderosa pine forests. *Canadian Journal of Forest Research, 42*, 593–604. https://doi.org/10.1139/x2012-010

Roebuck, J. A., Bladon, K. D., Donahue, D., Graham, E. B., Grieger, S., Morgenstern, K., et al. (2022). Spatiotemporal controls on the delivery of dissolved organic matter to streams following a wildfire. *Geophysical Research Letters, 49*, e2022GL099535. https://doi.org/10.1029/2022GL099535

Roering, J. J., & Gerber, M. (2005). Fire and the evolution of steep, soil-mantled landscapes. *Geology, 33*, 349–352. https://doi.org/10.1130/G21260.1

Rosso, R., Rulli, M. C., & Bocchiola, D. (2007). Transient catchment hydrology after wildfires in a Mediterranean basin: runoff, sediment and woody debris. *Hydrology and Earth System Sciences, 11*, 125–140. https://doi.org/10.5194/hess-11-125-2007

Roth, D. L., Doane, T. H., Roering, J. J., Furbish, D. J., & Zettler-Mann, A. (2020). Particle motion on burned and vegetated hillslopes. *Proceedings of the National Academy of Sciences, 117*, 25335–25343. https://doi.org/10.1073/pnas.1922495117

Ruan, Y., Mohtadi, M., Dupont, L. M., Hebbeln, D., van der Kaars, S., Hopmans, E. C., et al. (2020). Interaction of fire, vegetation, and climate in tropical ecosystems: a multiproxy study over the past 22,000 years. *Global Biogeochemical Cycles, 34*, e2020BG006677. https://doi.org/10.1029/2020GB006677

Ruckman, E. M., Schwinning, S., & Lyons, K. G. (2012). Effects of phenology at burn time on post-fire recovery in an invasive C4 grass. *Restoration Ecology*, *20*, 756–763. https://doi.org/10.1111/j.1526-100X.2011.00830.x

Russell, K. R., Van Lear, D. H., & Guynn, D. C. (1999). Prescribed fire effects on herpetofauna: review and management implications. *Wildlife Society Bulletin*, *27*, 374–384. https://www.jstor.org/stable/3783904

Ryan KC, Noste NV. 1985. *Evaluating prescribed fires*. USDA Forest Service Intermountain Forest and Range Experiment Station, General Technical Report INT-182, p 230-238.

Saab, V. A., Russell, R. E., & Dudley, J. G. (2007). Nest densities of cavity-nesting birds in relation to postfire salvage logging and time since wildfire. *The Condor*, *109*, 971–908. https://doi.org/10.1093/condor/109.1.97

Saab, V. A., & Vierling, K. (2001). Reproductive success of Lewis's Woodpecker in burned pine and cottonwood riparian forests. *The Condor*, *103*, 491–501. https://doi.org/10.1093/condor/103.3.491

Salmon, V. G., Schadel, C., Bracho, R., Pegoraro, E., Celis, G., Mauritz, M., et al. (2018). Adding depth to our understanding of nitrogen dynamics in permafrost soils. *Journal of Geophysical Research Biogeosciences*, *123*, 2497–2512. https://doi.org/10.1029/2018JG004518

Sanderfoot, O. V., Bassing, S. B., Brusa, J. L., Emmet, R. L., Gillman, S. J., Swift, K., & Gardner, B. (2021). A review of the effects of wildfire smoke on the health and behavior of wildlife. *Environmental Research Letters*, *16*, 123003.

Sankey, J. B., Germino, M. J., & Glenn, N. F. (2009). Aeolian sediment transport following wildfire in sagebrush steppe. *Journal of Arid Environments*, *73*, 912–919. https://doi.org/10.1016/j.jaridenv.2009.03.016

Santi, P., Cannon, S., & DeGraff, J. (2013). Wildfire and landscape change. In L. A. James, C. P. Harden, & J. J. Clague (Eds.), *Treatise on geomorphology, v 13, geomorphology of human disturbances, climate change, and natural hazards* (pp. 262–287). San Diego, CA: Academic Press.

Santi, P. M., deWolfe, V. G., Higgins, J. D., Cannon, S. H., & Gartner, J. E. (2008). Sources of debris flow material in burned areas. *Geomorphology*, *96*, 310–321. https://doi.org/10.1016/j.geomorph.2007.02.022

Santi, P. M., & Rengers, F. K. (2022). Wildfire and landscape change. In J. Shroder (Ed.), *Treatise on geomorphology* (2nd ed., Vol. 13, pp. 262–287). Amsterdam: Elsevier. https://doi.org/10.1016/B978-0-12-818234-5.00017-1

Santín, C., & Doerr, S. H. (2016). Fire effects on soils: the human dimension. *Philosophical Transactions Royal Society B*, *371*, 20150171. https://doi.org/10.1098/rstb.2015.0171

Santín, C., Doerr, S. H., Jones, M. W., Merino, A., Warneke, C., & Roberts, J. M. (2020). The relevance of pyrogenic carbon for carbon budgets from fires: insights from the FIREX experiment. *Global Biogeochemical Cycles*, *34*, e2020GB006647. https://doi.org/10.1029/2020GB006647

Santos, F., Wymore, A. S., Jackson, B. K., Sullivan, S. M. P., McDowell, W. H., & Berhe, A. A. (2019). Fire severity, time since fire, and site-level characteristics influence streamwater chemistry at baseflow conditions in catchments of the

Sierra Nevada, California, USA. *Fire Ecology*, *15*, 3. https://doi.org/10.1186/s42408-018-0022-8

Schieck, J., & Song, S. J. (2006). Changes in bird communities throughout succession following fire and harvest in boreal forests of western North America: literature review and meta-analyses. *Canadian Journal of Forest Research*, *36*, 1299–1318. https://doi.org/10.1139/x06-017

Schoennagel, T., Smithwick, E. A. H., & Turner, M. G. (2008). Landscape heterogeneity following large fires: insights from Yellowstone National Park, USA. *International Journal of Wildland Fires*, *17*, 742–753. https://doi.org/10.1071/WF07146

Schwilk, D. W., Keeley, J. E., Knapp, E. E., McIver, J., Bailey, J. D., Fettig, C. J., et al. (2009). The national fire and fire surrogate study: effects of fuel reduction methods on forest vegetation structure and fuels. *Ecological Applications*, *19*, 285–304. https://doi.org/10.1890/07-1747.1

Seedre, M., Shrestha, B. M., Chen, H. Y. H., Colombo, S., & Jogiste, K. (2011). Carbon dynamics of North American boreal forest after stand replacing wildfire and clearcut logging. *Journal of Forest Research*, *16*, 168–183. https://doi.org/10.1007/s10310-011-0264-7

Semenova-Nelsen, T. A., Platt, W. J., Patterson, T. R., Huffman, J., & Sikes, B. A. (2019). Frequent fire reorganizes fungal communities and slows decomposition across a heterogeneous pine savanna landscape. *New Phytologist*, *224*, 916–927. https://doi.org/10.1111/nph.16096

Shakesby, R. A. (2011). Post-wildfire soil erosion in the Mediterranean: review and future research directions. *Earth-Science Reviews*, *105*, 71–100. https://doi.org/10.1016/j.earscirev.2011.01.001

Shakesby, R., & Doerr, S. (2006). Wildfire as a hydrological and geomorphological agent. *Earth-Science Reviews*, *74*, 269–307. https://doi.org/10.1016/j.earscirev.2005.10.006

Sharkey, T. D., Chen, X., & Yeh, S. (2001). Isoprene increases thermotolerance of fosmidomycin-fed leaves. *Plant Physiology*, *125*, 2001–2006. https://doi.org/10.1104/pp.125.4.2001

Sharma, U., Sharma, J., & Devi, M. (2017). Effect of forest fire on soil nitrogen mineralization and microbial biomass: a review. *Journal of Pharmacognosy and Phytochemistry*, *6*, 682–685.

Shatford, J., Hibbs, D. E., & Puettmann, K. (2007). Conifer regeneration after forest fire in the Klamath-Siskiyous: how much, how soon? *Journal of Forestry*, *105*, 139–146. https://doi.org/10.1093/jof/105.3.139

Sherson, L. R., Van Horn, D. J., Gomez-Velez, J. D., Crossey, L. J., & Dahm, C. N. (2015). Nutrient dynamics in an alpine headwater stream: use of continuous water quality sensors to examine responses to wildfire and precipitation events. *Hydrological Processes*, *29*, 3193–3207. https://doi.org/10.1002/hyp.10426

Shi, Y. F., Shi, S. H., Jiang, Y. S., & Liu, J. (2022). A global synthesis of fire effects on soil seed banks. *Global Ecology and Conservation*, *36*, e02132. https://doi.org/10.1016/j.gecco.2022.e02132

Shibata, H., Petrone, K. C., Hinzman, L. D., & Boone, R. D. (2003). Effect of fire on dissolved organic carbon and inorganic solutes in spruce forest in the permafrost region of interior Alaska. *Soil Science and Plant Nutrition*, *49*, 25–29. https://doi.org/10.1080/00380768.2003.10409975

Shlisky, A., Alencar, A., Manta, M., & Curran, L. M. (2009). Overview: global fire regime conditions, threats, and opportunities for fire management in the tropics. In M. A. Cochrane (Ed.), *Tropical fire ecology* (pp. 65–83). Chichester: Praxis.

Shlisky A, Waugh J, Gonzalez P, Gonzalez M, Manta M, Santoso H, Alvarado E, Nuruddin AA, Rodriguez-Trejo DA, Swaty R, et al. 2007. *Fire, Ecosystems and People: Threats and Strategies for Global Biodiversity Conservation*. Global Fire Initiative Technical Report 2007-2, The Nature Conservancy, Arlington, VA.

Skorbiłowicz, M., & Ofman, P. (2014). Seasonal changes of nitrogen and phosphorus concentration in Supraśl River. *Journal of Ecological Engineering*, *15*, 26–31.

Smith, H. G., Blake, W. H., & Owens, P. N. (2013). Discriminating fine sediment sources and the application of sediment tracers in burned catchments: a review. *Hydrological Processes*, *27*, 943–958. https://doi.org/10.1002/hyp.9537

Smith, H. G., Sheridan, G. J., Lane, P. N. J., Nyman, P., & Haydon, S. (2011). Wildfire effects on water quality in forest catchments: a review with implications for water supply. *Journal of Hydrology*, *396*, 170–192. https://doi.org/10.1016/j.jhydrol.2010.10.043

Smith, N. R., Kishchuk, B. E., & Mohn, W. W. (2008). Effects of wildfire and harvest disturbances on forest soil bacterial communities. *Applied Environmental Microbiology*, *74*, 216–224. https://doi.org/10.1128/AEM.01355-07

Smucker, K. M., Hutto, R. L., & Steele, B. M. (2005). Changes in bird abundance after wildfire: importance of fire severity and time since fire. *Ecological Applications*, *15*, 1535–1549. https://doi.org/10.1890/04-1353

Song, Z., Tian, H., Li, Z., Luo, Y., & Liu, Y. (2021). Changes in plant nutrient utilization during ecosystem recovery after wildfire. *Journal of Environmental Management*, *295*, 112994. https://doi.org/10.1016/j.jenvman.2021.112994

Sosa-Pérez, G., & MacDonald, L. (2017). Wildfire effects on road surface erosion, deposition, and road-stream connectivity. *Earth Surface Processes and Landforms*, *42*, 735–748. https://doi.org/10.1002/esp.4018

Southon, G. E., Green, E. R., Jones, A. G., Barker, C. G., & Power, S. A. (2012). Long-term nitrogen additions increase likelihood of climate stress and affect recovery from wildfire in a lowland heath. *Global Change Biology*, *18*, 2824–2837. https://doi.org/10.1111/j.1365-2486.2012.02732.x

Spies T. A., Cline S. P. 1988. *Coarse woody debris in forests and plantations of coastal Oregon. In, From the Forest to the Sea: A Story of Fallen Trees*. USDA Forest Service General Technical Report PNW-GTR-229, Portland, OR, 5-24.

Spies, T. A., Franklin, J. F., & Thomas, T. B. (1988). Coarse woody debris in Douglas-fir forests of western Oregon and Washington. *Ecology*, *69*, 1689–1702. https://doi.org/10.2307/1941147

Staley, D. M., Negri, J. A., Kean, J. W., Laber, J. L., Tillery, A. C., & Youberg, A. M. (2017). Prediction of spatially explicit rainfall intensity-duration thresholds for post-fire debris-flow generation in the western United States. *Geomorphology*, *278*, 149–162. https://doi.org/10.1016/j.geomorph.2016.10.019

Stavi, I. (2019). Wildfires in grasslands and shrublands: a review of impacts on vegetation, soil, hydrology, and geomorphology. *Water*, *11*, 1042. https://doi.org/10.3390/w11051042

Steers, R. J., & Allen, E. B. (2011). Fire effects on perennial vegetation in the western Colorado desert, USA. *Fire Ecology*, *7*, 59–74. https://doi.org/10.4996/fireecology.0703059

Stephens, S. L., Bernal, A. A., Collins, B. M., Finney, M. A., Lautenberger, C., & Saah, D. (2022). Mass fire behavior created by extensive tree mortality and high tree density not predicted by operational fire behavior models in the southern Sierra Nevada. *Forest Ecology and Management*, *518*, 120258. https://doi.org/10.1016/j.foreco.2022.120258

Stephens, S. L., Kobziar, L. N., Collins, B. M., Davis, R., Fule, P. Z., Gaines, W., et al. (2019). Is fire "for the birds"? How two rare species influence fire management across the US. *Frontiers in Ecology and Environment*, *17*, 391–399. https://doi.org/10.1002/fee.2076

Stevens-Rumann, C., Morgan, P., & Hoffman, C. (2015). Bark beetles and wildfire: How does forest recovery change with repeated disturbances in mixed conifer forests? *Ecosphere*, *6*, 1–17. https://doi.org/10.1890/ES14-00443.1

Stockmann, U., Minasny, B., & McBratney, A. B. (2014). How fast does soil grow? *Geoderma*, *216*, 48–61. https://doi.org/10.1016/j.geoderma.2013.10.007

Stocks, B. J. (1987). Fire potential in the spruce budworm-damaged forests of Ontario. *The Forestry Chronicle*, *63*, 8–11. https://doi.org/10.5558/tfc63008-1

Stoddard, J. L. (1994). *Long-term changes in watershed retention of nitrogen*. Environmental chemistry of lakes and reservoirs (Vol. 237, pp. 223–284). Washington DC: American Chemical Society.

Stuart, J. D., Agee, J. K., & Gara, R. I. (1989). Lodgepole pine regeneration in an old, self-perpetuating forest in south Oregon. *Canadian Journal of Forest Research*, *19*, 1096–1104. https://doi.org/10.1139/x89-166

Suryanayanan, T. S., Govindarajulu, M. B., Thirumalai, E., Reddy, M. S., & Money, N. P. (2011). Agni's fungi: heat-resistant spores from the Western Ghats, southern India. *Fungal Biology*, *115*, 833–838. https://doi.org/10.1016/j.funbio.2011.06.011

Swanson, F. J. (1981). Fire and geomorphic processes. In H. A. Mooney, T. M. Bonnicksen, N. L. Christiansen, J. E. Lotan, & W. A. Reiners (Eds.), *Fire Regime and Ecosystem Properties*. USDA Forest Service General Technical Report WO, 26, Washington, DC (pp. 401–421).

Swanson, M. E., Franklin, J. F., Beschta, R. L., Crisafulli, C. M., DellaSala, D. A., Hutto, R. L., et al. (2011). The forgotten stage of forest succession: early successional ecosystems on forest sites. *Frontiers in Ecology and Environment*, *9*, 117–125. https://doi.org/10.1890/090157

Swindle, C., Shankin-Clarke, P., Meyerhof, M., Carlson, J., & Melack, J. (2021). Effects of wildfires and ash leaching on stream chemistry in the Santa Ynez Mountains of southern California. *Water*, *13*, 2402. https://doi.org/10.3390/w13172402

Tahovská, K., Choma, M., Kastovska, E., Oulehle, F., Barta, J., Santruckova, H., & Moldan, F. (2020). Positive response of soil microbes to long-term nitrogen input in spruce forest: results from Gardsjon whole-catchment N-addition experiment. *Soil Biology and Biochemistry*, *143*, 107732. https://doi.org/10.1016/j.soilbio.2020.107732

Tallmon, D., Jules, E. S., Radke, N. J., & Mills, L. S. (2003). Of mice and men and trillium: cascading effects of forest fragmentation. *Ecological Applications*, *13*, 1193–1203. https://doi.org/10.1890/02-5111

Tang, H., McGuire, L. A., Rengers, F. K., Kean, J. W., Staley, D. M., & Smith, J. B. (2019). Developing and testing physically based triggering thresholds for runoff-generated debris flows. *Geophysical Research Letters*, *46*, 8830–8839. https://doi.org/10.1029/2019GL083623

Tangney, R., Merritt, D. J., Fontaine, J. B., & Miller, B. P. (2019). Seed moisture content as a primary trait regulating the lethal temperature thresholds of seeds. *Journal of Ecology*, *107*, 0193–1105. https://doi.org/10.1111/1365-2745.13095

Tank, J. L., Rosi-Marshall, E. J., Griffiths, N. A., Entrekin, S. A., & Stephen, M. L. (2010). A review of allochthonous organic matter dynamics and metabolism in streams. *Journal of the North American Benthological Society*, *29*, 118–146. https://doi.org/10.1899/08-170.1

Taylor, A. R., Gao, B., & Chen, H. Y. H. (2020). The effect of species diversity on tree growth varies during forest succession in the boreal forest of central Canada. *Forest Ecology and Management*, *455*, 117641. https://doi.org/10.1016/j.foreco.2019.117641

Thomas, M. A., Rengers, F. K., Kean, J. W., McGuire, L. A., Staley, D. M., Barnhart, K. R., & Ebel, B. A. (2021). Postwildfire soil-hydraulic recovery and the persistence of debris flow hazards. *Journal of Geophysical Research Earth Surface*, *126*, e2021JF006091. https://doi.org/10.1029/2021JF006091

Thwaites, R. N., Brooks, A. P., Pietsch, T. J., & Spencer, J. R. (2022). What type of gully is that? The need for a classification of gullies. *Earth Surface Processes and Landforms*, *47*, 109–128. https://doi.org/10.1002/esp.5291

Tinker, D. B., & Knight, D. H. (2000). Coarse woody debris following fire and logging in Wyoming lodgepole pine forests. *Ecosystems*, *3*, 472–483. https://doi.org/10.1007/s100210000041

Tracy, B. F., & McNaughton, S. J. (1997). Elk grazing and vegetation responses following a late season fire in Yellowstone National Park. *Plant Ecology*, *130*, 111–119. https://doi.org/10.1023/A:1009794226291

Tubbesing, C. L., Young, D. J. N., York, R. A., Stephens, S. L., & Battles, J. J. (2022). Incorporating shrub neighborhood dynamics to predict forest succession trajectories in an altered fire regime. *Ecosystems*, *25*, 136–154. https://doi.org/10.1007/s10021-021-00645-5

Ucitel, D., Christian, D. P., & Graham, J. M. (2003). Vole use of coarse woody debris and implications for habitat and fuel management. *The Journal of Wildlife Management*, *67*, 65–72. https://doi.org/10.2307/3803062

Uecker, T. M., Kaspari, S. D., Musselman, K. N., & Skiles, S. M. (2020). The post-wildfire impact of burn severity and age on black carbon snow deposition and implications for snow water resources, Cascade Range, Washington. *Journal of Hydrometeorology*, *21*, 1777–1792. https://doi.org/10.1175/JHM-D-20-0010.1

Uzun, H., Dahlgren, R. A., Olivares, C., Erdem, C. U., Karanfil, T., & Chow, A. T. (2020). Two years of post-wildfire impacts on dissolved organic matter, nitrogen, and precursors of disinfection by-products in California stream waters. *Water Research*, *181*, 115891. https://doi.org/10.1016/j.watres.2020.115891

Valtera, M., & Schaetzl, R. J. (2017). Pit-mound microrelief in forest soils: review of implications for water retention and hydrologic modelling. *Forest Ecology and Management*, *393*, 40–51. https://doi.org/10.1016/j.foreco.2017.02.048

Violle, C., Navas, M. L., Vile, D., Kazakou, E., Fortunel, C., Hummel, I., & Garnier, E. (2007). Let the concept of trait be functional! *Oikos*, *116*, 882–892. https://doi.org/10.1111/j.0030-1299.2007.15559.x

Wagenbrenner, J. W., Robichaud, P. R., & Elliot, W. J. (2010). Rill erosion in natural and disturbed forests: 2. *Modeling approaches. Water Resources Research*, *46*, W10507. https://doi.org/10.1029/2009WR008315

Wagner, S., Cawley, K. M., Rosario-Ortiz, F. L., & Jaffe, R. (2015). In-stream sources and links between particulate and dissolved black carbon following a wildfire. *Biogeochemistry*, *124*, 145–161. https://doi.org/10.1007/s10533-015-0088-1

Wang, Q., Zhong, M., & Wang, S. (2012). A meta-analysis on the response of microbial biomass, dissolved organic matter, respiration, and N mineralization in mineral soil to fire in forest ecosystems. *Forest Ecology and Management*, *271*, 91–97. https://doi.org/10.1016/j.foreco.2012.02.006

Warren, S. D., Scifres, C. J., & Teel, P. D. (1987). Response of grassland arthropods to burning: a review. *Agriculture, Ecosystems and Environment*, *19*, 105–130. https://doi.org/10.1016/0167-8809(87)90012-0

Watson, S. J., Taylor, R. S., Nimmo, D. G., Kelly, L. T., Clarke, M. F., & Bennett, A. F. (2012). The influence of unburnt patches and distance from refuges on post-fire bird communities. *Animal Conservation*, *15*, 499–507. https://doi.org/10.1111/j.1469-1795.2012.00542.x

Weirich, F., Neumann, W., Brady, J., & Weirich, F. A. (2024). Progress in assessing the impact of fire and water repellency on hillslopes. In J. L. Florsheim, A. P. O'Dowd, & A. Chin (Eds.), *Biogeomorphic Responses to Wildfire in Fluvial Ecosystems*. Geological Society of America Special Paper 562, Boulder, CO (pp. 43–55).

Wester, T., Wasklewicz, T., & Staley, D. (2014). Functional and structural connectivity within a recently burned drainage basin. *Geomorphology*, *206*, 326–373. https://doi.org/10.1016/j.geomorph.2013.10.011

Whelan, R. J., Rodgerson, L., Dickman, C. R., & Sutherland, E. F. (2002). Critical life processes of plants and animals: developing a process-based understanding of population changes in fireprone landscapes. In R. A. Bradstock, J. E. Williams, & A. M. Gill (Eds.), *Flammable Australia: the fire regimes and biodiversity of a continent* (pp. 94–124). United Kingdom: Cambridge University Press.

Whicker, J. J., Breshears, D. D., Wasiolek, P. T., Kirchner, T. B., Tavani, R. A., Schoep, D. A., & Rodgers, J. C. (2002). Temporal and spatial variation of episodic wind erosion in unburned and burned semiarid shrubland. *Journal of Environmental Quality*, *31*, 599–612. https://doi.org/10.2134/jeq2002.5990

White A. 2015. *Examining the impacts of wildfire on throughfall and stemflow chemistry and flux at plot and catchment scales*. MS thesis, University of Arizona, Tucson, AZ.

Whitman, T., Woolet, J., Sikora, M., Johnson, D. B., & Whitman, E. (2022). Resilience in soil bacterial communities of the boreal forest from one to five years after wildfire across a severity gradient. *Soil Biology and Biochemistry*, *172*, 108755. https://doi.org/10.1016/j.soilbio.2022.108755

Wilkinson, S. N., Wallbrink, P. J., Hancock, G. J., Blake, W. H., Shakesby, R. A., & Doerr, S. H. (2009). Fallout radionuclide tracers identify a switch in sediment sources and transport-limited sediment yield following wildfire in a eucalypt forest. *Geomorphology*, *110*, 140–151. https://doi.org/10.1016/j.geomorph.2009.04.001

Wills, S. A., Williams, C. O., Duniway, M. C., Veenstra, J., Seybold, C., & Presley, D. (2016). Human land-use and soil change. In L. West, M. Singer, & A. Hartemink (Eds.), *The soils of the USA*, World Soils Book Series (pp. 351–371). Cham: Springer. https://doi.org/10.1007/978-3-319-41870-4_18

Wilson, C., Kampf, S. K., Ryan, S., Covino, T., MacDonald, L. H., & Gleason, H. (2021). Connectivity of post-fire runoff and sediment from nested hillslopes and watersheds. *Hydrological Processes*, *35*, e13975. https://doi.org/10.1002/hyp.13975

Wilson, C., Kampf, S. K., Wagenbrenner, J. W., & MacDonald, L. H. (2018). Rainfall thresholds for post-fire runoff and sediment delivery from plot to watershed scales. *Forest Ecology and Management*, *430*, 346–356. https://doi.org/10.1016/j.foreco.2018.08.025

Wohl, E. (2013a). Migration of channel heads following wildfire in the Colorado Front Range, USA. *Earth Surface Processes and Landforms*, *38*, 1049–1053. https://doi.org/10.1002/esp.3429

Wohl, E. (2013b). Wilderness is dead: whither critical zone studies and geomorphology in the Anthropocene? *Anthropocene*, *2*, 4–15. https://doi.org/10.1016/j.ancene.2013.03.001

Wohl, E., Brierley, G., Cadol, D., Coulthard, T. J., Covino, T., Fryirs, K. A., et al. (2019). Connectivity as an emergent property of geomorphic systems. *Earth Surface Processes and Landforms*, *44*, 4–26. https://doi.org/10.1002/esp.4434

Wohl, E., & Scott, D. N. (2017). Transience of channel head locations following disturbance. *Earth Surface Processes and Landforms*, *42*, 1132–1139. https://doi.org/10.1002/esp.4124

Woinarski, J. C. Z., & Recher, H. F. (1997). Impact and response: a review of the effects of fire on the Australian avifauna. *Pacific Conservation Biology*, *3*, 183–205. https://doi.org/10.1071/PC970183

Wolf, E. C., Cooper, D. J., & Hobbs, N. T. (2007). Hydrologic regime and herbivory stabilize an alternative state in Yellowstone National Park. *Ecological Applications*, *17*, 1572–1587. https://doi.org/10.1890/06-2042.1

Wolf, K. D., Higuera, P. E., Davis, K. T., & Dobrowski, S. Z. (2021). Wildfire impacts on forest microclimate vary with biophysical context. *Ecosphere*, *12*, e03467. https://doi.org/10.1002/ecs2.3467

Wondzell, S. M., & King, J. G. (2003). Post-fire erosional processes in the Pacific Northwest and Rocky Mountain regions. *Forest Ecology and Management*, *178*, 75–87. https://doi.org/10.1016/S0378-1127(03)00054-9

Wright, H. A., & Bailey, H. W. (1982). *Fire ecology, United States and Southern Canada*. NY: Wiley.

Xenopoulos, M. A., Barnes, R. T., Boodoo, K. S., Butman, D., Catalan, N., D'Amario, S. C., et al. (2021). How humans alter dissolved organic matter composition in freshwater: relevance for the Earth's biogeochemistry. *Biogeochemistry*, *154*, 323–348. https://doi.org/10.1007/s10533-021-00753-3

Yanagiya, K., & Furuya, M. (2020). Post-wildfire surface deformation near Batagay, eastern Siberia, detected by L-band and C-band InSAR. *Journal of Geophysical Research: Earth Surface*, *125*, e2019JF005473. https://doi.org/10.1029/2019JF005473

Yanagiya, K., Furuya, M., Danilov, P., & Iwahana, G. (2023). Transient freeze–thaw deformation responses to the 2018 and 2019 fires near Batagaika megaslump, northeast Siberia. *Journal of Geophysical Research: Earth Surface*, *128*, e2022JF006817. https://doi.org/10.1029/2022JF006817

Yang, T., Tedersoo, L., Lin, X., Fitzpatrick, M. C., Jia, Y., Liu, X., et al. (2020). Distinct fungal successional trajectories following wildfire between soil horizons in a cold-temperate forest. *New Phytologist*, *227*, 572–587. https://doi.org/10.1111/nph.16531

Yang, X., Liu, R.-T., Shao, M.-A., Wei, X.-R., Li, T.-C., Chen, M.-Y., et al. (2022). Short-term effects of wildfire on soil arthropods in a semi-arid grassland on the Loess Plateau. *Frontiers in Microbiology*, *13*, 989351. https://doi.org/10.3389/fmicb.2022.989351

Yoshikawa, K., Bolton, W. R., Romanovsky, V. E., Fukuda, M., & Hinzman, L. D. (2002). Impacts of wildfire on the permafrost in the boreal forests of interior Alaska. *Journal of Geophysical Research Atmospheres*, *107*. FFR 4-1-FFR 4-14. https://doi.org/10.1029/2001JD000438

Young, J. M., Alvarez, A., van der Sluijs, J., Kokelj, S. V., Rudy, A., McPhee, A., et al. (2022). Recent intensification (2004-2020) of permafrost mass-wasting in the central Mackenzie Valley foothills is a legacy of past forest fire disturbances. *Geophysical Research Letters*, *49*, e2022GL100559. https://doi.org/10.1029/2022GL100559

Zak, J. C., & Wicklow, D. (1980). Structure and composition of a post-fire ascomycete community: role of abiotic and biotic factors. *Canadian Journal of Botany*, *58*, 1915–1922. https://doi.org/10.1139/b80-222

Zhai, J., Ning, Z., Dahal, R., & Yang, S. (2023). Wildfire susceptibility of land use and topographic features in the western United States: implications for the landscape management. *Forests*, *14*, 807. https://doi.org/10.3390/f14040807

Zwolak, R. (2009). A meta-analysis of the effects of wildfire, clearcutting, and partial harvest on the abundance of North American small mammals. *Forest Ecology and Management*, *258*, 539–545. https://doi.org/10.1016/j.foreco.2009.05.033

4

FIRES AND FRESHWATER ENVIRONMENTS

Abstract

First- and second-order headwater channels constitute most of the total channel length in a river network. They are closely coupled to adjacent uplands and are typically transport-dominated areas that effectively transmit post-fire fluxes from uplands to downstream areas. This transmission can be attenuated by unburned riparian zones. The middle and lower portions of river networks are typically the response reaches where more extensive floodplains retain post-fire inputs of materials and logjams, beaver dams, and spatially complex channels store materials after fires, although post-fire alterations of process and form can also persist longer in higher-order river corridors. Macroinvertebrates and fishes may experience greater post-fire environmental fluctuation in headwaters, but fire effects can be more prolonged in higher-order river corridors. Sedimentary records preserved in alluvial fans can be used to infer long-term fire recurrence intervals. Fluxes of matter and energy through river networks create the template in which organisms live, but the organisms actively adjust that template. Wildfire can alter magnitudes and pathways of fluxes into lakes. The effects on the lake

Landscapes on Fire: Impacts on Uplands, Rivers, and Communities, Special Publications 80,
First Edition. Ellen Wohl.

DOI: 10.1002/9781394235186.ch4

ecosystem reflect parameters such as the magnitude and duration of inputs, as well as the pre-fire water chemistry, lake-water residence time, and the ratio of lake area to catchment area.

4.1. Introduction

This chapter focuses on post-fire fluxes, forms, and biota within freshwater environments, primarily river corridors. A river corridor includes the active channel(s), the adjacent riparian zone and floodplain, and the underlying hyporheic zone. River corridors receive much of the accelerated fluxes that occur in steep uplands following severe wildfire. Some of this material is passed downstream rapidly but most of it, including water, is stored for varying lengths of time within individual river corridors and the river network as a whole. Storage times may vary from minutes and hours for water moving between surface and hyporheic flow, to millennia for sediment and large wood stored in depositional features such as the floodplain and alluvial fans. Transit and storage times influence the response of river forms and the habitat, food resources, and connectivity for diverse organisms.

The chapter first considers riparian zones. Riparian zones are treated separately from channels and floodplains because they are typically delineated based on hydrology and vegetation communities. Riparian zones can overlap with floodplains and can create many of the same effects during and after fire, but they do not necessarily correspond to the floodplain, which is a fluvially constructed landform.

The chapter then examines headwater stream corridors (first- and second-order streams), followed by higher-order streams in the river network. I use "stream" and "river" interchangeably because there is no consistently adopted differentiation between these terms in river-science literature. Headwater and higher-order stream sections are each divided into fluxes, forms, and biota. Discussion of river corridor forms includes living organisms as these influence forms, whereas the sections on biota focus on the effects of fire and post-fire changes on living organisms. The chapter ends with a brief discussion of lakes in the context of wildfire. Lakes are treated separately because they may or may not have surface connectivity with a river network.

In the preceding chapter, I referred to the movement of diverse materials downslope within the terrestrial environment as the first part of the wildfire disturbance cascade. The subsequent movement of these materials through river networks and/or into lakes is the second part of the wildfire disturbance cascade. As with the terrestrial portion of the cascade, diverse materials that move through river networks exhibit a wide range of magnitude, duration, frequency, and spatial extent of movement. Movement of materials can be altered by fire for many decades after the fire because of continuing interactions between river processes, forms, and biota, as well as feedbacks between headwaters and higher-order rivers in the same river network.

4.2. Riparian Zones

The riparian zone encompasses the river channel between low- and highwater marks and that portion of the terrestrial landscape from the highwater mark toward the uplands where vegetation may be influenced by elevated water tables or flooding and by the ability of the soils to hold water (Naiman & Decamps, 1997, p. 623). The riparian zone may overlap or coincide with the floodplain, which is a low-relief sedimentary surface adjacent to the active channel that is constructed by fluvial processes and frequently inundated (Nanson & Croke, 1992). In some cases, however, the two may not coincide, such as where steep valley walls preclude the development of a floodplain but the toe of each slope adjacent to the active channel is a riparian zone. In the context of wildfire fluxes, riparian zones can burn and experience fluxes of water, solutes, sediment, organic matter, and large wood analogous to the fluxes described for uplands. In many wildfires, however, riparian zones are critically important because they can buffer streams by retaining some of the accelerated fluxes of materials that occur following wildfire. Pettit and Naiman (2007) provide a comprehensive overview of wildfire and riparian zones.

Wildfires within riparian zones are strongly influenced by the size of the active channel and the moisture regime of the riparian zone (Benda et al., 1998; Agee et al., 2002). Riparian zones, especially those along perennial channels, may have more extensive but lower severity burns than upland areas (McMahon et al., 2023). Riparian fire regimes typically involve less frequent fires of lower intensity than in surrounding uplands. Riparian zones are less likely to burn severely because of factors that help to keep fuel moisture levels high, including higher relative humidity; shading and lower air temperatures; greater soil moisture; dense vegetation that reduces wind speeds; and greater moisture levels within vegetation (Dwire & Kauffman, 2003; Pettit & Naiman, 2007). Beaver-modified riparian zones in the northern hemisphere are less likely to be burnt than otherwise similar riparian zones because of the greater abundance of standing water and floodplain wetlands, as well as the higher riparian water table and moisture within vegetation (Fairfax & Whittle, 2020; Fairfax et al., 2024).

The riparian zone acts as a buffer against fire and provides a refuge for fire-sensitive species (Meave et al., 1991; Kobziar & McBride, 2006; Pettit & Naiman, 2007). Intense upland fires can penetrate limited portions of moist riparian zones along small and large streams, however, creating burnt and unburnt patches.

Intense wildfires can also completely burn riparian zones. This scenario is more likely when prolonged droughts have lowered the riparian water table and dried riparian vegetation; when the stream is narrow; when prior fires have created canopy gaps that dry the riparian zone and permit the establishment of fire-adapted species such as grasses (Jackson & Sullivan, 2009); and when fluvial deposition of floodplain logjams has created ladder fuels that can allow fire to reach the riparian canopy (Ellis, 2001; Pettit et al., 2005). In some cases, riparian zones can provide a conduit for fire into unburnt uplands (Agee, 1988), as when steep stream banks funnel winds and increase the rate of fire spread (Dwire & Kauffman, 2003). The riparian disturbances of flood and fire may have a nonlinear relationship, with fires occurring more frequently when flood frequency is intermediate (Pettit & Naiman, 2007). Frequent flooding inhibits vegetation growth and accumulation of fuel. In contrast, occasional floods create potentially flammable logjams, as well as destroying portions of the riparian forest and thus opening the riparian canopy to greater radiant heat and drying of the surviving vegetation.

As in uplands, the effects of riparian fire on nutrient fluxes depend strongly on factors such as fire severity. Low-severity fires can increase plant-available nutrients whereas high-severity fires can reduce nutrients through volatilization and loss via ash, erosion, and leaching (Aranibar et al., 2003).

Fire within the riparian zone can increase stream bank erosion by (i) removing vegetation and the associated hydraulic roughness created by the vegetation during overbank flow, thus facilitating higher values of flow velocity and shear stress exerted against the banks, (ii) increasing surface runoff and reducing infiltration, and (iii) reducing the ability of plant roots to increase soil cohesion. Bank erosion, combined with fire-induced tree mortality, leads to greater recruitment of large wood into channels (Bragg, 2000; Bêche et al., 2005).

Materials flowing downslope from surrounding uplands can be deposited in riparian zones because of the decrease in topographic gradient that is commonly present adjacent to a stream (Scott, 1993; Carroll et al., 2007). The degree to which the riparian zone buffers the adjacent stream channel from these upland inputs depends on the magnitude of the inputs, the width of the riparian zone, and the vegetation density or burn status of the riparian zone. Riparian zones can be particularly effective in buffering increased post-fire nitrogen inputs, for example, because riparian zones are hotspots for denitrification due to greater soil moisture (Gustine et al., 2022). Even if riparian vegetation is consumed during fire and plant uptake of nitrogen is reduced, riparian soil microbial communities may continue denitrification after the fire (Gustine et al., 2022).

If riparian zone vegetation is unburned or recovers quickly, it can limit post-fire increases in stream flow by taking up water and releasing it through transpiration (Lane et al., 2006). Bart and Tague (2017) examined this effect in detail by quantifying how fire decreases baseflow recession rates in California catchments. Lower post-fire baseflow recession rates can be associated with higher levels of antecedent groundwater storage. Groundwater storage reflects characteristics such as size, hydraulic conductivity, porosity, and connectivity of aquifers, as well as recharge and losses through evapotranspiration from plants that directly access groundwater. Wildfire can increase recharge by increasing the soil water content above the water table via reduced vegetation and evapotranspiration (Silberstein et al., 2013), reduced upland interception, and increased infiltration. Wildfire can also decrease recharge because of soil hydrophobicity (Letey, 2001), as discussed in detail in Chapter 3. Post-fire reductions in vegetation and associated decreased interception and evapotranspiration more strongly influence baseflow recession rate than does upland soil hydrophobicity in California catchments (Bart & Tague, 2017). Burn severity in the riparian zone is a primary control on baseflow recession rate because of riparian plants that access groundwater (phreatophytes). Changes in post-fire evapotranspiration losses can persist for years, depending on how fast the vegetation recovers after fire (Figure 4.1).

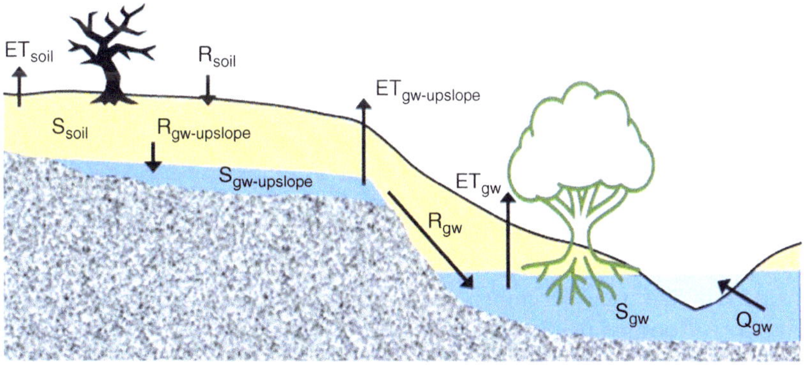

Figure 4.1 Conceptual model of fluxes affecting base flow recession rates. ET is evapotranspiration; Q_{gw} is groundwater discharge to a stream; S_{gw} is groundwater storage; R_{gw} is recharge to groundwater; S_{soil} is unsaturated soil matrix; R_{soil} is soil infiltration. (Adapted from Bart et al., 2017.)

In summary, the degree to which riparian processes and forms are altered during and after wildfire depends strongly on whether and how much of the riparian vegetation is killed/consumed during the fire. Similarly, the degree to which a riparian zone buffers river corridors from enhanced upland fluxes after a wildfire depends on the severity of the fire within the riparian zone. Riparian zones that are hydrologically connected to the active channel(s) can have sufficient moisture levels to resist burning. Moisture within riparian plant tissues in ponded surface water or floodplain wetlands and in a floodplain aquifer with a high-water table can limit fire severity within the riparian zone. This fire-resistant area can then store excess water and sediment entering from uplands; biologically process excess solutes; maintain erosional resistance of stream banks and litterfall inputs to the stream; and create fire refugia for diverse organisms. Alternatively, an intense fire or a narrow and relatively dry riparian zone can result in high fire severity within the riparian zone, exacerbating post-fire changes in the river corridor precipitated by accelerated inputs from adjacent uplands and from upstream portions of the catchment.

4.3. Headwater Stream Corridors

Headwaters are commonly defined as first- and second-order streams in the Strahler (1952) stream ordering system (Wohl, 2017b). First-order streams have no tributaries and second-order streams have only first-order tributaries. These small streams make up 70–80% of the total channel length of river networks (Leopold et al., 1964; Downing et al., 2012). Although these smallest streams are likely to be ephemeral or intermittent even in wet climates (Meyer et al., 2007), headwater streams influence hydrology, nutrient and sediment dynamics, and biota in downstream portions of the river network, as well illustrated by processes occurring within headwater streams following wildfire.

A defining characteristic of headwater streams is their close connection to adjacent uplands. In high-relief environments, headwater streams can have steep channels narrowly confined between valley walls. These are transport-dominated reaches (Montgomery & Buffington, 1997) with minimal sediment storage. Headwater streams, especially in high-relief catchments, can lack well-developed or extensive floodplains and the buffering that such floodplains can provide in terms of peak-flow attenuation or storage of sediment inputs from adjacent uplands (Wohl, 2021). Headwater streams in low-relief terrains can be weakly channeled marshy or swampy environments that are more capable of storing inputs from adjacent uplands (Wohl et al., 2021), especially when modified by beavers

(*Castor* spp.) (Larsen et al., 2021). Across diverse terrains, however, headwater streams typically lack broad valley floors and wide floodplains or extensive alluvial aquifers and hyporheic zones (McGlynn et al., 2004).

The smaller contributing area, shorter upland-channel surface and subsurface flow paths, and less water storage in the river corridor can create greater spatial and temporal variability of flow in headwater streams (Gomi et al., 2002; Richardson & Danehy, 2007). Disturbances such as convective storms or wildfire are more likely to affect the entire contributing area of headwaters. Characteristics such as channel morphology, presence of a floodplain, and riparian vegetation strongly influence the degree to which sediment and nutrients are retained and biochemically processed or passed directly downstream in headwater stream corridors (Peterson et al., 2001; Richardson et al., 2005; Alexander et al., 2007). Headwater streams are also more likely than larger rivers to undergo substantial change in channel form via alternating incision and aggradation through time in response to disturbances such as wildfire (e.g., Wohl & Pearthree, 1991; Benda et al., 2003; Leonard et al., 2017).

4.3.1. Headwater Stream Corridor Fluxes

Water. As might be expected, the increased surface runoff following wildfire typically creates both greater magnitude peak flow and a flashier hydrograph with more rapid rise and recession in headwater streams, particularly in steep terrain with high fire severity (e.g., Lane et al., 2006; Shakesby & Doerr, 2006) (Figure 4.2). A review by Paul et al. (2022) of 16 studies indicated some burn sites with no change in stream flow (24% of the studies surveyed), but a median increase of about 60% and a median duration of increased stream flow over five years. Some streams experienced an increase of nearly 10,000% (Paul et al., 2022). Peak flow is typically more sensitive to post-fire changes than total flow (Moody & Martin, 2001; Shakesby & Doerr, 2006) and changes are also likely to be greater in humid ecosystems with high pre-fire evapotranspiration that declines substantially immediately after severe fires (Shakesby & Doerr, 2006).

Solutes. Substantial impacts on stream solute concentrations can follow severe wildfires in catchments with: strong hydrologic connectivity and gradients from steep uplands; heavy precipitation following the fire; strong winds that can transport ash directly to streams; and upland soils with low cation-exchange capacity (Gustine et al., 2022). These conditions make N and other solutes more available and mobile. Catchments that are N-saturated before a fire can also experience substantially higher post-fire N export to streams (Gustine et al., 2022).

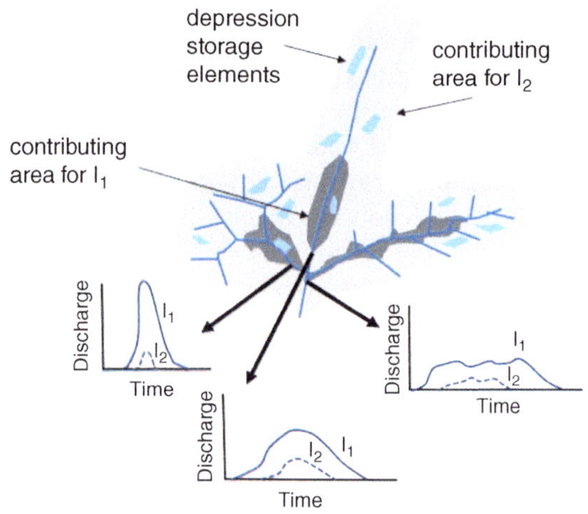

Figure 4.2 Conceptual illustration of how the morphological pattern of drainage networks in channels might affect the timing and peak values of runoff hydrographs after fire. Three sub-catchments with different morphologies are shown. Two hypothetical rainfall intensities, I_1 and I_2, and the corresponding contributing areas are labeled. The inset hydrographs are schematic illustrations of how differences in characteristics of the sub-catchments and rainfall of differing intensities can produce different runoff hydrographs. Depression storage refers to surface concavities without surface hydrologic connectivity and includes burnt-out stump and root holes in burned areas. Proportion of area burned and severity of burn will also affect runoff. (Moody et al., 2013 / with permission of ELSEVIER.)

A meta-analysis of 105 burned catchments, mostly in North America and Australia, found a consistent post-fire increase (75% of sites) in inorganic and organic N and P, as well as suspended sediment (Hampton et al., 2022). Even sites at which the mean concentrations did not change after fire had an increase in the high percentile concentrations. Not all solutes exhibit the same magnitude or even direction of change, however. Following wildfire in a boreal forest underlain by permafrost, Betts and Jones (2009) documented a threefold increase in stream nitrate concentration, whereas dissolved organic carbon and dissolved organic nitrogen decreased after the fire. These types of conflicting trends reflect the numerous feedbacks and thresholds involved in most nutrient dynamics.

Nitrogen levels in streams are not as persistently elevated following low-severity fire as they are following high-severity fires (Gustine et al., 2022). High-severity fires can create stream nitrate concentrations that are an order of magnitude higher (Stephan et al., 2012) or 50 times greater (Hauer & Spencer, 1998) than adjacent unburned catchments.

Nitrogen dynamics in headwaters and downstream portions of a river network can also be altered by feedbacks between the riparian zone and the stream. Destruction of the riparian canopy during fire can allow more sunlight to reach the stream, elevating water temperatures for a decade or longer (Rhoades et al., 2019). Warmer water and greater nutrient concentrations in runoff can increase algal biomass. Alternatively, algal biomass can remain lower than in unburned catchments when riparian vegetation is not consumed and is able to buffer excess nutrient inputs from burned uplands (Klose et al., 2015). Biomass of algae and instream moss can reduce downstream nitrate delivery (Stephan et al., 2015). In addition, post-fire increases in stream nitrate levels change the composition of DOM and the ratios of stream DOM (Roebuck et al., 2022), dissolved organic N, and dissolved organic C relative to nitrate. Because nitrate uptake depends on water biogeochemistry, changes in relative concentrations can reduce instream nitrate uptake efficiency and subsequently increase downstream N delivery (Gustine et al., 2022).

Over longer time periods, post-fire changes in upland and riparian vegetation can influence solute inputs to streams. In boreal forest underlain by permafrost, for example, soils underlying black spruce (*Picea mariana*) stands are generally wetter and more acidic than soil underlying stands of deciduous trees (Betts & Jones, 2009). This keeps organic soil oxidation low under black spruce stands and promotes maintenance of a thick organic layer and underlying permafrost. Wildfire that combusts organic soils and disrupts the soil thermal regime can lead to establishment of deciduous forest that can alter soil nutrient chemistry, soil active layer depth, and catchment hydrologic flow paths, leading to changes in solute inputs to streams (Betts & Jones, 2009).

The percentage of C and N in upland sediment can be good predictors of the leachate dissolved organic carbon concentration, as demonstrated by work in the Poudre River catchment in Colorado, USA (Cawley et al., 2018). Burned and unburned soil and biomass transported from uplands can accumulate near a stream channel and dissolved organic matter leached from these sediments can increase DOC concentrations in stream water (Cawley et al., 2018).

Most constituents are only sampled for five years or less following wildfire, which limits the ability to estimate recovery timescales for solute concentrations (Hampton et al., 2022). For at least some burned catchments, dissolved fluxes return to pre-fire levels within five years or less (e.g., Granath et al., 2021). Exceptions to this short post-fire monitoring exist, however. Monitoring of N export for 14 years after a wildfire in Colorado, USA indicated that stream nitrate and total dissolved N remained elevated throughout this period because of much lower retention of atmospheric N inputs (Rhoades et al., 2019). Increased N levels in these streams corresponded to greater autotrophic biomass and primary productivity (Rhea et al., 2021). Remotely sensed estimates of upland and riparian vegetation cover suggested that burned forests could require several more decades before forest cover and nutrient demand return to pre-fire levels (Rhoades et al., 2019). Rodriguez-Cardona et al. (2020) examined 17 catchments in the Central Siberian Plateau that included a range from 3 to >100 years since wildfire. Fire increased nitrate concentrations for a decade but decreased the concentrations of dissolved organic carbon and nitrogen and DOM for five decades.

In addition to nutrients, wildfires can affect the mobilization of pollutants into and through headwater stream corridors. Ash provides a complex, heterogeneous matrix that can contain toxic compounds such as polycyclic aromatic hydrocarbons (PAHs) and metals including cadmium, chromium, copper, and lead. These toxins are mutagens that cause genetic mutations and carcinogens that cause cancer (Wohl, 2015). They are known for persistence in the environment and the ability to bioaccumulate across the food chain (Nordberg et al., 2007). Consequently, transport of ash can significantly increase pollutant levels in headwater streams draining burned catchments. Wildfire can also substantially alter water quality by releasing contaminants such as lead, mercury, and other trace metals into the air. These contaminants are then deposited on soils and vegetation, remobilized by increased runoff and erosion after the fire, and carried into streams (Burke et al., 2013; Kristensen et al., 2014).

Sediment. Increased sediment yields from uplands and greater stream flow and associated transport capacity translate to increased suspended (e.g., Troendle and Bevenger 1996; Smith et al., 2011; Warrick et al., 2012; Hampton et al., 2022) and bedload (Beaty, 1994; Robinne et al., 2020) transport in headwater streams following wildfire. Again, these trends are most pronounced in steep terrain and high-severity wildfires.

Suspended materials can include ash, particularly during the first year after fire, and aggregates of clay particles, microbial communities, and organic particles promoted by soil heating in uplands (Smith et al., 2011). The magnitude of the increase in sediment yields corresponds with topographic relief and burn severity and ranges from negligible in low-relief environments with rapid vegetation regrowth after fire, through relatively minor increases in sediment that may create local effects on landforms and infrastructure, to substantial effects of large decreases in reservoir storage space (Bladon et al., 2014) and fatalities and extensive damage to infrastructure caused by post-fire debris flows (e.g., Kean et al., 2019).

Suspended sediment export during the first year after fire is especially variable, with reported increases of 1–1459 times unburned exports across

Figure 4.3 Sources and sinks (italicized text) in the sediment budget for a burned catchment.

catchment sizes from 0.02 to 1655 km^2 (Smith et al., 2011). (These increases can be comparable to those resulting in small catchments from human alterations such as removal of native land cover, which can increase sediment yield by up to an order of magnitude, although the proportional increase in sediment yield declines with increasing catchment size (Walling & Fang, 2003).) Suspended sediment yields then generally decline with time as vegetation cover regrows and fire effects on soil water repellency and surface roughness decrease. Storage of fire-mobilized sediment within river corridors, however, can result in lagged responses in suspended sediment yield when large flows subsequently mobilize the stored sediment (Wilkinson et al., 2009; Smith et al., 2011; Murphy et al., 2019) (Figure 4.3). Such storage is particularly likely at channel junctions and in lower gradient stream reaches (Nyman et al., 2020). Some headwater stream networks, however, can very effectively pass sediment downstream to larger river corridors (e.g., Kean et al., 2019).

Channel instability also results in greater sediment supply to downstream reaches as greater post-fire peak flows form headcuts that migrate upstream or into the floodplain, erode banks and colluvial deposits impinging on channels, and incise the channel (Moody & Martin, 2009). The magnitudes of sediment transport and channel changes depend

(a)

(b)

Figure 4.4 Examples of headcuts in the channel (a) and floodplain (b) following a post-fire flood on Little Beaver Creek, Colorado, USA. In each case, the headcut is 0.5–1 m tall.

strongly on post-fire precipitation patterns (Smith et al., 2011; Warrick et al., 2012), just as for upland post-fire responses (Figure 4.4). Local excess sediment supply can result in channel aggradation and braiding in higher-order rivers (Shakesby & Doerr, 2006), which can also occur in headwater channels (e.g., Florsheim et al., 2017).

In general, the magnitudes and sources of sediment vary through time following fire with changing interactions among precipitation, sediment available to be mobilized, and vegetation. A scenario observed in multiple burned areas in California, USA for example, involves initial post-fire mobilization of hillslope sediment into channels via dry ravel, followed by mobilization during debris flows and floods of pre-fire sediment stored along channels during channel incision, and then continued sediment supply from lateral sources including upland sheetwash and rilling, erosion along stream cutbanks, and shallow landslides (Florsheim et al., 2017; Guilinger et al., 2023).

Silt and clay-sized particles can also provide a vector for dispersing contaminants into streams. Trace metals and synthetic chemicals such as PAHs can adsorb, or physically attach, to sediment particles and move with the particles. When contaminated sediment accumulates in channel and floodplain environments, it creates a reservoir of toxins that may be released when the sediments are subsequently remobilized during a flood or accessed by burrowing or benthic organisms (Bosch et al., 2009).

POM and large wood. Loss or alteration of riparian vegetation can alter inputs of particulate organic matter (POM) and large wood to streams (Betts & Jones, 2009). POM inputs are likely to decrease because of combustion of upland POM and destruction of riparian vegetation that when unburned creates litterfall directly into streams.

Fire can consume large wood within very small or ephemeral streams (Pettit & Naiman, 2007) or debris flows can transport instream large wood farther down the river network (Burton, 2005). The more common scenario, however, is that large wood inputs increase post-fire as a result of bank erosion that recruits wood to the stream, tree topple in burned riparian zones, and mass movements that bring upland wood into the stream (Minshall et al., 1997; Benda et al., 2003; Burton, 2005; Chen et al., 2005; Jones & Daniels, 2008).

This is a transient increase in large wood load at the scale of years to decades, followed by a decline in large wood inputs as trees regrow (Robinne et al., 2020): the patterns of large wood inputs to streams through time after a fire are analogous to those exhibited by coarse woody debris in the uplands. The duration of increased wood loads in streams depends in part on the sequence of flows capable of mobilizing wood pieces and logjams through time after the fire (Comiti et al., 2008; Marcus et al., 2011), as well as the rate of forest regrowth. Although the burned wood pieces are commonly larger in diameter and length than unburned wood (e.g., Young, 1994; Vaz et al., 2011; Praskievicz & Sigdel, 2021), fires result in increased wood mobility in streams because of higher stream flows and increased stream bank erosion (Minshall & Robinson, 1992; Young, 1994; Zelt & Wohl, 2004). Large wood introduced by fire may also be more mobile because it has a simpler shape (less branching complexity), is more decayed, and is less likely to be partly buried in the bed or bank than unburned wood (Vaz et al., 2015).

Increased large wood loads and wood piece mobility in streams create secondary effects such as formation of logjams (Pettit & Naiman, 2007; Iskin & Wohl, 2021). These effects are most likely in second-order and larger streams but can occasionally occur during the higher peak flows occurring in first-order streams after fire. Logjams and burned wood pieces that retain their shape complexity (Vaz et al., 2011) can enhance in-stream sediment storage, formation of pools, geomorphic heterogeneity, and habitat diversity and structural complexity, all of which can benefit fish (Rieman et al., 1997; Burton, 2005; Flitcroft et al., 2016; Robinne et al., 2020; Jager et al., 2021). In forested catchments, large wood exerts substantial influences on process and form in river corridors (Gurnell et al., 2002; Wohl, 2017a), so fluctuations in wood abundance and spatial distribution create corresponding fluctuations in river corridor dynamics.

The timespan over which these channel adjustments occur reflects the rate at which large wood enters channels after a fire. This varies regionally in relation to the half-life of standing dead trees, which can be 15 years in conifer forests of the northwestern United States but only a few years in *Eucalyptus* forests of Australia (Pinto et al., 2008). Development of an old-growth forest can require 200 years after stand-killing wildfires and studies in diverse regions of the world suggest that instream wood loads may also require circa 200 years to return to pre-disturbance levels (Bragg, 2000; Livers et al., 2018; Stout et al., 2018) (Text Box 4.1). Thus, the delayed impacts of wildfire on stream process and form can last for well over a century (Minshall et al., 1997). This long lag time may reflect the decline in wood recruitment to streams while trees regrow after stand-killing wildfire, and the export of large wood from the active channel into the floodplain

Text Box 4.1 How Do we Know? Recovery Times for Instream Large Wood Loads

Different lines of evidence suggest that circa 200 years are required before wood loads—cubic meters of large wood per area of stream channel—return to pre-disturbance levels following a disturbance that completely removes wood. This recovery time is based on (i) comparison of wood loads in streams with old-growth forest and/or only natural disturbances such as fires and floods versus wood loads in streams with a history of timber harvest (e.g., Livers et al., 2018), (ii) Monte Carlo simulation based on observed mass balances of instream wood with time following wood removal (Stout et al., 2018), and (iii) numerical models of forest growth and wood yield (Bragg, 2000; Meleason et al., 2003). These studies come from the U.S. Southern Rockies, southern Australia, and the Pacific Northwest region of North America, suggesting the general applicability of the inferred time necessary to restore instream wood loads.

Figure 4.5 Examples of channel and floodplain logjams pinned against standing vegetation after a large flood in July 2022, 2 years after a wildfire in Little Beaver Creek, Colorado, USA. Arrows indicate flow direction. Field notebook (oval outline) in two of the photos is 18 cm long.

during flashy, high-magnitude peak flows after wildfire (e.g., Lininger et al., 2021; Wohl et al., 2024). Floodplain logjams created by post-fire peak flows can form on the upstream side of standing trees that remain on the floodplain after the fire. The large wood in these jams is unlikely to be redistributed to the active channel by subsequent floods that are of much lower magnitude and transport capacity than the anomalously high flows occurring immediately after fire (Figure 4.5). These persistent logjams can provide habitat for riparian organisms and provide moisture and nutrients to the soil as the wood decays (e.g., Fetherston et al., 1995; Pettit & Naiman, 2005).

In a review of wildfire effects on organic carbon stored in floodplains in the form of soil organic carbon and large wood, Wohl et al. (2020) noted that the net effect of fire in steep, confined valley segments within temperate latitudes is to reduce carbon storage via floodplain erosion, whereas fire can increase floodplain carbon in laterally unconfined valley segments by promoting floodplain aggradation and wood deposition. The effects in partly confined valley segments depend on site-specific interactions among floodplain aggradation and erosion, and, in high latitudes, permafrost degradation. In unconfined valleys underlain by permafrost, wildfire is likely in the short-term to decrease floodplain organic carbon via permafrost degradation and reduced organic-layer thickness. As permafrost degradation reduces stream bank erosional resistance, carbon fluxes into channels increase (Wohl et al., 2020).

In summary, the first- and second-order stream channels that constitute the headwaters are closely coupled to adjacent uplands. They are less likely to be buffered than are higher-order stream by the presence of floodplains and wide riparian zones. Headwaters in burned catchments typically experience higher flood peaks and flashier hydrographs, especially in steep terrain following high-severity wildfire. Nitrogen and other solutes are more available and mobile after wildfire and N, P, and suspended sediment concentrations increase in headwater streams. Contaminants can be mobilized and enter streams through atmospheric, surface, and subsurface transport pathways. Increased inputs from adjacent uplands can be reduced by interactions between the stream and the riparian zone, especially where the riparian zone was resistant to wildfire or where riparian vegetation regrows rapidly. In general, though, headwater streams efficiently transmit increased upland inputs down through the river network to higher-order streams. Headwater channel instability after a fire can further increase sediment supply to downstream areas. Large wood inputs commonly increase after fire, then decrease after most fire-killed trees have toppled and while the upland and riparian forest regrow, before stabilizing over a timespan of a century or longer. The most common scenario following severe wildfire is that headwater channels become at least temporarily unstable and rapidly transmit increased inputs from adjacent uplands to downstream portions of the river network.

4.3.2. Headwater Stream Corridor Forms

Both headwaters and stream corridors farther down the river network can exhibit nonlinear behavior involving spatially and temporally complex feedbacks following wildfire. The primary drivers of these nonlinear adjustments in channel and floodplain form are (i) altered water and sediment yields from uplands and associated alterations in force exerted on the channel and floodplain as the water and sediment move downstream and (ii) altered boundary conditions within the river corridor because of floodplain vegetation and large wood consumed during fire and redistributed after fire.

Adjustments to changed inputs and boundary conditions include altered channel and floodplain morphology. Channel morphologic change can be expressed through planform, lateral and vertical stability, cross-sectional area, bedform type and dimensions, and substrate grain size and patchiness (e.g., Florsheim et al., 1991; Benda et al., 2003; Eaton et al., 2010; Chin et al., 2019) (Figure 4.6). Excessive sediment supply from uplands

Figure 4.6 Schematic illustration of different degrees of post-fire channel morphologic adjustment. Asterisks indicate approximately the same point in paired photos. In the upper and lower pairs of photographs, a logjam failed and the backwater pool and stored fine sediment disappeared, creating changes in bedform type and dimensions and substrate grain size and patchiness. In the pair of photographs at left, the channel alternated through time after the fire between incision to bedrock (upper photograph) and aggradation (lower photograph), creating changes in lateral and vertical stability and cross-sectional area and shape.

following fire, for example, can create channel aggradation and subsequent incision when sediment inputs decline, leaving fill terraces (Florsheim et al., 1991; Moody, 2017). Channel aggradation can bury bedforms, creating a relatively planar bed with more uniform grain size (Florsheim et al., 2017). Channels can also change planform to braiding or anastomosing, which can then increase bank erosion and recruitment of large wood, as well as increasing lateral connectivity between channel(s) and floodplain and vertical connectivity via hyporheic exchange flows (e.g., Benda et al., 2003). Excess sediment can also create or enlarge alluvial fans where headwaters join higher-order streams (Rathburn et al., 2018). Increased post-fire water yields, flashier peak flows, and burned floodplain vegetation can change the ratio of hydraulic driving force to erosional resistance in streams, leading to greater bed and bank erosion.

The post-fire scenario inferred for Wangi Creek in Australia by Larsen et al. (2016) provides an example of feedbacks that can moderate post-fire enhanced erosion. Episodic headward retreat of alluvial knickpoints leads to a drop in the floodplain water table, desiccating the floodplain soils that support a wet monsoon forest and peat formation in the midst of eucalypt savanna. Fires can then burn into the stream corridor and this, combined with the water table decline, destroys the forest. As the dead trees fall, they create logjams that form local baselevels, trapping sediment and helping to raise the floodplain water table sufficiently to once more support wet monsoon forest (Larsen et al., 2016).

The Wangi Creek example illustrates a fundamental aspect of post-fire response in headwater and higher-order river corridors: the response of the river corridor to altered fluxes from adjacent uplands is not linear or consistent through time. Feedbacks can cause reversals in stream corridor form, such as the incision followed by wood recruitment and then aggradation in Wangi Creek. Scaling these interactions up to a river network, first- and second-order channels can respond to post-fire fluxes at different rates and magnitudes, and even in different directions, than portions of the river network farther downstream. A scenario for debris-flow initiation described in the preceding chapter, for example, involves mobilization of sediment stored along headwater channels to create a debris flow downstream. The headwater channels incise in this scenario, even while the debris flow likely causes substantial aggradation downstream in the river network. Following the decline in runoff and stream peak flows as vegetation recovers after the fire, the headwater channels may begin to once more accumulate sediment, whereas the downstream portions of the river network incise in response to declines in sediment supply.

Complex response, in which a single initial perturbation can cause continuing alternations in channel stability and form, and the response to perturbation varies through time and space across a river network, was initially described for dryland channels experiencing base level fall (Schumm, 1973). The diverse scenarios that play out in river networks, especially following severe fire in steep terrain, also exemplify complex response.

Post-fire floodplain morphologic change can occur in floodplain topography, alluvial volume, stratigraphy, and sediment residence time (e.g., Sutfin & Wohl, 2019). Direct erosion and sediment deposition, or creation of floodplain logjams that then induce local erosion and deposition, can alter floodplain topography (Brogan et al., 2019). Increasing water and sediment inputs, combined with reduced erosional resistance where floodplain vegetation has burned, can drive lateral channel mobility and floodplain erosion, creating shorter residence time of floodplain sediment (Sutfin & Wohl, 2019). Increased sediment inputs can also cause floodplain aggradation and increased alluvial volume (e.g., Meyer et al., 1992; Malmon et al., 2007). The greater erosional force and transport capacity of post-fire floods can strip finer sediment or deposit very coarse (cobble to boulder) sediment on floodplains, changing floodplain stratigraphy and hyporheic and groundwater exchange flows.

The details of these diverse floodplain responses reflect the balance between factors such as the magnitude of post-fire water and sediment fluxes as this magnitude governs spatial extent, duration, and erosive energy during floodplain inundation, versus the erosional resistance of the channel banks and floodplain as governed by vegetation, surface hydraulic roughness, and grain-size distribution. This balance can be very site-specific and can vary through time, creating a diverse array of potential outcomes for headwater floodplains following wildfire.

Wildfires also change biotic influences on stream corridor form and process, including aquatic and riparian plants, large wood, and beaver in the northern hemisphere (e.g., Dwire & Kauffman, 2003; Chen et al., 2005). Post-fire alterations in stream corridor processes influenced by biota include (i) hyporheic exchanges that are affected by changes in streambed and floodplain surface topography and grain-size distribution mediated by the presence or absence of large wood (e.g., Dunham et al., 2007; Ader et al., 2021), (ii) biogeochemical uptake of nutrients such as N and P mediated by regrowth of riparian vegetation (e.g., Hauer & Spencer, 1998), and (iii) retention of organic carbon in floodplain soil mediated by the ability of floodplain vegetation to increase erosional resistance and contribute organic matter to the floodplain (e.g., Wohl et al., 2020). Post-fire alterations in stream corridor forms influenced by biota include (i) channel bank and floodplain stability as mediated by aboveground hydraulic roughness created by vegetation and belowground erosional resistance created by plant roots (e.g., Eaton et al., 2010), (ii) channel-bed morphology as mediated by the presence of large wood (e.g., Wohl et al., 2022), and (iii) channel planform as mediated by the presence or absence of large wood and beaver dams (e.g., Short et al., 2015).

The magnitude of post-fire adjustments in headwater stream corridors reflects interactions among multiple factors, including:

- the ratios of pre- to post-fire water and sediment yields—a relatively minor increase in sediment yield, for example, may cause minor or temporary secondary effects;
- the stream corridor's conveyance for water and transport capacity for sediment relative to the magnitude of post-fire increases (e.g., Murphy et al., 2019; Nyman et al., 2020);
- the erosional resistance of the stream corridor relative to the magnitude of increased flow energy or sediment supply (Constantine et al., 2014)—erosional resistance in this context may reflect whether the river corridor itself burned and, if so, how rapidly riparian vegetation regrows;
- position in the river network—headwaters or narrow valley segments that are closely coupled to adjacent uplands may experience greater secondary responses than wide valley segments in which broad floodplains and other stream corridor landforms create longitudinal disconnectivity and thus buffer increased water and sediment yields (Fryirs et al., 2007);
- pre-fire stability of the stream corridor—an incising or aggrading portion of the river network might experience lesser secondary change in response to post-fire increases in sediment yield, for example, than a stable reach (Schumm et al., 1984); and
- spatial heterogeneity and connectivity of the stream corridor as these influence the degree to which post-fire increases in water and sediment yield are retained within a reach or transmitted downstream (Wohl et al., 2022, 2023, 2024).

Post-fire responses in Little Beaver Creek, Colorado, USA, illustrate the influences of heterogeneity and connectivity on stream corridor adjustment (Wohl et al., 2022). Prior to the Cameron Peak Wildfire in 2020, this stream corridor draining a 40 km^2 catchment had numerous abandoned beaver dams that, although breached by the creek, remained in place across much of the stream corridor as meter-high berms stabilized by densely growing, multi-stemmed willows (*Salix* spp.) (Figure 4.7). The beaver berms effectively trapped mobile large wood and many of the logjams that spanned the active channel prior to the fire were associated with the berms. The stream corridor alternates longitudinally between beads—segments of relatively wide floodplain in which beaver berms and logjams promote

Figure 4.7 An abandoned beaver dam (crest outlined in white dashed line) in Little Beaver Creek, Colorado, USA just before (a) and after (b) a 2020 wildfire. Note the large burned log deposited along the beaver berm in (b).

lateral and vertical connectivity and multithread channel planform—and steep, laterally, confined single channels.

The 2020 Cameron Peak Wildfire burned 86% of the catchment at high severity, including much of the length of the stream corridor. Higher water and sediment yields from the burned uplands during the 2021 runoff season created a series of small, flashy floods that redistributed logjams along the active channel(s), mobilized the gravel and finer sediment stored upstream from the logjams, and caused local deposition of fine sediment on the floodplain along the channel margins. Although substantial amounts of water and sediment moved downstream, the basic stream corridor morphology did not change, in large part because of the attenuation of hydraulic force and at least transient storage of sediment in the beads (Wohl et al., 2022). Multiple tributary debris flows associated with a major flood during the 2022 runoff season caused more substantial channel bed and bank erosion and coarse floodplain deposition. This flood moved most of the large wood from the channel onto the floodplain and initiated new segments of multithread channel planform (Wohl et al., 2024). The massive influxes of sediment from the uplands did not create an alluvial fan or excessive sediment inputs to the larger channel downstream, however, because of the ability of the beads along Little Beaver Creek to attenuate forces and fluxes during and after the flood.

In summary, the morphology of headwater channels and floodplains can exhibit nonlinear behavior involving spatially and temporally complex feedbacks following wildfire. Both channels and floodplains have multiple degrees of freedom in which changes can occur, including channel planform, lateral and vertical stability, cross-sectional area, bedform type and dimensions, substrate grain size and patchiness, large wood abundance and distribution, and floodplain topography, alluvial volume, stratigraphy, and sediment residence time. The nature of post-fire changes in headwater stream corridors reflects factors such as the magnitude of increased fluxes from uplands and the ratio between increased erosional energy and erosional resistance in the stream corridor; the pre-fire stability of the stream corridor; the existence of spatial heterogeneity within the stream corridor; and recovery of floodplain and channel biota such as woody plants or beaver after the fire.

4.3.3. Headwater Stream Biota

This section focuses on aquatic food webs, macroinvertebrates, and fishes. Other organisms, including terrestrial insects, amphibians, reptiles, birds, and mammals use headwater stream corridors and are affected by wildfire and post-fire changes in fluxes and forms, but the responses of these organisms are discussed in the chapter on terrestrial environments.

Headwater streams that experience substantial increases in flow, sediment supply and mobility, and water temperature (due to decreased riparian shading) following fire can present challenges to survival and post-fire recolonization by macroinvertebrates, fish, and other stream biota. Multiple studies indicate that increased water temperatures, for example, can persist for more than a decade after fire if vegetation capable of shading the channel slowly recovers and post-fire erosion and deposition in the channel create wide, shallow cross-sectional geometry that facilitates solar heating of the stream water (e.g., Moore et al., 2005; Dunham et al., 2007).

The effects of fire start with the base of a stream's food web, whether that base is primarily autochthonous (based on instream photosynthesis by algae) or allochthonous (based primarily on organic matter such as plant litter dropped from the adjacent riparian vegetation) (Figure 4.8). Autochthonous production can be increased by both the flush of solutes coming from uplands and the greater availability of sunlight if the riparian vegetation is consumed by the fire (Betts & Jones, 2009; Leach & Moore, 2010; Cooper et al., 2015; Klose et al., 2015; Rhea et al., 2021). These changes can lead to excessive growth of algae and multiple studies describe post-fire algal blooms (e.g., Mishra et al., 2021). Even when fire does not alter algal or periphyton biomass, streams can exhibit a shift in diatom assemblage within the periphyton, for example (Earl & Blinn, 2003). Increased nutrients and organic matter after fires can enrich the entire stream food web, from phytoplankton and microbial communities to fish, for several years after a fire (Robinne et al., 2020).

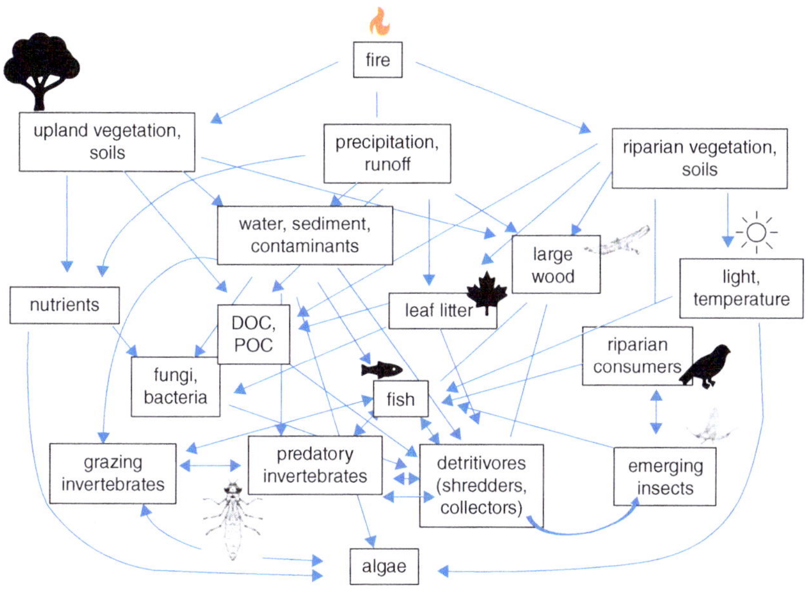

Figure 4.8 Flow diagram showing probable cause-effect relationships leading from fire to stream communities. Lines without arrows indicate factors associated with each other; unidirectional arrows point from driver to response variable; double-headed arrows indicate consumer-resource interactions in which consumers depress and benefit from the consumption of their resources. DOC is dissolved organic carbon; POC is particulate organic carbon. (Adapted from Bixby et al., 2015.)

Allochthonous production can be lowered if burning consumes vegetation that would otherwise provide plant litter (Cooper et al., 2015) and if charred organic matter replaces unburned material, but rates of organic matter input gradually recover with regrowth of vegetation. Altered species composition of riparian vegetation and overall inputs of particulate organic matter through time can also influence stream food webs. The chemical and physical structure of particulate organic matter (POM) affects the rate at which this material breaks down. Different plant species and plant parts have varying palatability and nutritional value that influence colonization preferences and community composition in microbial and macroinvertebrate stream communities (Musetta-Lambert et al., 2020). Rate of POM breakdown correlates positively with N concentration in the POM, for example, and negatively with lignin concentration (Bruder et al., 2014). Increased stream temperatures caused by decreased riparian shading can create faster POM processing by both microbial and macroinvertebrate communities (Rodríguez-Lozano et al., 2015).

Benthic macroinvertebrates. Benthic macroinvertebrates live on or within the stream bed for at least some portion of their lifecycle and are visible to the naked eye. Such organisms include mussels and clams, worms, crayfish, snails, and a diverse array of insects, from beetles to larval mayflies, caddisflies, stoneflies, and dragonflies.

Stream ecologists commonly differentiate four functional groups among benthic macroinvertebrates based on feeding strategies (Allan et al., 2021). Shredders feed on particulate detritus, submerged wood, and living plants. Grazers or scrapers feed on the periphyton—attached algae and associated microflora and microfauna. Collectors eat smaller particulate detritus and the microbes living on the detritus. Collectors can either filter their food from the passing water or gather their food from the channel substrate. Predators eat other macroinvertebrates.

The proportions of different functional groups within the benthic macroinvertebrate community can change in response to changes in resources, such as greater exposure to sunlight and autochthonous production that favors grazers after fire. These changes can then influence other portions of the stream food web, as illustrated by field experiments along the Horonai Stream in northern Japan (Baxter et al., 2004). Terrestrial invertebrates falling into streams provide about half of the food eaten by native Dolly Varden charr in streams of northern Japan. Emerence of adult aquatic insects from the stream provide about a quarter of the food eaten by forest birds and much of the food for web-building spiders along the stream. When nonnative rainbow trout invaded the Horonai Stream in northern Japan, they ate so many of the terrestrial insects that fell into the stream that the Dolly Vardens changed their diet to focus on insects grazing algae on the streambed. Removal of the insect grazing pressure caused algae to become more abundant. Fewer insects emerged from the stream as winged adults and the populations of spiders along the streambanks declined by 65%. Although this example does not involve wildfire, it illustrates how changes in a top predator or the base of the food web can influence other aspects of the river ecosystem.

The influence of wildfires on macroinvertebrates varies from minimal to substantial, depending on factors such as magnitude of post-fire fluxes and channel disturbance and the species composition of the pre-fire community. Fire typically has little direct effect on benthic macroinvertebrates (Minshall, 2003), but the physical disturbances associated with increased water and sediment yields after fire can cause short-term declines in macroinvertebrate abundance (Rinne, 1996; Monaghan et al., 2019). These physical disturbances include greater substrate mobility, abrasion or dislodgement by mobile sediment, bed coarsening or fining, channel aggradation or incision, and greater turbidity.

Post-fire physical disturbances can also cause shifts in the abundance of different functional groups within the macroinvertebrate community toward a greater proportion of disturbance-adapted taxa such as Chironomidae (non-biting midges) and Baetis mayflies (Jackson & Füreder, 2006; Rugenski & Minshall, 2014; Martens et al., 2019). Some macroinvertebrate taxa, such as Ephemeroptera and Diptera, are expected to recover more rapidly after fire-related disturbances because of their rapid dispersal and early maturity (Vieira et al., 2004). Other post-fire investigations document an increased abundance of predatory macroinvertebrates specialized for soft sediments, with reduced abundances of shredding and algal-grazing caddisflies (Robson et al., 2018), or an increased abundance of collector-gatherers that scavenge for dead organisms and detritus (Williams-Subiza & Brand, 2021).

The details of which functional groups do best after fire depend on the continuing disturbance regime (e.g., greater sunlight that favors algae and grazers vs soft sediment deposition) and the degree to which vegetation in the riparian zone burned. Densities of macroinvertebrate predators can also increase where fish are extirpated by fire (Cooper et al., 2015). Reviewing 30 studies of post-fire benthic invertebrate responses, Paul et al. (2022) found that most studies document a decline in abundance and diversity, averaging 45%, but also note recovery within 5, or at most 10, years. Fire can also result in greater invertebrate productivity and abundance because of increased food resources (Malison & Baxter, 2010; Rugenski & Minshall, 2014; Martens et al., 2019).

Minshall (2003) predicted that abundance of shredder macroinvertebrates should track the loss and subsequent recovery of leaf litter, whereas scraper macroinvertebrates should reflect changes in periphyton associated with greater sunlight reaching the stream and greater nutrient inputs (Minshall et al., 2001a). However, Minshall (2003, p. 157) cautioned that "strict adherence to a patterned sequence of feeding group replacements generally has not been observed . . . because physical factors, particularly turbidity, sedimentation, and scouring, have an overriding influence on macroinvertebrate occurrence, and furthermore, most stream invertebrates are not narrow food specialists." In other words, reality is more complex than scientists might expect and predictions of post-fire macroinvertebrate response include substantial uncertainty.

The rate of macroinvertebrate community recovery also varies widely between studies. Macroinvertebrate species richness and diversity were greater in burned headwater streams than in unburned streams a decade after wildfire within dry conifer forests in Arizona, USA, despite post-fire flooding, channel incision, and changes in substrate grain size (Leonard et al., 2017). Within 20 years, macroinvertebrate communities in the two sets of streams did not differ. Canopy closure over the burned streams remained lower 20 years after the fire, but stream temperatures were higher, and concentrations of key nutrients such as nitrate and phosphate were lower. Leonard et al. (2017) concluded that the process of stream recovery would extend beyond the 21-year time period of the study.

In contrast, other studies indicate that macroinvertebrate densities and community structure can recover within a year (e.g., Earl & Blinn, 2003; Monaghan et al., 2019) or 2–3 years (Robson et al., 2018; O'Dowd et al., 2023). Addition of ash to burned streams can create a rate of macroinvertebrate drift over 10-fold that in unburned reaches, but the community can still recover within a year (Earl & Blinn, 2003). (Drift is a sort of lottery for macroinvertebrates. By drifting downstream, the organisms can disperse and colonize new areas, but they can also fall prey to drift-feeding trout.) In other burned areas in the Boise River catchment of central Idaho, detectable changes in macroinvertebrate drift and diet of trout feeding on the macroinvertebrates persisted for a decade after wildfire, although these changes correlated best with stream-specific characteristics such as water temperature and canopy cover (Rosenberger et al., 2011).

Benthic macroinvertebrates can also be exposed to contaminants adsorbed to sediment or dissolved in pore water in the hyporheic zone even after post-fire water and sediment yields decline (Harkey et al., 1994). Exposure to water and sediments affected by wildfire can reduce feeding in macroinvertebrates following the exposure, which reduces the energy available to the organisms for detoxification, growth, and reproduction (Ré et al., 2021).

Fishes. Although fishes are more likely to be present in higher-order streams, they can be present in headwaters. Populations of individual species or communities in a stream can be described via resistance and resilience. Resistance is the ability of individuals to survive disturbance such as wildfire. Resilience is a population's capacity to restore its abundance to levels similar to those before the disturbance (Gido et al., 2019).

Species traits, the habitat template of the stream—including the magnitude and duration of post-fire disturbance—and availability of, and connectivity to, refuge habitats interact to govern how fish populations and communities respond to disturbance (Townsend & Hildrew, 1994). Among species traits, mobility can confer resistance and resilience (McCluney et al., 2014). Large body size might add resilience because of greater mobility, capacity for compensatory reproduction, and physiological ability to survive extreme anoxic conditions for longer periods (Nilsson & Östlund-Nilsson, 2008). Small-bodied fish species, however, might be more resilient because they can be broadly distributed and have large population sizes and short generation times (Shurin et al., 2009).

The effects of post-fire changes in sediment transport, bed substrate, channel morphology, and other variables can be distinguished with respect to life-stage of fishes. Fine sediment on the channel bed that limits interstitial flow and clogs pore spaces, for example, can reduce spawning and larval habitat for fishes (e.g., Bowerman et al., 2014), whereas loss of pool volume may affect summer and overwinter habitat for adult fishes (Cunjak, 1996).

Organisms in headwater streams should be more likely to experience extreme disturbance because wildfires are more likely to spread across the entire catchment and post-fire water and sediment yields are greater per unit area of catchment, especially in catchments with steep slopes and shallow soils that experienced high-severity fire (Minshall et al., 1989). Larger streams are likely to have greater habitat complexity and thermal refugia in deep pools than are headwater streams (Magoulick & Kobza, 2003; Roberts & Hitt, 2010). Habitats in larger streams are also likely to be more connected within stream networks (i.e., not disconnected by dry channel reaches or barriers such as waterfalls) and more resilient following disturbance relative to headwater habitats because dispersal of stream organisms is facilitated in a downstream direction (Brown & Swan, 2010).

The response of salmonid fishes in North American streams following fire has received a great deal of attention. Salmonids are coldwater-adapted species that typically prefer water temperatures below 20 °C (Warren et al., 2022). Severe wildfires that burn much of a catchment, including the riparian zone, can create elevated water temperatures that regularly exceed 20 °C. Progressive declines in spawning and rearing habitat for salmonids in the U.S. Northern Rockies have been attributed to both warming climate and wildfire (Isaak et al., 2010). However, more than one study in the western United States demonstrates persistence of native salmonids and coldwater frogs (Dunham et al., 2007; Warren et al., 2022). This resilience of coldwater vertebrates may reflect several factors, including greater thermal tolerance than previously recognized; the presence of cooler microhabitats created by groundwater discharge into the stream; physiological recovery during cooler overnight temperatures; and an increase in food availability due to increased autotrophs at the base of stream food webs, allowing consumption to increase with metabolic demands at higher temperatures (Beakes et al., 2014; Railsback, 2022). Alternatively, fish may be moving from warmer to cooler portions of the stream network, if connectivity allows (Warren et al., 2022) (Text Box 4.2).

Text Box 4.2 How Do We Know? Fish Movements After Fire

Among the "tags" that fish carry within their bodies are molecular genetic markers and otoliths. An otolith is a calcium carbonate structure found inside the head of bony fish (i.e., ear bones). As a fish grows, trace elements from the surrounding environment are absorbed into the bloodstream via the gills. These elements move into the endolymph that surrounds otoliths and are accreted into otoliths, remaining in their original precipitation point throughout the fish's lifespan (Olsson et al., 1998; Campana, 1999). In other words, otoliths are like tree rings in recording change through time. Variation in elemental concentrations within an otolith can thus be used to determine an individual's dispersal across chemically heterogeneous habitats (Zeigler & Whitledge, 2011), including how fish recolonize burned catchments (Whitney et al., 2017). Surveys of genetic diversity of fish in numerous streams within a burned area can reveal the presence of founder effects or population bottlenecks associated with wildfire, post-fire floods or debris flows, or infrastructure such as culverts that disconnect habitat (Neville et al., 2009).

Following immediate post-fire increases in sediment input to streams, the degree to which habitat recovers may depend on factors such as the ability of peak flows to export fine sediments and the supply and redistribution of instream large wood that can create habitat structure and diversity. Even the most severely impacted streams in conifer-forested mountains of central Idaho, USA had substantial improvements in habitat and trout populations within 5–10 years of fire. These trends may reflect the presence of wood and large boulders introduced by post-fire debris flows that enhanced habitat structure, as well as removal of excess fine sediment by peak flows (Burton, 2005).

In addition to natural disturbances during and after fire, aquatic communities can be affected by the use of fire retardants, which are sometimes applied aerially during fire. Retardants are designed to lower the temperature of fuel so that it will burn slowly by glowing combustion rather than by flaming. Retardants are mixtures of water and selected inorganic salts (e.g., ammonium sulfates, diammonium phosphates) with additives such as clays to thicken the mixture, bactericides to inhibit corrosion and spoilage, and dyes to color the liquid (Kalabokidis, 2000). Retardants can encourage eutrophication of aquatic systems and can sometimes contribute to fish kill, especially when they are applied directly to water bodies (Kalabokidis, 2000). Many of the ions in retardants—and particularly ammonia—are lethal to aquatic life at the concentrations found in modern fire chemicals (Norris et al., 1978).

In summary, where the riparian zone burns, wildfire alters headwater stream food webs that begin with algae and those that start with plant litter from the riparian zone. Greater inputs of solutes, more sunlight reaching the stream, and reduction in plant litterfall can each ripple through the food web of a headwater stream. Stream macroinvertebrates are affected to a greater or lesser extent after the fire, depending on the magnitude of upland inputs and channel disturbance. Fishes in headwaters can be resistant or resilient to wildfire effects. Headwater stream organisms are more likely than organisms in high-order channels to experience extreme post-fire disturbance because of the greater likelihood of wildfire across the entire catchment; the lack of storage and buffering areas in the typically narrow stream corridors of headwaters; the simpler habitat and absence of thermal refugia such as deep pools; and the presence of features such as intermittent stream reaches or waterfalls that limit connectivity with downstream portions of the river network. These are generalizations, however, and wildfire effects on fluxes, forms, and biota in headwater river corridors vary through time in a manner that can be nonlinear because of interactions between water and sediment fluxes through upland, headwater, and higher-order river corridors in a catchment.

4.4. River Corridors in Higher-Order Portions of River Networks

Downstream from the headwaters, river corridors are more likely to be response reaches (Montgomery & Buffington, 1997) that retain and store sediment coming from uplands and headwaters. These higher-order rivers are more likely to have longitudinally continuous floodplains that can act as a safety valve during peak flows by storing and gradually releasing overbank flow; dissipating flow energy in a hydraulically rough floodplain; and facilitating hyporheic exchange and microbial uptake of excess nutrients. Higher-order rivers also receive the cumulative effects of increased fluxes from uplands and headwaters. Because many headwaters can effectively pass excess inputs downstream, the junction of a tributary and higher-order river can be an important response point in a river network, especially following large, punctuated increases in water and sediment supply (Benda et al., 2003) (Figure 4.9).

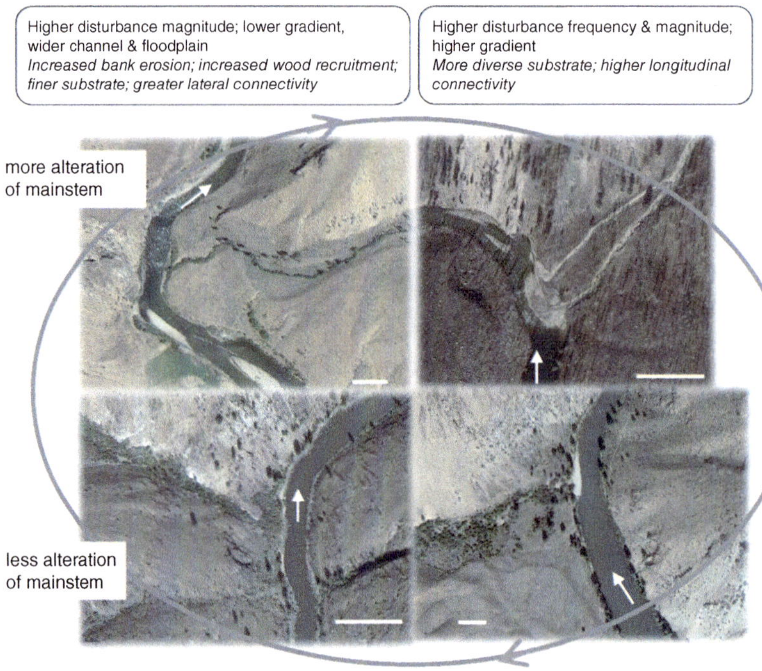

Figure 4.9 Schematic illustration of how large influxes of sediment from burned tributary catchments along the Middle Fork Salmon River, Idaho, USA can affect the morphology of the receiving channel in proximity to rejuvenated alluvial fans. Inset photos clockwise from upper left (Pole Creek, Kotch Creek, Sheep Creek, unnamed tributary) show different degrees of mainstem response to tributary inputs, from greatest at upper left to least at lower left. Arrows indicate flow direction on mainstem. White bar indicates 100 m. Plain font text in boxes indicates disturbance characteristics and river corridor geometry; italicized font indicates mainstem response. Text boxes are not associated with a particular inset photo. (After Benda et al., 2003, Figure 8.)

4.4.1. River Corridor Fluxes

Although the magnitude of increases in discharge post-fire may decline moving down a river network because of the potential for attenuation of water fluxes via floodplain and hyporheic storage, Kinoshita and Hogue (2011, 2015) documented increased stream flow for seven years and elevated dry season flow for over a decade after fire in both headwater and higher-order catchments in southern California, USA. In other words, there are exceptions to the general relations described in the introduction to this section.

Analogously, Ortega-Becerril et al. (2024) described temporal variations in the response of a higher-order river corridor in the mountainous Sistema Central of Spain. Although the contributing catchment burned at generally low severity, a post-fire increase in flow rates for similar rainfall scenarios, combined with steep slopes that included high-severity fire patches and substantial volumes of regolith, resulted in debris flows along the river network, followed by channel erosion (including erosion into bedrock).

Post-fire responses in headwaters and third and fourth order rivers during the five years after a wildfire in New Mexico, USA indicate less pronounced changes in dissolved oxygen and turbidity in the higher-order rivers, which Reale et al. (2015) attributed to groundwater contributions and retention of post-fire debris flow sediments in the headwaters. On the other hand, retention of very fine sand, silt, and clay in higher-order, gravel-bed rivers for several years after wildfire can provide a significant, long-term source of bioavailable P within the channel (Emelko et al., 2016).

Measurements before and after a wildfire in Little Granite Creek, Wyoming, USA revealed suspended sediment transport about five times pre-fire levels that gradually declined to baseline levels within a few years (Ryan & Dixon, 2008). Bedload transport, in contrast, did not increase relative to pre-burn values. The first few years after the fire were warm and dry, which likely limited the magnitude of post-fire response, although flows with substantial suspended sediment, ash, and charcoal occurred during the first year after the fire (Ryan et al., 2011).

Robinson et al. (2005) compared burned and unburned tributaries of the Middle Fork Salmon River, Idaho, U.S. Tributaries with drainage areas in the range of 5–213 km^2 (2nd to 5th order rivers) exhibited the greatest change in channel longitudinal profiles during the first 10 years post-fire, with little additional change during the succeeding decade. Twenty years after the fire, instream wood loads were mostly comparable in burned and unburned catchments. Large wood in higher-order rivers resulted from downstream transport, as well as recruitment from adjacent riparian or floodplain forests (Minshall et al., 1997).

In summary, the limited number of studies that explicitly compare post-wildfire fluxes in headwater streams and higher-order channels in the same network indicate that spatial and temporal trends are highly variable. Individual constituents—N versus P or suspended vs bedload sediment—exhibit different patterns through time and space. In general, concentrations of material in transport can decline downstream because of greater storage capacity in the more extensive floodplains and spatially heterogeneous channels of higher-order streams.

The repeated use of phrases such as "more likely to" and words such as 'can' in my summaries of post-fire changes in headwaters and higher-order rivers reflects the existence of exceptions. There are higher-order rivers in narrow bedrock canyons that very effectively transmit post-fire fluxes downstream, for example, and that exhibit minimal changes in channel form after fire. The generalized summaries, however, are intended to characterize the most common patterns of response across the case studies thus far described in peer-reviewed publications.

4.4.2. River Corridor Forms

As in headwater streams, post-fire alterations in water and sediment inputs to higher-order river corridors can result in changes to channel and floodplain forms. Channel substrate, bedform type and dimensions, channel cross-sectional geometry, planform, and floodplain topography, alluvial volume, and stratigraphy can all change over varying timespans following fire (Moody, 2017; Florsheim & Chin, 2022). The magnitude and type of changes will largely reflect the balance between increased erosional force and sediment supply versus erosional resistance and accommodation space for water and sediment storage in the higher-order river corridor. A complicating factor in higher-order rivers is that post-fire increases in water and sediment can be staggered through time as individual headwater catchments respond differently to precipitation after the fire. In other words, some tributary catchments may transmit large volumes of water and sediment downstream to higher-order rivers in the first runoff season following the fire, whereas other tributaries may transmit peak water and sediment volumes two or more years after the fire. The primary sediment sources to higher-order river corridors can also change with time after fire, from alluvium stored along headwater channels to continuing hillslope erosion (e.g., Guilinger et al., 2020) or from widespread hillslope rilling and sheetwash to debris flows that evacuate sediment stored in tributary channels (Wohl et al., 2024).

As an example of these interactions among sediment inputs and river corridor morphology, changes over 15 years after the massive 1988 wildfires in Yellowstone National Park, Wyoming, USA took the form of sequential changes in sediment transport and channel morphology (Legleiter et al., 2003). Increased sediment supply to river corridors continued for 5–10 years after the fire, causing channels to aggrade. First- to third-order streams retained substantial fine sediment for about four years after the fire (Minshall et al., 1997), but sediment storage by alluvial fans likely reduced the volume of sediment entering high-order rivers (Meyer & Wells, 1997). Morphology in first-order streams did not differ between burned and unburned catchments within a decade of the fire (Ernstrom, 1999). In higher-order rivers, however, continuing post-fire increase in water yields, combined with decreased upland sediment supply, induced channel incision over the next few years. Twelve to 13 years after the fire, higher-order channels were narrower and deeper in more extensively burned catchments (Legleiter et al., 2003).

Post-fire feedbacks in river corridor form involving beaver dams can vary with stream size and the associated ability to support beaver across climatic fluctuations occurring over millennia. Stratigraphic records from the Yellowstone region indicate decreasing beaver activity and coarse sediment aggradation in smaller streams during periods of increased drought and wildfire (Persico & Meyer, 2009, 2013), whereas beaver activity continued on nearby larger rivers (Levine & Meyer, 2014).

The feedbacks between riparian vegetation and channel morphology described for headwater streams also have analogs in higher-order rivers. A 2003 wildfire in the Fishtrap Creek catchment ($158\,km^2$) of British Columbia, Canada burned the riparian forest. During the next four years the creek widened substantially despite the absence of intense rainfall, major sediment pulses from uplands, and significant increases in peak flows. Eaton et al. (2010) attributed this widening to loss of bank erosional resistance as the roots of burned trees decayed.

Feedbacks between large wood and channel morphology and substrate grain-size are more common in higher-order rivers than in headwaters because of the typically greater mobility of large wood in higher-order rivers. Formation of logjams can cause upstream bed aggradation and fining, but subsequent destruction of the logjams can result in channel incision and coarsening in the vicinity of the logjam (Wohl et al., 2022, 2023; Chin et al., 2024).

Many of the adjustments in higher-order rivers following wildfire involve alluvial fans that form at channel junctions. Alluvial fans can attenuate post-fire fluxes as excess stream flows infiltrate the fan alluvium (e.g., Houston, 2002; Huang et al., 2021). The fan also stores sediment coming into a higher-order river corridor (Benda et al., 2003; Riley et al., 2015) and influences mainstem river corridor morphology and longitudinal sediment connectivity (Fryirs et al., 2007). In the decade following a 1995 wildfire in the Boise River catchment of Idaho, USA, alluvial fans impinged on higher-order channels. Each fan created a knickpoint with decreased channel gradient for up to 4 km upstream from the fan and increased channel gradient downstream from the fan (Benda et al., 2003).

The effect of post-fire fan deposition depends partly on the geometry of the receiving valley and partly on the grain-size distribution of fan deposits. A massive post-fire debris flow crossed and temporarily dammed the South Fork Payette River, Idaho at a narrow canyon stretch of the river, but the predominantly pebble and finer size sediments within the fan deposit were largely removed within a few months (Meyer et al., 2001).

Another post-fire debris-flow fan along a different portion of the river included numerous large boulders and this fan created a steep rapid on the river, with a finer-grained, low gradient reach upstream (Meyer et al., 2001) that has now persisted for 20 years (G. Meyer, pers. comm., 2022).

The broad glacially eroded valleys of the Yellowstone region of the United States facilitate storage of large volumes of sediment on alluvial fans. Much of this sediment is not accessible to mainstem channels. The fans thus buffer post-fire sediment inputs to axial channels because the coarse sediment mostly remains on the fans, with predominantly sand and finer sediment reaching axial streams (Meyer & Wells, 1997). Stratigraphic records in the region indicate that alluvial fans aggrade during periods of relatively frequent fire-related sediment inputs (Meyer et al., 1992). The fans impinge on axial streams, limiting their lateral mobility and floodplain extent, and the axial streams incise and create terraces. When wildfire activity declines during wetter climatic phases, greater river flow removes sediment from alluvial fans via incision of channels on the fan and carries the sediment into the axial river, which also undercuts the toe of the fan, causing floodplain widening. These alternations in stream corridor morphology occur over timespans of about 1300 years (Meyer et al., 1995).

Following a 2000 wildfire in Montana's Sleeping Child Creek catchment, debris flows from 1st and 2nd order streams created fans that spanned the entire mainstem river corridor and locally buried the creek (Hoffman & Gabet, 2007). Five years after the fire, the creek had aggraded upstream from each fan, incised across the fan, and then aggraded downstream. After 11 years, however, substantial wood loads in the creek, along with a newly constructed beaver dam, had caused continuous aggradation in the creek (Short et al., 2015). Sediment-trapping by the dam and logjams and subsequent vegetation colonization of the sediment is likely to retain the sediment within the river corridor over longer timespans. The 2000 wildfire burned only 3.5% of the catchment, but intense rains during summer 2001 triggered the debris flows (Hoffman & Gabet, 2007).

In summary, higher-order rivers are more likely to be response reaches than are the transport-dominated headwaters. Consequently, higher-order rivers can store sediment and take longer to stabilize after wildfire than headwater streams, with lag times that reflect interactions between continuing supplies of sediment and precipitation abundance and intensity over multiple years. Morphological adjustments in higher-order rivers are likely to be influenced by the presence and characteristics of alluvial fans that form at tributary confluences. Alluvial fans can store excess water and sediment after wildfire. Fans can also influence mainstem river corridor morphology and longitudinal sediment connectivity. The manner in which alluvial fans influence post-fire river corridor morphology depends on characteristics including fan volume, grain-size distribution, the erosional and transport capacity of the axial channel, and feedbacks with large wood and beaver.

4.4.3. River Corridor Biota

As noted for the headwaters section, other types of biota beyond aquatic insects and fishes use river corridors and are affected by fire, but this is discussed in the terrestrial environments chapter.

As mentioned in the discussion of biota in headwaters, aquatic organisms in higher-order rivers may have more options to find refugia up- or downstream during and after a fire if the organisms are sufficiently mobile (e.g., Whitney et al., 2017). Higher-order rivers can also provide greater habitat diversity and potential refugia post-fire. Fish communities in higher-order rivers can thus be more resistant to post-fire disturbance (e.g., Reale et al., 2021). Substantial negative effects of post-fire disturbance on benthic macroinvertebrate communities are unlikely in rivers of sixth order or larger (Mihuc et al., 1996; Minshall et al., 2001b). Taxonomic richness and biomass of many individual macroinvertebrate taxa decreased post-fire in a higher-order river in New Mexico, USA, for example, yet total macroinvertebrate community biomass decreased only slightly (Webster et al., 2022).

Higher-order rivers, however, may be more likely than headwaters to retain excess fine sediment inputs post-fire, with concomitant loss of pool volume, channel-marginal habitat, and interstitial habitat and spawning gravel in coarse-grained channels (Capelli, 2023). Greater mobility of sand-sized sediment following fire may also abrade benthic organisms and limit the stability of their substrate. In this context, stable features such as large wood or boulders can provide critically important stable substrate (Wallace & Benke, 1984; Pilotto et al., 2016).

Arid-region rivers such as the Rio Grande in New Mexico, USA illustrate some of the local interactions that influence macroinvertebrate responses to fire (Webster et al., 2022). Rivers in drylands may have limited longitudinal connectivity and fewer perennial tributaries and associated refuges than rivers in wetter regions (Bogan et al., 2015). High flow variability in dryland rivers is a primary influence on benthic macroinvertebrate community richness (Leigh & Datry, 2017). Ash flows during wildfire cause extensive mortality by changing stream morphology, water quality, habitat availability and quality, and food web dynamics (Bixby et al., 2015). Post-fire recovery of benthic macroinvertebrate communities can be rapid if sources of colonizers are nearby and population growth rates are high, but post-fire flooding can delay recovery (Leonard et al., 2017). Interactions among such limited conditions can delay macroinvertebrate recovery after fire: caddisflies in montane grassland streams in New Mexico affected by a 2011 wildfire continued to exhibit reduced size at maturity seven years after the fire (Kremer & Caldwell, 2022). In conditions where drought stresses macroinvertebrate communities by reducing habitat abundance and diversity, and drought increases the risk of fires that are then followed by flashier runoff where vegetation has been denuded by fire, the macroinvertebrate community may be subjected to everything everywhere all at once.

The response of fishes following wildfire can be spatially heterogeneous, analogous to the spatial heterogeneity of changes such as higher water temperature, greater substrate mobility, and elevated turbidity that may stress the fishes. Severe wildfires in New Mexico, USA during 2011–2013, for example, caused declines that ranged from 0 to 70% for different common species and across locations, but by 2017 all but 2 of 13 species had recovered at one or more sites (Gido et al., 2019). Resistance of fish communities correlated best with spatial factors such as intensity of disturbance. Resilience was driven by immigration or species ability to rebound from severely depressed numbers (Gido et al., 2019). Analogously, extremely high turbidity levels persisted for three years after a fire in Colorado, USA and may have contributed to significant fish kills, but insect diversity and fish populations nonetheless recovered to pre-fire levels within three years (Rust et al., 2019). In another example, although fishes were absent from moderate- and high-intensity burn areas in the John Day River of Oregon, USA immediately after fire, movement into these areas had returned fish numbers to background levels within seven years of the fire (Howell, 2006). Some of this recolonization was facilitated by elimination of a culvert near the mouth of one of the study streams that was likely a barrier to fish movement (Howell, 2006).

Thus, although fish mortality immediately post-fire can be high, recolonization from refugia can allow rapid recovery of fish communities if connectivity is present within the river network (Verkaik et al., 2013). Fisheries management can deliberately include population redundancy and spatial separation—establishing populations of a target fish species in widely dispersed catchments—as a form of insurance against all viable populations of that species being extirpated in one catastrophic event or series of events (Capelli, 2023).

Among post-fire stressors for fish, even short-duration exposure to ash-laden runoff can alter fish behavior and physiology. Post-fire ash flows can increase stream water turbidity, pH, and conductivity and deplete dissolved oxygen (Silva et al., 2016). This can kill or displace fishes over stream lengths of many kilometers, although the effects can be patchy

(Rieman et al., 2012). Indirect effects after the ash-laden flow can include reduced abundance and spatial distribution of fishes, loss of habitat connectivity, and loss of habitat and food resources (Rinne & Carter, 2008; Rieman et al., 2012). Adverse effects of ash-laden flows have been documented for North American salmonids and for Iberian barbel (*Luciobarbus bocagei*), a cyprinid species in the Mediterranean (Gonino et al., 2019). Debris flows and hyperconcentrated flows that persist as they move down a river network into higher-order rivers can also cause abrupt, massive fish kills, as documented in the news media for the Klamath River, California during the 2022 McKinney Fire or the Poudre River, Colorado after the 2020 Cameron Peak Fire.

Fire can be a primary driver of spatial and temporal heterogeneity in floodplain communities. The concept of a floodplain shifting habitat mosaic proposes that floods with sufficient power to geomorphically alter floodplain topography and stratigraphy via erosion and deposition strongly influence habitat patch dynamics on floodplains (Arscott et al., 2002; Stanford et al., 2005). Wildfire is also an important landscape-scale disturbance in floodplain communities and, in combination with flood-related disturbances, strongly affects floodplain habitat patch composition in the North American Rocky Mountains (Kleindl et al., 2015). The fire followed by flood scenario described earlier for Little Beaver Creek, Colorado, for example, created extensive disruption of ground cover and understory communities on the forested floodplain and changed the soil (via aggradation and erosion) and ground cover (via changed vegetation and deposition of logjams) resources available to floodplain organisms.

Examining stream-riparian food webs at multiple locations one year after a high-intensity wildfire in Colorado, USA, Preston et al. (2023) found decreased benthic macroinvertebrate density, flux of emerging aquatic insects, and riparian spider abundance. Fluxes of insects entering the stream did not change despite the loss of riparian vegetation during the fire. Trout were less abundant but larger, suggesting mortality or displacement of smaller trout. Burned sites generally showed greater variation in community characteristics than unburned sites. The results indicate that fire strongly altered some food web responses (e.g., riparian spiders) whereas other aspects of food webs (e.g., prey composition consumed by trout) were resilient.

Although beaver (*Castor canadensis* in North America and *C. fiber* in Eurasia) can build dams in headwater streams, they are more likely to occupy moderate-sized channels. Beaver may not be able to survive during or after severe fires that burn the riparian zone, largely because the fire destroys their key dam-building materials and winter food of deciduous riparian trees such as *Salix, Alnus, Populus*, and *Betula* species. However, beaver receive much greater attention in the context of wildfire for the ability of their ecosystem engineering to influence burn characteristics and post-fire fluxes.

Beaver build multiple dams through time in valley floors with sufficient width, creating a mosaic of ponds in different stages of infilling with sediment (Laurel & Wohl, 2019). Beaver also dig small canals across the floodplain, facilitating formation of a multithread system of subparallel channels that branch and rejoin downstream (John & Klein, 2004). These portions of river corridor, known as beaver meadows (Ives, 1942; Polvi & Wohl, 2012), have strong lateral connectivity between active channel(s) and the floodplain; strong vertical connectivity via hyporheic exchange; and limited longitudinal connectivity (Burchsted et al., 2010). This creates a river-wetland corridor (Wohl et al., 2021) with extensive areas of saturated soils and vegetation with high moisture levels. Multiple studies in burned areas indicate that the presence of ponded water behind beaver dams and the greater soil and fuel moisture in beaver-modified riparian zones equate to greater resistance to burning than other parts of the river network (Fairfax & Whittle, 2020; Weirich, 2021; Fairfax et al., 2024).

Beaver also increase post-fire physical and biotic resilience. Beaver-modified river corridors in burn areas exhibit reduced stream nutrient transport (Roth et al., 2022) and pH, increased channel-floodplain connectivity, greater floodplain vegetation diversity, and fewer non-native species than other burned areas in the river network (Whipple, 2019). Beaver ponds and meadows can also effectively store excess post-fire sediment (Figure 4.10).

In a synthesis of work on fire effects on stream biota, Bixby et al. (2015) highlighted research needs, including (i) additional geographic areas and biomes such as tropical savannas, (ii) other aquatic habitats such as lakes, ponds, and wetlands, (iii) fires with different characteristics of severity, extent, and frequency, (iv) additional response variables such as sublethal effects of fire on fish distributions, food availability, growth, reproductive potential, and population structure, (v) monitoring of effects over longer time frames beyond five years post-fire, and (vi) the effects on stream biota of different fire management practices, such as use of fire retardant, construction of instream structures to retain post-fire sediment, and pre- and post-fire vegetation removal via prescribed burns and salvage logging.

A uniquely integrated case study of post-fire interactions between phosphorus (P) dynamics, algae, benthic invertebrates, and fish communities in river ecosystems comes from the Rocky Mountains of Alberta, Canada (Silins et al., 2014). Mean concentrations of all forms of P were 2–13 times greater after the fire and significant differences persisted for at least five years, likely because of continuing release of P from sediment. P

Figure 4.10 Post-fire sediment filling off-channel, abandoned beaver ponds, 2 years after a 2020 wildfire in Little Beaver Creek, Colorado, USA. Active channel is at rear of photo, with arrow indicating flow direction. The upland grass in the foreground burned during the wildfire, but the valley-floor vegetation around the abandoned beaver ponds did not burn.

loading created annual algal productivity 5–71 times greater in rivers within burned catchments and these differences also persisted five years after the fire. Greater algal production correlated with greater invertebrate densities, biomass, species diversity, and shifts in species composition. More insect food corresponded to increased weight, length, and growth rate for cutthroat trout (*Oncorhynchus clarkia*). These cascading effects of wildfire on river primary production, secondary invertebrate consumers, and fish are persistent legacies of the fire because of the slow recovery of P regimes (Silins et al., 2014).

In summary, macroinvertebrates and fishes in higher-order rivers are likely to have more options for refugia during and after wildfire because of greater habitat diversity and longitudinal connectivity relative to headwater streams. The longer retention of fine sediment in higher-order rivers after fire, however, can cause prolonged loss of important habitat and abrasion and lack of substrate stability for benthic organisms. In the northern hemisphere, beaver in the river corridor can increase resistance to fire by increasing the extent of surface water and the water content of floodplain vegetation. Beaver ponds can also store substantial amounts of excess post-fire sediment and create refugia for herpetofauna and other organisms.

4.5. Stratigraphic Records of Long-Term Fire History and River Corridor Response

In addition to their influences on post-fire response in river corridors, alluvial fans can preserve records of sediment fluxes from source catchments and indirect records of fire occurrence in the form of pyrogenic carbon (Power et al., 2008). These records can be used to infer wildfire recurrence intervals and to understand river corridor response to wildfire. The substantial field data and interpretations related to this topic come almost entirely from the western United States, although related work has been done in other regions such as Central America (Anderson & Wahl, 2016) and Australia (Romanin et al., 2016) using sediment cores from lakes, particularly in the context of the millennia-long history of human use of fire.

Stratigraphic records and chronologies largely based on radiocarbon dating can be combined with independent climate indicators, such as pollen abundance or tree rings, to examine potential correlations between climate, wildfire, and sedimentation. The consistent pattern is that alluvial fans aggrade during warmer, drier periods with frequent, fire-related sedimentation events (e.g., Pierce et al., 2004; Pierce & Meyer, 2008; Frechette & Meyer, 2009). Sediment is removed from alluvial fan storage during wetter periods. As Meyer et al. (1992, p. 147) described it, "The dominant alluvial activity is strongly modulated by climate, with fire acting as a drought-actuated catalyst for sediment transport."

Nuances in this general pattern appear in relation to elevation and aspect. Wetter, cooler conditions during the Little Ice Age (circa 1450–1800 AD) probably promoted grass growth and frequent, low-severity fires in Idaho, USA, whereas colder temperatures at higher elevations in the nearby Yellowstone region limited overall fire activity during this time. Stand-replacing wildfires and large debris flows occurred in both Idaho and Yellowstone during warmer conditions with severe, multidecadal droughts circa 950–1150 AD (Pierce et al., 2004; Pierce & Meyer, 2008). Examining multiple tributary-catchment alluvial fans in the Middle Fork Salmon River of Idaho, Riley et al. (2015) found that fire-related deposits constitute nearly 70% of fan volume in more densely forested upper catchments versus about 33% in shrub-steppe-dominated catchments at lower elevations.

Within a single catchment in central Idaho, north-facing slopes can have similar vegetation and cooler temperatures analogous to higher elevations, whereas south-facing slopes are more similar to lower elevations (Poulos & Pierce, 2018). Sub-catchments with different aspects may achieve similar erosion rates because those with southern aspects have more frequent fires but less severe erosional responses in the form of sheetfloods, whereas those with northern aspects burn less frequently but erode more severely via debris flows (Poulos & Pierce, 2018).

In the Jemez Mountains of New Mexico, USA, south-facing slopes have sparser vegetation and more exposed bedrock, generating runoff and sediment even when unburned, whereas soil-mantled north-facing slopes produce little sediment unless severely burned (Fitch & Meyer, 2016). Consequently, fire-related deposits constitute nearly 80% of fan sediments from north-facing catchments but only about 40% from south-facing catchments (Fitch & Meyer, 2016).

Wildfire recurrence intervals estimated from alluvial fan stratigraphy and chronology vary from 150 to 400 years in central Idaho (Riley et al., 2015) and 250–400 years in the Jemez Mountains of New Mexico (Fitch & Meyer, 2016), to 800–1200 years in east-central British Columbia, Canada (Sanborn et al., 2006), and ~900–1000 years in northern Colorado, USA (Elliott & Parker, 2001). These are approximations because the inferred intervals reflect large erosional responses to fire, such as debris flows. These responses may be less frequent than fire itself because debris flows (i) require intense precipitation as a trigger and (ii) deplete sediment stored in colluvial hollows and ephemeral channels, which then require time to refill before mass movements can again occur (Wohl & Pearthree, 1991; Poulos & Pierce, 2018; Morell et al., 2021).

An intriguing addition to the sediment-charcoal record of past fires involves the use of Fourier Transform Infrared (FTIR) Spectroscopy to assess fire-affected chemical bonds within sediment. Distinctive FTIR spectra may allow approximation of fire intensity and temperature. Ryan et al. (2023) analyzed sediment cores from a burned catchment in Australia and found a notable shift in both the FTIR spectra and the boron isotope composition of stream-corridor sediment in response to the fire. These results suggest the potential of FTIR spectra and boron isotopes to provide a proxy for identifying past wildfires.

4.6. Summary: Wildfire Effects in River Networks

Headwaters are typically the transport-dominated portions of a river network. These first- and second-order channels constitute most of the total channel length in any river network. Closely coupled to adjacent uplands and relatively narrow and simple in form, headwater stream corridors are more likely to effectively transmit post-fire fluxes from the uplands to downstream portions of the river network. This transmission can be attenuated by unburned riparian zones and by feedbacks such as large wood recruitment and accumulation in logjams, but the formation of logjams and modification of the stream corridor by beaver is more widespread in higher-order rivers.

The middle and lower portions of river networks are typically the response reaches. More extensive floodplains retain post-fire inputs of water, solutes, sediment, organic matter, and large wood. Logjams, beaver dams, and spatially complex channels help to at least temporarily store diverse materials moving downstream after fires, but post-fire alterations of process and form can also persist longer in higher-order river corridors. Macroinvertebrates and fishes may experience greater post-fire environmental fluctuation in headwaters, but fire effects can be more prolonged in higher-order river corridors. The severity of wildfire in the riparian zone influences the nature and magnitude of physical and biotic responses to fire in both headwaters and higher-order river corridors. Sedimentary records, especially those preserved in alluvial fans, can be used to infer long-term fire recurrence intervals and to understand the responses of river corridors to fire-related fluctuations in water and sediment inputs over timespans up to millennia.

As in terrestrial environments of the uplands, the physical, chemical, and biotic responses of freshwater environments to wildfire are so closely linked that it is unrealistic to regard them in isolation. The fluxes of matter and energy through river networks create the template in which organisms live, but the organisms actively adjust that template, from biofilm communities and silk-spinning macroinvertebrates that alter the stability of very fine streambed sediment (e.g., Albertson et al., 2019), to salmon and other fish species modifying portions of the streambed in which they lay eggs (e.g., Buxton et al., 2015), to aquatic macrophytes and riparian vegetation influencing energy dissipation, substrate stability, and channel and floodplain form (e.g., Camporeale et al., 2013; Maji et al., 2020). These interactions are reflected in conceptualizations such as the stream evolution triangle (Castro & Thorne, 2019) and the Green New Balance (Bywater-Reyes et al., 2022), which emphasize interactions among hydrology, geology, and biology as primary controls on river form. Feedbacks between matter, energy, and organisms change during fire and with ongoing time after fire, but these feedbacks ultimately govern the processes, forms, and functions of river networks.

4.7. Lake Ecosystems

Lakes can serve as natural firebreaks (Nielsen et al., 2016). Lakes are also distinctive environments that may or may not have surface hydrologic connectivity to river networks. Regardless, fire affects exchanges of nutrients and other materials between terrestrial and lake ecosystems (Scrimgeour et al., 2001).

Fewer studies have examined the effects of wildfire on lake ecosystems than on terrestrial or river ecosystems. Post-fire inputs to lakes can occur via surface and subsurface water or via atmospheric deposition. Fire-related changes in water chemistry can be more persistent or cumulative throughout the year in lakes receiving river inputs relative to isolated lakes (McCullough et al., 2023).

For lakes in boreal forests, some studies document post-fire increases in N, P, and dissolved organic carbon (e.g., Allen et al., 2005; Garcia & Carignan, 2005; Wagner et al., 2014), whereas others find few changes in water quality variables (Pretty et al., 2021). Evans et al. (2017) documented large, sustained increases in nitrate concentrations and reductions in concentrations of dissolved organic carbon following fire in a lake draining a moorland catchment in the United Kingdom. The biogeochemical responses in this moorland catchment were broadly similar to those observed in forest ecosystems elsewhere, but historically high nitrogen deposition made the lake ecosystem especially susceptible to nitrate leaching and acidification (Evans et al., 2017).

In contrast, Olefeldt et al. (2013) discussed how regional characteristics of climate, surface geology, and lake morphometry prevent wildfire from causing pulse perturbations in dissolved organic carbon in a peatland lake of western Canada's Boreal Plains region. In upland Boreal Plain lakes, for example, wildfire can increase total P, inorganic N, and algal biomass, whereas wildfire in wetland- and permafrost-dominated catchments on the northern Boreal Plain causes increases in dissolved organic carbon and water color, causing light-limitation of algae (Pinel-Alloul et al., 2002). The number of water quality indicators affected by wildfire is greater in the low-nutrient lakes of the Boreal Shield (Pinel-Alloul et al., 2002).

Studies from southern Norway also emphasize the importance of site-specific factors such as pre-fire water chemistry, lake-water residence time, and the ratio of lake area to catchment area as controlling factors on the responses of lake chemistry to wildfire (Lydersen et al., 2014). Tundra wildfire reduces organics available for hydrologic transport and microbial respiration and mobilizes nitrogen into freshwaters (Zolkos et al., 2022), which could drive persistent shifts in aquatic inorganic nutrient cycling.

For lakes at lower latitudes, post-fire increases in dissolved black carbon can enhance the water's absorption of incident radiation, leading to increased water temperatures and associated effects on aquatic organisms, such as thermal stresses for native fishes (Kruger et al., 2023).

Lake sediments can also preserve records of response to wildfire. Pelletier et al. (2020) used sediment cores from subarctic lakes to document small but significant increases in atmospheric deposition of trace metals, metalloids, or major ions within two years of local wildfires. Charette and Prepas (2003) used quantification of sedimentary pigments that record changes in lake phytoplankton communities to examine phytoplankton response across 3 lakes with 50-year sediment chronologies. Based on these records, cyanobacterial biomass is most likely to increase after fire in shallow headwater lakes. A 2500-year sediment record from three subalpine lakes in Colorado, USA indicated that, on average, C, N, and S concentrations in lake sediments increased after fire (Pompeani et al., 2020). Post-fire C and N increased due to both erosion of upland soils and organic matter and increased primary productivity within the lake.

Increased nutrient inputs in runoff can create increases in primary producers such as macrophytes (Pretty et al., 2021) and algae, leading to short-term increases in zooplankton biomass (Patoine et al., 2000). Benthic macroinvertebrate groups may exhibit more persistent changes after fire and some studies indicate that increases in opportunistic feeders such as amphipods can fuel an increase in predators for at least 30 years post-fire (Scrimgeour et al., 2001; Bladon et al., 2014). Examining lakes in northern Alberta, Canada that lay within catchments that had burned between 1 and 26 years earlier, Scrimgeour et al. (2001) found that benthic macroinvertebrate biomass can remain elevated for 15–20 years after fire.

Wildfire smoke can also affect lake ecosystems (Williamson et al., 2016). Polycyclic aromatic hydrocarbons introduced to lakes via atmospheric deposition during wildfires are toxic and may impact lake biota (Gorshkov et al., 2021). Ultraviolet radiation is an influence on zooplankton vertical migration within lake waters. Examining zooplankton in Lake Tahoe, California, USA, Urmy et al. (2016) found that zooplankton position themselves closer to the surface when wildfire smoke reduces incident ultraviolet radiation reaching the lake surface. Diel vertical migration of zooplankton then affects predators and structures aquatic ecosystems. Smoke can also reduce photosynthetically active radiation and lake heat content, increase primary production in shallow offshore habitat by releasing photoinhibition, and decrease deep-water primary production (Scordo et al., 2021). The effects of fire on smoke plumes likely also affect incident ultraviolet radiation on large areas of the world's oceans (Ohneiser et al., 2022).

Vaz et al. (2015) focused on fire-derived wood in a lake in Minnesota, USA. Wood provides substrate for periphyton and invertebrates, as well as habitat and attached food for fish (Smokorowski et al., 2006). Wildfire created an area of lake shoreline with disproportionately more wood than areas unaffected by fire and the burned wood had greater piece shape complexity, which provided more abundant habitat.

As in other environments, feedbacks and cumulative effects can influence lake responses to wildfire. Following wildfires in 1986 and 2004, much of the vegetation in the catchment of a small, subarctic lake in Alaska, USA changed from spruce-dominated forest to deciduous forest, which likely indicates permafrost degradation (Larsen et al., 2023). The fire influenced lake chemistry, but about 15 years after the second fire, lake chemistry changed again in response to the development of a groundwater seep near the lake that likely formed as a result of permafrost thaw and new subsurface flowpaths in response to long-term warming and fire effects (Larsen et al., 2023). As Larsen et al. (2023) noted, source waters and flow paths contributing to subarctic lakes will continue to evolve uniquely in response to climate change and thawing permafrost, illustrating the cascading effects of warming climate and wildfires on catchment characteristics, terrestrial vegetation, and aquatic vegetation and lake water chemistry.

In summary, although fewer studies have examined the effects of wildfire on lake ecosystems than on river ecosystems, a few general points emerge from existing work. Wildfire can alter fluxes of water, solutes, sediment, particulate organic matter, and large wood into lakes and can affect surface and subsurface flow paths as well as atmospheric deposition. The magnitude of effects on the lake ecosystem reflects parameters such as the magnitude of inputs and the duration of these inputs through time, with duration reflecting rate of vegetation regrowth and, in some lakes, continuing river inputs. The response of the lake ecosystem to post-fire inputs also depends on the pre-fire water chemistry, lake-water residence time, and the ratio of lake area to catchment area. One distinctive aspect of lake responses is the potential effect of wildfire smoke, which can introduce contaminants to the lake via atmospheric deposition, as well as influencing solar radiation reaching the lake and thus the lake food web. Feedbacks and cumulative effects influence a lake's prolonged response to wildfire, as illustrated by the fire-accelerated permafrost degradation and changes in groundwater influxes and lake chemistry in a subarctic Alaskan lake.

References

Ader, E., Wohl, E., McFadden, S., & Singha, K. (2021). Logjams as a driver of transient storage in a mountain stream. *Earth Surface Processes and Landforms*, *46*, 701–711. https://doi.org/10.1002/esp.5057

Agee, J. K. (1988). Successional dynamics in forest riparian zones. In K. J. Raedeke (Ed.), *Streamside management: Riparian wildlife and forestry interactions* (pp. 31–43). Seattle, WA: University of Washington Institute of Forest Resources, Contrib. No. 59.

Agee, J. K., Wright, C. S., Williamson, N., & Huff, M. H. (2002). Foliar moisture content of Pacific northwest vegetation and its relation to wildland fire behavior. *Forest Ecology and Management*, *167*, 57–66. https://doi.org/10.1016/S0378-1127(01)00690-9

Albertson, L. K., Sklar, L. S., Cooper, S. D., & Cardinale, B. J. (2019). Aquatic macroinvertebrates stabilize gravel bed sediment: a test using silk net-spinning caddisflies in semi-natural river channels. *PLoS ONE, 14*, e020987. https://doi.org/10.1371/journal.pone.0209087

Allan, J. D., Castillo, M. M., & Capps, K. A. (2021). *Stream Ecology: Structure and Function of Running Waters* (3rd ed.). Cham, Switzerland: Springer Nature.

Alexander, R. B., Boyer, E. W., Smith, R. A., Schwarz, G. E., & Moore, R. B. (2007). The role of headwater streams in downstream water quality. *Journal of the American Water Resources Association, 43*, 41–59. https://doi.org/10.1111/j.1752-1688.2007.00005.x

Allen, E. W., Prepas, E. E., Gabos, S., Strachan, W. M., & Zhang, W. (2005). Methyl mercury concentrations in macroinvertebrates and fish from burned and undisturbed lakes on the boreal plain. *Canadian Journal of Fisheries and Aquatic Sciences, 62*, 1963–1977. https://doi.org/10.1139/f05-103

Anderson, L., & Wahl, D. (2016). Two Holocene paleofire records from Peten, Guatemala: Implications for natural fire regime and prehispanic Maya land use. *Global and Planetary Change, 138*, 82–92. https://doi.org/10.1016/j.gloplacha.2015.09.012

Aranibar, J. N., Macko, S. A., Anderson, I. C., Potgieter, A. L., Sowry, R., & Shugart, H. H. (2003). Nutrient cycling responses to fire frequency in the Kruger National Park (South Africa) as indicated by stable isotope analysis. *Isotopes Environ Health Studies, 39*, 141–158. https://doi.org/10.1080/1025601031000096736

Arscott, D. B., Tockner, K., van der Nat, D., & Ward, J. V. (2002). Aquatic habitat dynamics along a braided alpine river ecosystem (Tagliamento River, Northeast Italy). *Ecosystems, 5*, 802–814. https://doi.org/10.1007/s10021-002-0192-7

Bart, R. R., & Tague, C. L. (2017). The impact of wildfire on baseflow recession rates in California. *Hydrological Processes, 31*, 1662–1673. https://doi.org/10.1002/hyp.11141

Baxter, C. V., Fausch, K. D., Murakami, M., & Chapman, P. L. (2004). Fish invasion restructures stream and forest food webs by interrupting reciprocal prey subsidies. *Ecology, 85*, 2656–2663. https://doi.org/10.1890/04-138

Beakes, M. P., Moore, J. W., Hayes, S. A., & Sogard, S. M. (2014). Wildfire and the effects of shifting stream temperature on salmonids. *Ecosphere, 5*, 63. https://doi.org/10.1890/ES13-00325.1

Beaty, K. G. (1994). Sediment transport in a small stream following two successive forest fires. *Canadian Journal of Fisheries and Aquatic Sciences, 51*, 2723–2733. https://doi.org/10.1139/f94-273

Bêche, L. A., Stephens, S. L., & Resh, V. H. (2005). Effects of prescribed fire on a Sierra Nevada (California, USA) stream and its riparian zone. *Forest Ecology and Management, 218*, 37–59. https://doi.org/10.1016/j.foreco.2005.06.010

Benda, L., Miller, D., Bigelow, P., & Andras, K. (2003). Effects of post-wildfire erosion on channel environments, Boise River, Idaho. *Forest Ecology and Management, 178*, 105–119. https://doi.org/10.1016/S0378-1127(03)00056-2

Benda, L. E., Miller, D. J., Dunne, T., Reeves, G. H., & Agee, J. K. (1998). Dynamic landscape systems. In R. J. Naiman & R. E. Bilby (Eds.), *River ecology and management: Lessons from the Pacific coastal ecoregion* (pp. 261–288). NR: Springer.

Betts, E. F., & Jones, J. B. (2009). Impact of wildfire on stream nutrient chemistry and ecosystem metabolism in boreal forest catchments of interior Alaska. *Arctic, Antarctic, and Alpine Research*, *41*, 407–417. https://doi.org/10.1657/1938-4246-41.4.407

Bixby, R. J., Cooper, S. D., Gresswell, R. E., Brown, L. E., Dahm, C. N., & Dwire, K. A. (2015). Fire effects on aquatic ecosystems: an assessment of the current state of the science. *Freshwater Science*, *34*, 1340–1350. https://doi.org/10.1086/684073

Bladon, K. D., Emelko, M. B., Silins, U., & Stone, M. (2014). Wildfire and the future of water supply. *Environmental Science and Technology*, *48*, 8936–8943. https://doi.org/10.1021/es500130g

Blake, W. H., Wallbrink, P. J., Wilkinson, S. N., Humphreys, G. S., Doerr, S. H., Shakesby, R. A., & Tomkins, K. M. (2009). Deriving hillslope sediment budgets in wildfire-affected forests using fallout radionuclide tracers. *Geomorphology*, *104*, 105–116. https://doi.org/10.1016/j.geomorph.2008.08.004

Bogan, M. T., Boersma, K. S., & Lytle, D. A. (2015). Resistance and resilience of invertebrate communities to seasonal and supraseasonal droughts in arid-land headwater streams. *Freshwater Biology*, *60*, 2547–2558. https://doi.org/10.1111/fwb.12522

Bosch, C., Olivares, A., Faria, M., Navas, J. M., del Olmo, I., Grimalt, J. O., et al. (2009). Identification of water soluble and particle bound compounds causing sublethal toxic effects: a field study on sediments affected by a chlor-alkali industry. *Aquatic Toxicology*, *94*, 16–27. https://doi.org/10.1016/j.aquatox.2009.05.011

Bowerman, T., Neilson, B. T., & Budy, P. (2014). Effects of fine sediment, hyporheic flow, and spawning site characteristics on survival and development of bull trout embryos. *Canadian Journal of Fisheries and Aquatic Sciences*, *71*, 1059–1071. https://doi.org/10.1139/cjfas-2013-0372

Bragg, D. C. (2000). Simulating catastrophic and individualistic large woody debris recruitment for a small riparian system. *Ecology*, *81*, 1383–1394. https://doi.org/10.1890/0012-9658(2000)081[1383:SCAILW]2.0.CO;2

Brogan, D. J., Nelson, P. A., & MacDonald, L. H. (2019). Spatial and temporal patterns of sediment storage and erosion following a wildfire and extreme flood. *Earth Surface Dynamics*, *7*, 563–590. https://doi.org/10.5194/esurf-7-563-2019

Brown, B. L., & Swan, C. M. (2010). Dendritic network structure constrains metacommunity properties in riverine ecosystems. *Journal of Animal Ecology*, *79*, 571–580. https://doi.org/10.1111/j.1365-2656.2010.01668.x

Bruder, A., Schindler, M. H., Moretti, M. S., & Gessner, M. O. (2014). Litter decomposition in a temperate and a tropical stream: the effects of species mixing, litter quality and shredders. *Freshwater Biology*, *59*, 438–449. https://doi.org/10.1111/fwb.12276

Burchsted, D., Daniels, M., Thorson, R., & Vokoun, J. (2010). The river discontinuum: applying beaver modifications to baseline conditions for restoration of forested headwaters. *BioScience*, *60*, 908–922. https://doi.org/10.1525/bio.2010.60.11.7

Burke, M. P., Hogue, T. S., Kinoshita, A. M., Barco, J., Wessel, C., & Stein, E. D. (2013). Pre- and post-fire pollutant loads in an urban fringe watershed in southern California. *Environmental Monitoring and Assessment*, *185*, 10131–10145. https://doi.org/10.1007/s10661-013-3318-9

Burton, T. A. (2005). Fish and stream habitat risks from uncharacteristic wildfire: Observations from 17 years of fire-related disturbances on the Boise National Forest, Idaho. *Forest Ecology and Management*, *211*, 140–149. https://doi.org/10.1016/j.foreco.2005.02.063

Buxton, T. H., Buffington, J. M., Yager, E. M., Hassan, M. A., & Fremier, A. K. (2015). The relative stability of salmon redds and unspawned streambeds. *Water Resources Research*, *51*, 6074–6092. https://doi.org/10.1002/2015WR016908

Bywater-Reyes, S., Diehl, R. M., Wilcox, A. C., Stella, J. C., & Kui, L. (2022). A green new Balance: interactions among riparian vegetation plant traits and morphodynamics in alluvial rivers. *Earth Surface Processes and Landforms*, *47*, 2410–2436. https://doi.org/10.1002/esp.5385

Campana, S. E. (1999). Chemistry and composition of fish otoliths: pathways, mechanisms, and applications. *Marine Ecology Progress Series*, *188*, 263–297.

Camporeale, C., Perucca, E., Ridolfi, L., & Gurnell, A. M. (2013). Modeling the interactions between river morphodynamics and riparian vegetation. *Reviews of Geophysics*, *51*, 379–414. https://doi.org/10.1002/rog.20014

Capelli, M. H. (2023). The role of wildfires in the recovery strategy for the endangered southern California steelhead. In J. L. Florsheim, A. P. O'Dowd, & A. Chin (Eds.), *Biogeomorphic responses to wildfire in fluvial ecosystems* (pp. 95–129). Boulder, CO: Geological Society of America Special Paper 562.

Carroll, E. M., Miller, W. W., Johnson, D. W., Saito, L., Qualls, R. G., & Walker, R. F. (2007). Spatial analysis of a large magnitude erosion event following a Sierran wildfire. *Journal of Environmental Quality*, *36*, 1105–1111. https://doi.org/10.2134/jeq2006.0466

Castro, J. M., & Thorne, C. R. (2019). The stream evolution triangle: Integrating geology, hydrology, and biology. *River Research and Applications*, *35*, 315–326. https://doi.org/10.1002/rra.3421

Cawley, K. M., Hohner, A. K., McKee, G. A., Borch, T., Omur-Ozbek, P., Oropeza, J., & Rosario-Ortiz, F. L. (2018). Characterization and spatial distribution of particulate and soluble carbon and nitrogen from wildfire-impacted sediments. *Journal of Soils and Sediments*, *18*, 1314–1326. https://doi.org/10.1007/s11368-016-1604-1

Charette, T., & Prepas, E. E. (2003). Wildfire impacts on phytoplankton communities of three small lakes on the boreal plain, Alberta, Canada: a paleolimnological study. *Canadian Journal of Fisheries and Aquatic Sciences*, *60*, 584–593. https://doi.org/10.1139/f03-049

Chen, X., Wei, X., & Scherer, R. (2005). Influence of wildfire and harvest on biomass, carbon pool, and decomposition of large woody debris in forested streams of southern interior British Columbia. *Forest Ecology and Management*, *208*, 101–114. https://doi.org/10.1016/j.foreco.2004.11.018

Chin, A., Burton, J. W., Humphreys, K. M., Florsheim, J. L., Kinoshita, A. M., Andreano, E. C., & Eckermann, T. K. (2024). Vegetation and channel recovery ten years following the Waldo canyon fire of Colorado. In J. L. Florsheim, A. P. O'Dowd, & A. P. Chin (Eds.), *Biogeomorphic responses to wildfire in fluvial ecosystems* (pp. 27–42). Boulder, CO: Geological Society of America Special Paper 562.

Chin, A., Solverson, A. P., O'Dowd, A. P., Florsheim, J. L., Kinoshita, A. M., Nourbakhshbeidokhti, S., et al. (2019). Interacting geomorphic and ecological response of step-pool streams after wildfire. *Geological Society of America Bulletin*, *131*, 1480–1500. https://doi.org/10.1130/B35049.1

Comiti, F., Andreoli, A., Mao, L., & Lenzi, M. A. (2008). Wood storage in three mountain streams of the southern Andes and its hydro-morphological effects. *Earth Surface Processes and Landforms*, *33*, 244–262. https://doi.org/10.1002/esp.1541

Constantine, J. A., Dunne, T., Ahmed, J., Legleiter, C., & Lazarus, E. D. (2014). Sediment supply as a driver of river meandering and floodplain evolution in the Amazon Basin. *Nature Geoscience*, *7*, 899–903. https://doi.org/10.1038/ngeo2282

Cooper, S. D., Page, H. M., Wiseman, S. W., Klose, K., Bennett, D., Even, T., et al. (2015). Physicochemical and biological responses of streams to wildfire severity in riparian zones. *Freshwater Biology*, *60*, 2600–2619. https://doi.org/10.1111/fwb.12523

Cunjak, R. A. (1996). Winter habitat of selected stream fishes and potential impacts from land-use activity. *Canadian Journal of Fisheries and Aquatic Sciences*, *53*, 267–282. https://doi.org/10.1139/f95-275

Downing, J. A., Cole, J. J., Duarte, C. M., Middelburg, J. J., Melack, J. M., Prairie, Y. T., et al. (2012). Global abundance and size distribution of streams and rivers. *Inland Waters*, *2*, 229–236. https://doi.org/10.5268/IW-2.4.502

Dunham, J. B., Rosenberger, A. E., Luce, C. H., & Rieman, B. E. (2007). Influences of wildfire and channel reorganization on spatial and temporal variation in stream temperature and the distribution of fish and amphibians. *Ecosystems*, *10*, 335–346. https://doi.org/10.1007/s10021-007-9029-8

Dwire, K. A., & Kauffman, J. B. (2003). Fire and riparian ecosystems in landscapes of the western USA. *Forest Ecology and Management*, *178*, 61–74. https://doi.org/10.1016/S0378-1127(03)00053-7

Earl, S. R., & Blinn, D. W. (2003). Effects of wildfire ash on water chemistry and biota in South-Western USA streams. *Freshwater Biology*, *48*, 1015–1030. https://doi.org/10.1046/j.1365-2427.2003.01066.x

Eaton, B. C., Andrews, C. A. E., Giles, T. R., & Phillips, J. C. (2010). Wildfire, morphologic change and bed material transport at Fishtrap Creek, British Columbia. *Geomorphology*, *118*, 409–424. https://doi.org/10.1016/j.geomorph.2010.02.008

Elliott, J. G., & Parker, R. S. (2001). Developing a post-fire chronology and recurrence probability from alluvial stratigraphy in the Buffalo Creek watershed, Colorado, USA. *Hydrological Processes*, *15*, 3039–3051. https://doi.org/10.1002/hyp.390

Ellis, L. M. (2001). Short-term response of woody plants to fire in a Rio Grande riparian forest, Central New Mexico, USA. *Biological Conservation*, *97*, 159–170. https://doi.org/10.1016/S0006-3207(00)00108-7

Emelko, M. B., Stone, M., Silins, U., Allin, D., Collins, A. L., Williams, C. H. S., et al. (2016). Sediment-phosphorus dynamics can shift aquatic ecology and cause downstream legacy effects after wildfire in large river systems. *Global Change Biology*, *22*, 1168–1184. https://doi.org/10.1111/gcb.13073

Ernstrom, K. (1999). Effects of wildfire on first order stream channel morphology, Yellowstone National Park, USA (MS thesis). Bozeman, MT: Montana State University.

Evans, C. D., Malcolm, I. A., Shilland, E. M., Rose, N. L., Turner, S. D., Crilly, A., et al. (2017). Sustained biogeochemical impacts of wildfire in a mountain lake catchment. *Ecosystems*, *20*, 813–829. https://doi.org/10.1007/s10021-016-0064-1

Fairfax, E., & Whittle, A. (2020). Smokey the beaver: beaver-dammed riparian corridors stay green during wildfire throughout the western United States. *Ecological Applications*, *30*, e02225. https://doi.org/10.1002/eap.2225

Fairfax, E., Whipple, A., Wheaton, J. M., Osorio, B., Miller, J., Kirksey, K., et al. (2024). Impacts of beaver dams on riverscape burn severity during megafires in the Rocky Mountain region, western United States. In J. L. Florsheim, A. P. O'Dowd, & A. Chin (Eds.), (pp. 131–151). Boulder, CO: Biogeomorphic Responses to Wildfire in Fluvial Ecosystems, Geological Society of America Special Paper 562.

Fetherston, K. L., Naiman, R. J., & Bilby, R. E. (1995). Large woody debris, physical process, and riparian forest development in montane river networks of the Pacific northwest. *Geomorphology*, *13*, 133–144. https://doi.org/10.1016/0169-555X(95)00033-2

Fitch, E. P., & Meyer, G. A. (2016). Temporal and spatial climatic controls on Holocene fire-related erosion and sedimentation, Jemez Mountains, New Mexico. *Quaternary Research*, *85*, 75–86. https://doi.org/10.1016/j.yqres.2015.11.008

Flitcroft, R. L., Falke, J. A., Reeves, G. H., Hessburg, P. F., McNyset, K. M., & Benda, L. E. (2016). Wildfire may increase habitat quality for spring Chinook salmon in the Wenatchee River subbasin, WA, USA. *Forest Ecology and Management*, *359*, 126–140. https://doi.org/10.1016/j.foreco.2015.09.049

Florsheim, J. L., & Chin, A. (2022). Geomorphic responses to wildfire in fluvial systems. In *Treatise on geomorphology* (2nd ed., Vol. v 9, pp. 478–503). San Diego, CA: Academic Press. https://doi.org/10.1016/B978-0-12-818234-5.00045-6

Florsheim, J. L., Chin, A., Kinoshita, A. M., & Nourbakhshbeidokhti, S. (2017). Effect of storms during drought on post-wildfire recovery of channel sediment dynamics and habitat in the southern California chaparral, USA.

Earth Surface Processes and Landforms, *42*, 1482–1492. https://doi.org/10.1002/esp.4117

Florsheim, J. L., Keller, E. A., & Best, D. W. (1991). Fluvial sediment transport in response to moderate storm flows following chaparral wildfire, Ventura County, southern California. *Geological Society of America Bulletin*, *103*, 504–511. https://doi.org/10.1130/0016-7606(1991)103%3C0504:FSTIRT%3E2.3.CO;2

Frechette, J. D., & Meyer, G. A. (2009). Holocene fire-related alluvial-fan deposition and climate in ponderosa pine and mixed-conifer forests, Sacramento Mountains, New Mexico, USA. *The Holocene*, *19*, 639–651. https://doi.org/10.1177/0959683609104031

Fryirs, K. A., Brierley, G. J., Preston, N. J., & Kasai, M. (2007). Buffers, barriers and blankets: the (dis)connectivity of catchment-scale sediment cascades. *Catena*, *70*, 49–67. https://doi.org/10.1016/j.catena.2006.07.007

Garcia, E., & Carignan, R. (2005). Mercury concentrations in fish from forest harvesting and fire-impacted Canadian boreal lakes compared using stable isotopes of nitrogen. *Environmental Toxicology and Chemistry*, *24*, 685–693. https://doi.org/10.1897/04-065R.1

Gido, K. B., Propst, D. L., Whitney, J. E., Hedden, S. C., Turner, T. F., & Pilger, T. J. (2019). Pockets of resistance: Response of arid-land fish communities to climate, hydrology, and wildfire. *Freshwater Biology*, *64*, 761–777. https://doi.org/10.1111/fwb.13260

Gomi, T., Sidle, R. C., & Richardson, J. S. (2002). Understanding processes and downstream linkages of headwater systems. *Bioscience*, *52*, 905–916. https://doi.org/10.1641/0006-3568(2002)052[0905:UPADLO]2.0.CO;2

Gonino, G., Branco, P., Benedito, E., Ferreira, M. T., & Santos, J. M. (2019). Short-term effects of wildfire ash exposure on behaviour and hepatosomatic condition of a potamodromous cyprinid fish, the Iberian barbel *Luciobarbus bocagei* (Steindacher, 1864). *Science of the Total Environment*, *665*, 226–234. https://doi.org/10.1016/j.scitotenv.2019.02.108

Gorshkov, A. G., Izosimova, O. N., Kustova, O. V., Marinaite, I. I., Galachyants, Y. P., Sinyukovich, V. N., & Khodzher, T. V. (2021). Wildfires as a source of PAHs in surface waters of background areas (Lake Baikal, Russia). *Water*, *13*, 2636. https://doi.org/10.3390/w13192636

Granath, G., Evans, C. D., Strengbom, J., Folster, J., Grelle, A., Stromqvist, J., & Kohler, S. J. (2021). The impact of wildfire on biogeochemical fluxes and water quality in boreal catchments. *Biogeosciences*, *18*, 3243–3261. https://doi.org/10.5194/bg-18-3243-2021

Guilinger, J. J., Foufoula-Georgiou, E., Gray, A. B., Randerson, J. T., Smyth, P., Barth, N. C., & Goulden, M. L. (2023). Predicting postfire sediment yields of small steep catchments using airborne lidar differencing. *Geophysical Research Letters*, *50*, e2023GL104626. https://doi.org/10.1029/2023GL104626

Guilinger, J. J., Gray, A. B., Barth, N. C., & Fong, B. T. (2020). The evolution of sediment sources over a sequence of postfire sediment-laden flows revealed through repeat high-resolution change detection. *Journal of Geophysical Research Earth Surface*, *125*, e2020JF005527. https://doi.org/10.1029/2020JF005527

Gurnell, A. M., Piegay, H., Swanson, F. J., & Gregory, S. V. (2002). Large wood and fluvial processes. *Freshwater Biology*, *47*, 601–619. https://doi.org/10.1046/j.1365-2427.2002.00916.x

Gustine, R. N., Hanan, E. J., Robichaud, P. R., & Elliot, W. J. (2022). From burned slopes to streams: how wildfire affects nitrogen cycling and retention in forests and fire-prone watersheds. *Biogeochemistry*, *157*, 51–68. https://doi.org/10.1007/s10533-021-00861-0

Hampton, T. B., Lin, S. G., & Basu, N. B. (2022). Forest fire effects on stream water quality at continental scales: a meta-analysis. *Environmental Research Letters*, *17*, 064003. https://iopscience.iop.org/article/10.1088/1748-9326/ac6a6c/meta

Harkey, G. A., Landrum, P. F., & Klaine, S. J. (1994). Comparison of whole-sediment, elutriate and pore-water exposures for use in assessing sediment-associated organic contaminants in bioassays. *Environmental Toxicology and Chemistry*, *13*, 1315–1329. https://doi.org/10.1002/etc.5620130814

Hauer, F. R., & Spencer, C. N. (1998). Phosphorus and nitrogen dynamics in streams associated with wildfire: a study of intermediate and longterm effects. *International Journal of Wildland Fire*, *8*, 183–198. https://doi.org/10.1071/WF9980183

Hoffman, D., & Gabet, E. J. (2007). Effects of sediment pulses on channel morphology in a gravel bed river. *Geological Society of America Bulletin*, *119*, 116–125. https://doi.org/10.1130/B25982.1

Houston, J. (2002). Groundwater recharge through an alluvial fan in the Atacama Desert, northern Chile: Mechanisms, magnitudes and causes. *Hydrological Processes*, *16*, 3019–3035. https://doi.org/10.1002/hyp.1086

Howell, P. J. (2006). Effects of wildfire and subsequent hydrologic events on fish distribution and abundance in tributaries of North Fork John Day River. *North American Journal of Fisheries Management*, *26*, 983–994. https://doi.org/10.1577/M05-114.1

Huang, X., Ping, J., Leng, W., Yu, Y., Zhang, M., & Zhu, Y. (2021). A study on groundwater recharge in the Anyanghe River alluvial fan, North China plain, based on hydrochemistry, stable isotopes and tritium. *Hydrogeology Journal*, *29*, 2149–2170.

Isaak, D. J., Luce, C. H., Rieman, B. E., Nagel, D. E., Peterson, E. E., Horan, D. L., et al. (2010). Effects of climate change and wildfire on stream temperatures and salmonid thermal habitat in a mountain river network. *Ecological Applications*, *20*, 1350–1371. https://doi.org/10.1890/09-0822.1

Iskin, E. P., & Wohl, E. (2021). Wildfire and the patterns of floodplain large wood on the Merced River, Yosemite National Park, California, USA. *Geomorphology*, *389*, 107805. https://doi.org/10.1016/j.geomorph.2021.107805

Ives, R. L. (1942). The beaver-meadow complex. *Journal of Geomorphology*, *5*, 191–203.

Jackson, B. K., & Sullivan, S. M. P. (2009). Influence of wildfire severity on riparian plant community heterogeneity in an Idaho, USA wilderness. *Forest Ecology and Management*, *259*, 24–32. https://doi.org/10.1016/j.foreco.2009.09.036

Jackson, J. K., & Füreder, L. (2006). Long-term studies of freshwater macroinvertebrates: a review of the frequency, duration and ecological significance. *Freshwater Biology*, *51*, 591–603. https://doi.org/10.1111/j.1365-2427.2006.01503.x

Jager, H. I., Long, J. W., Malison, R. L., Murphy, B. P., Rust, A., Silva, L. G. M., et al. (2021). Resilience of terrestrial and aquatic fauna to historical and future wildfire regimes in western North America. *Ecology and Evolution*, *11*, 12259–12284. https://doi.org/10.1002/ece3.8026

John, S., & Klein, A. (2004). Hydrogeomorphic effects of beaver dams on floodplain morphology: avulsion processes and sediment fluxes in upland valley floors (Spessart, Germany). *Quaternaire*, *15*, 219–231. https://www.persee.fr/doc/quate_1142-2904_2004_num_15_1_1769

Jones, T. A., & Daniels, L. D. (2008). Dynamics of large woody debris in small streams disturbed by the 2001 Dogrib fire in the Alberta foothills. *Forest Ecology and Management*, *256*, 1751–1759. https://doi.org/10.1016/j.foreco.2008.02.048

Kalabokidis, K. D. (2000). Effects of wildfire suppression chemical on people and the environment – a review. *Global Nest: The International Journal*, *2*, 129–137.

Kean, J. W., Staley, D. M., Lancaster, J. T., Rengers, F. K., Swanson, B. J., Coe, J. A., et al. (2019). Inundation, flow dynamics, and damage in the 9 January 2018 Montecito debris-flow event, California, USA: Opportunities and challenges for post-wildfire risk assessment. *Geosphere*, *15*, 1140–1163. https://doi.org/10.1130/GES02048.1

Kinoshita, A. M., & Hogue, T. S. (2011). Spatial and temporal controls on post-fire hydrologic recovery in southern California watersheds. *Catena*, *87*, 240–252. https://doi.org/10.1016/j.catena.2011.06.005

Kinoshita, A. M., & Hogue, T. S. (2015). Increased dry season water yield in burned watersheds in southern California. *Environmental Research Letters*, *10*, 014003. https://iopscience.iop.org/article/10.1088/1748-9326/10/1/014003/meta

Kleindl, W. J., Rains, M. C., Marshall, L. A., & Hauer, F. R. (2015). Fire and flood expand the floodplain shifting habitat mosaic concept. *Freshwater Science*, *34*, 1366–1382. https://doi.org/10.1086/684016

Klose, K., Cooper, S. D., & Bennett, D. M. (2015). Effects of wildfire on stream algal abundance, community structure, and nutrient limitation. *Freshwater Science*, *34*, 1494–1509. https://doi.org/10.1086/683431

Kobziar, L. N., & McBride, J. R. (2006). Wildfire burn patterns and riparian vegetation response along two northern Sierra Nevada streams. *Forest Ecology and Management*, *222*, 254–265. https://doi.org/10.1016/j.foreco.2005.10.024

Kremer, L. B., & Caldwell, C. A. (2022). Wildfire effects on mass and thermal tolerance of *Hydropsyche oslari* (Trichoptera) in southwestern USA montane grassland streams. *Freshwater Science*, *41*, 62–76. https://doi.org/10.1086/718556

Kristensen, L., Taylor, M., Odigie, K., Hibdon, S., & Flegal, A. (2014). Lead isotopic compositions of ash sourced from Australian bushfires. *Environmental Pollution*, *190*, 159–165. https://doi.org/10.1016/j.envpol.2014.03.025

Kruger, B. R., Hausner, M. B., Chellman, N., Weaver, M., Samburova, V., & Khlystov, A. (2023). Dissolved black carbon as a potential driver of surface water heating dynamics in wildfire-impacted regions: A case study from pyramid Lake, NV, USA. *Science of the Total Environment*, *888*, 164141. https://doi.org/10.1016/j.scitotenv.2023.164141

Lane, P. N. J., Sheridan, G. J., & Noske, P. J. (2006). Changes in sediment loads and discharge from small mountain catchments following wildfire in southeastern Australia. *Journal of Hydrology*, *331*, 495–510. https://doi.org/10.1016/j.jhydrol.2006.05.035

Larsen, A., Larsen, J. R., & Lane, S. N. (2021). Dam builders and their works: Beaver influences on the structure and function of river corridor hydrology, geomorphology, biogeochemistry, and ecosystems. *Earth-Science Reviews*, *218*, 103623. https://doi.org/10.1016/j.earscirev.2021.103623

Larsen, A., May, J. H., Moss, P., & Hacker, J. (2016). Could alluvial knickpoint retreat rather than fire drive the loss of alluvial wet monsoon forest, tropical northern Australia? *Earth Surface Processes and Landforms*, *41*, 1583–1594. https://doi.org/10.1002/esp.3933

Larsen, A. S., Rupp, D. L., Swanson, D. K., & Hill, K. R. (2023). Cascading effects of climate change and wildfire on a subarctic lake: A 20-year case study of watershed change. *Ecosphere*, *14*, e4558. https://doi.org/10.1002/ecs2.4558

Laurel, D., & Wohl, E. (2019). The persistence of beaver-induced geomorphic heterogeneity and organic carbon stock in river corridors. *Earth Surface Processes and Landforms*, *44*, 342–353. https://doi.org/10.1002/esp.4486

Leach, J. A., & Moore, R. D. (2010). Above-stream microclimate and stream surface energy exchanges in a wildfire-disturbed riparian zone. *Hydrological Processes*, *24*, 2369–2381. https://doi.org/10.1002/hyp.7639

Legleiter, C. J., Lawrence, R. L., Fonstad, M. A., Marcus, W. A., & Aspinall, R. (2003). Fluvial response a decade after wildfire in the northern Yellowstone ecosystem: a spatially explicit analysis. *Geomorphology*, *54*, 119–136. https://doi.org/10.1016/S0169-555X(02)00332-X

Leigh, C., & Datry, T. (2017). Drying as a primary hydrological determinant of biodiversity in river systems: A broad-scale analysis. *Ecography*, *40*, 487–499. https://doi.org/10.1111/ecog.02230

Leonard, J. M., Magana, H. A., Bangert, R. K., Neary, D. G., & Montgomery, W. L. (2017). Fire and floods: The recovery of headwater stream systems following high-severity wildfire. *Fire Ecology*, *13*, 62–84. https://doi.org/10.4996/fireecology.130306284

Leopold, L. B., Wolman, M. G., & Miller, J. P. (1964). *Fluvial processes in geomorphology*. San Francisco, CA: WH Freeman & Co.

Letey, J. (2001). Causes and consequences of fire-induced soil water repellency. *Hydrological Processes*, *15*, 2867–2875. https://doi.org/10.1002/hyp.378

Levine, R., & Meyer, G. A. (2014). Beaver dams and channel sediment dynamics on Odell Creek, Centennial Valley, Montana, USA. *Geomorphology*, *205*, 51–64. https://doi.org/10.1016/j.geomorph.2013.04.035

Lininger, K. B., Scamardo, J. E., & Guiney, M. R. (2021). Floodplain large wood and organic matter jam formation after a large flood: investigating the influence of floodplain forest stand characteristics and river corridor morphology. *Journal of Geophysical Research Earth Surface*, *126*, e2020JF006011. https://doi.org/10.1029/2020JF006011

Livers, B., Wohl, E., Jackson, K. J., & Sutfin, N. A. (2018). Historical land use as a driver of alternative states for stream form and function in forested mountain watersheds of the southern Rocky Mountains. *Earth Surface Processes and Landforms*, *43*, 669–684. https://doi.org/10.1002/esp.4275

Lydersen, E., Høgberget, R., Moreno, C. E., Garmo, O. A., & Hagen, P. C. (2014). The effects of wildfire on the water chemistry of dilute, acidic lakes in southern Norway. *Biogeochemistry*, *119*, 109–124. https://doi.org/10.1007/s10533-014-9951-8

Magoulick, D. D., & Kobza, R. M. (2003). The role of refugia for fishes during drought: A review and synthesis. *Freshwater Biology*, *48*, 1186–1198. https://doi.org/10.1046/j.1365-2427.2003.01089.x

Maji, S., Hanmaiahgari, P. R., Balachandar, R., Pu, J. H., Ricardo, A. M., & Ferreira, R. M. L. (2020). A review on hydrodynamics of free surface flows in emergent vegetated channels. *Water*, *12*, 1218. https://doi.org/10.3390/w12041218

Malison, R. L., & Baxter, C. V. (2010). Effects of wildfire of varying severity on benthic stream insect assemblages and emergence. *Journal of the North American Benthological Society*, *29*, 1324–1338. https://doi.org/10.1899/09-022.1

Malmon, D. V., Reneau, S. L., Katzman, D., Lavine, A., & Lyman, J. (2007). Suspended sediment transport in an ephemeral stream following wildfire. *Journal of Geophysical Research Earth Surface*, *112*, F02006. https://doi.org/10.1029/2005JF000459

Marcus, W. A., Rasmussen, J., & Fonstad, M. A. (2011). Response of the fluvial wood system to fire and floods in northern Yellowstone. *Annals of the Association of American Geographers*, *101*, 21–44. https://doi.org/10.1080/00045608.2010.539154

Martens, A. M., Silins, U., Proctor, H. C., Williams, C. H. S., Wagner, M. J., Emelko, M. B., & Stone, M. (2019). Long-term impact of severe wildfire and post-wildfire salvage logging on macroinvertebrate assemblage structure in Alberta's Rocky Mountains. *International Journal of Wildland Fire*, *28*, 738–749. https://doi.org/10.1071/WF18177

McCluney, K. E., Poff, N. L., Palmer, M. A., Thorp, J. H., Poole, G. C., Williams, B. S., et al. (2014). Riverine macrosystems ecology: Sensitivity, resistance, and resilience of whole river basins with human alterations. *Frontiers in Ecology and the Environment*, *12*, 48–58. https://doi.org/10.1890/120367

McCullough, I. M., Brentrup, J. A., Wagner, T., Lapierre, J. F., Henneck, J., Paul, A. M., et al. (2023). Fire characteristics and hydrologic connectivity influence short-term responses of north temperate lakes to wildfire. *Geophysical Research Letters*, *50*, e2023GL103953. https://doi.org/10.1029/2023GL103953

McGlynn, B. L., McDonnell, J. J., Seibert, J., & Kendall, C. (2004). Scale effects on headwater catchment runoff timing, flow sources, and groundwater-streamflow relations. *Water Resources Research*, *40*. https://doi.org/10.1029/2003WR002494

McMahon, C., Cooper, S. D., & Wiseman, S. W. (2023). Postfire stream responses to spatial fire patterns in riparian and upland zones. In J. L. Florsheim, A. P. O'Dowd, & A. Chin (Eds.), *Biogeomorphic responses to wildfire in fluvial ecosystems* (pp. 1–25). Boulder, CO: Geological Society of America Special Paper 562.

Meave, J., Kellman, M., MacDougall, A., & Rosales, J. (1991). Riparian habitats as tropical forest refugia. *Global Ecology and Biogeography Letters*, *1*, 69–76. https://doi.org/10.2307/2997492

Meleason, M. A., Gregory, S. V., & Bolte, J. P. (2003). Implications of riparian management strategies on wood in streams of the Pacific northwest. *Ecological Applications*, *13*, 1212–1221. https://doi.org/10.1890/02-5004

Meyer, G. A., Pierce, J. L., Wood, S. H., & Jull, A. J. T. (2001). Fire, storms, and erosional events in the Idaho batholith. *Hydrological Processes*, *15*, 3025–3038. https://doi.org/10.1002/hyp.389

Meyer, G. A., & Wells, S. G. (1997). Fire-related sedimentation events on alluvial fans, Yellowstone National Park, USA. *Journal of Sedimentary Research*, *67*, 776–791. https://doi.org/10.1306/D426863A-2B26-11D7-8648000102C1865D

Meyer, G. A., Wells, S. G., Balling, R. C., & Jull, A. J. T. (1992). Response of alluvial systems to fire and climate change in Yellowstone National Park. *Nature*, *357*, 147–150. https://doi.org/10.1038/357147a0

Meyer, G. A., Wells, S. G., & Jull, A. J. T. (1995). Fire and alluvial chronology in Yellowstone National Park: climatic and intrinsic controls on Holocene geomorphic processes. *Geological Society of America Bulletin*, *107*, 1211–1230. https://doi.org/10.1130/0016-7606(1995)107%3C1211:FAACIY%3E2.3.CO;2

Meyer, J. L., Kaplan, L. A., Newbold, D., Woltemade, C. J., Zedler, J. B., Beilfuss, R., et al. (2007). *Where rivers are born: The scientific imperative for defending small streams and wetlands*. San Francisco, CA: Sierra Club.

Mihuc, T. B., Minshall, G. W., & Robinston, C. T. (1996). Response of benthic macroinvertebrate populations in Cache Creek, Yellowstone National Park to the 1988 wildfires. In J. M. Greenlee (Ed.), *The Ecological Implications of Fire in Greater Yellowstone* (pp. 83–94). Fairfield, WA: International Association of Wildland Fire.

Minshall, G. W. (2003). Responses of stream benthic macroinvertebrates to fire. *Forest Ecology and Management*, *178*, 155–161. https://doi.org/10.1016/S0378-1127(03)00059-8

Minshall, G. W., Brock, J. T., & Varley, J. D. (1989). Wildfires and Yellowstone's stream ecosystems. *BioScience*, *39*, 707–715. https://doi.org/10.2307/1311002

Minshall, G. W., & Robinson, C. T. (1992). *Annual report – effects of the 1998 wildfires on stream systems of Yellowstone National Park: 1991*. West Yellowstone, MT: Yellowstone National Park.

Minshall, G. W., Robinson, C. T., & Lawrence, D. E. (1997). Postfire responses of lotic ecosystems in Yellowstone National Park, USA. *Canadian Journal of Fisheries and Aquatic Sciences*, *54*, 2509–2525. https://doi.org/10.1139/f97-160

Minshall, G. W., Robinson, C. T., Lawrence, D. E., Andrews, D. A., & Brock, J. T. (2001a). Benthic macroinvertebrate assemblages in five central Idaho (USA) streams over a 10-year period following disturbance by wildfire. *International Journal of Wildland Fire*, *10*, 201–213. https://doi.org/10.1071/WF01018

Minshall, G. W., Royer, T. V., & Robinson, C. T. (2001b). Response of the Cache Creek macroinvertebrates during the first 10 years following disturbance by the 1988 Yellowstone wildfires. *Canadian Journal of Fisheries and Aquatic Sciences*, *58*, 1077–1088. https://doi.org/10.1139/f01-056

Mishra, A., Alnahit, A., & Campbell, B. (2021). Impact of land uses, drought, flood, wildfire, and cascading events on water quality and microbial communities: A review. *Journal of Hydrology*, *596*, 125707. https://doi.org/10.1016/j.jhydrol.2020.125707

Monaghan, K. A., Machado, A. L., Corado, M., Wrona, F. J., & Soares, A. M. V. M. (2019). Seasonal time-series reveal the impact and recovery in richness, abundance and community structure of benthic macroinvertebrates following catchment wildfire. *Science of the Total Environment*, *651*, 3117–3126. https://doi.org/10.1016/j.scitotenv.2018.10.176

Montgomery, D. R., & Buffington, J. M. (1997). Channel-reach morphology in mountain drainage basins. *Geological Society of America Bulletin*, *109*, 596–611. https://doi.org/10.1130/0016-7606(1997)109%3C0596:CRMIMD%3E2.3.CO;2

Moody, J. A. (2017). Residence times and alluvial architecture of a sediment superslug in response to different flow regimes. *Geomorphology*, *294*, 40–57. https://doi.org/10.1016/j.geomorph.2017.04.012

Moody, J. A., & Martin, D. A. (2001). Post-fire, rainfall intensity-peak discharge relations for three mountainous watersheds in the western USA. *Hydrological Processes*, *15*, 2981–2993. https://doi.org/10.1002/hyp.386

Moody, J. A., & Martin, D. A. (2009). Synthesis of sediment yields after wildland fire in different rainfall regimes in the western United States. *International Journal of Wildland Fire*, *18*, 96–115. https://doi.org/10.1071/WF07162

Moody, J. A., Shakesby, R. A., Robichaud, P. R., Cannon, S. H., & Martin, D. A. (2013). Current research issues related to post-wildfire runoff and erosion processes. *Earth-Science Reviews*, *122*, 10–37. https://doi.org/10.1016/j.earscirev.2013.03.004

Moore, R. D., Sunderland, P., Gomi, T., & Dhakal, A. (2005). Thermal regime of a headwater stream within a clearcut, coastal British Columbia, Canada. *Hydrological Processes*, *19*, 2591–2608. https://doi.org/10.1002/hyp.5733

Morell, K. D., Alessio, P., Dunne, T., & Keller, E. (2021). Sediment recruitment and redistribution in mountain channel networks by post-wildfire debris flows. *Geophysical Research Letters*, *48*, e2021GL095549. https://doi.org/10.1029/2021GL095549

Murphy, B. P., Czuba, J. A., & Belmont, P. (2019). Post-wildfire sediment cascades: A modeling framework linking debris flow generation and network-scale sediment routing. *Earth Surface Processes and Landforms*, *44*, 2126–2140. https://doi.org/10.1002/esp.4635

Musetta-Lambert, J. L., Kreutzweiser, D. P., & Sibley, P. K. (2020). Assessing the influence of wildfire on leaf decomposition and macroinvertebrate communities in boreal streams using mixed-species leaf packs. *Freshwater Biology*, *65*, 1047–1062. https://doi.org/10.1111/fwb.13488

Naiman, R. J., & Decamps, H. (1997). The ecology of interfaces: riparian zones. *Annual Review of Ecology and Systematics*, *28*, 621–658. https://doi.org/10.1146/annurev.ecolsys.28.1.621

Nanson, G. C., & Croke, J. C. (1992). A genetic classification of floodplains. *Geomorphology*, *4*, 459–486.

Neville, H., Dunham, J., Rosenberger, A., Umek, J., & Nelson, B. (2009). Influences of wildfire, habitat size, and connectivity on trout in headwater streams revealed by patterns of genetic diversity. *Transactions of the American Fisheries Society*, *138*, 1314–1327. https://doi.org/10.1577/T08-162.1

Nielsen, S. E., DeLancey, E. R., Reinhardt, K., & Parisien, M. A. (2016). Effects of lakes on wildfire activity in the boreal forests of Saskatchewan. *Canada. Forests*, *7*, 265. https://doi.org/10.3390/f7110265

Nilsson, G. E., & Östlund-Nilsson, S. (2008). Does size matter for hypoxia tolerance in fish? *Biological Reviews*, *83*, 173–189. https://doi.org/10.1111/j.1469-185X.2008.00038.x

Nordberg, G., Fowler, B., Nordberg, M., & Friberg, L. (2007). *Handbook on the toxicology of metals* (3rd ed.). Elsevier Academic Press.

Norris, L. A., Hawkes, C. L., Webb, W. L., Moore, D. G., Bollen, W. B., & Holcombe, E. (1978). *A report of research on the behavior and impact of chemical fire retardants in forest streams*. Corvallis, OR: Forestry Sciences Laboratory, PNW Forest and Range Experiment Station.

Nyman, P., Box, W. A. C., Stout, J. C., Sheridan, G. J., Keesstra, S. D., Lane, P. N. J., & Langhans, C. (2020). Debris-flow-dominated sediment transport through a channel network after wildfire. *Earth Surface Processes and Landforms*, *45*, 1155–1167. https://doi.org/10.1002/esp.4785

O'Dowd, A. P., Solverson, A. P., & Chin, A. (2023). Ecological response of step-pool streams to wildfire in the front range of Colorado (USA). In J. L. Florsheim, A. P. O'Dowd, & A. Chin (Eds.), *Biogeomorphic responses to wildfire in fluvial ecosystems* (pp. 79–94). Boulder, CO: Geological Society of America Special Paper 562.

Ohneiser, K., Ansmann, A., Kaifler, B., Chudnovsky, A., Barja, B., Knopf, D. A., et al. (2022). Australian wildfire smoke in the stratosphere: The decay phase in 2020/2021 and impact on ozone depletion. *Atmospheric Chemistry and Physics*, *22*, 7417–7442. https://doi.org/10.5194/acp-22-7417-2022

Olefeldt, D., Devito, K. J., & Turetsky, M. R. (2013). Sources and fate of terrestrial dissolved organic carbon in lakes of a Boreal Plains region recently affected by wildfire. *Biogeosciences*, *10*, 6247–6265. https://doi.org/10.5194/bg-10-6247-2013

Olsson, P. E., Kling, P., & Hogstrand, C. (1998). Mechanisms of heavy metal accumulation and toxicity in fish. In W. J. Langston & M. J. Bebianno (Eds.), *Metal metabolism in aquatic environments* (pp. 32–350). London: Chapman and Hall.

Ortega-Becerril, J. A., Suarez, C., Vazquez-Tarrio, D., Garrote, J., & Gomez-Heras, M. (2024). Sediment response after wildfires in mountain streams and their effects on cultural heritage: The case of the 2021 Navalcruz wildfire (Avila, Spain). *Fire*, *7*, 52. https://doi.org/10.3390/fire7020052

Patoine, A., Pinel-Alloul, B., Prepas, E. E., & Carignan, R. (2000). Do logging and forest fires influence zooplankton biomass in Canadian boreal Shield Lakes? *Canadian Journal of Fisheries and Aquatic Sciences*, *57*, 155–164. https://doi.org/10.1139/f00-105

Paul, M. J., LeDuc, S. D., Lassiter, M. G., Moorhead, L. C., Noyes, P. D., & Leibowitz, S. G. (2022). Wildfire induces changes in receiving waters: A review with considerations for water quality management. *Water Resources Research*, *58*, e2021WR030699. https://doi.org/10.1029/2021WR030699

Pelletier, N., Chételat, J., Blarquez, O., & Vermaire, J. C. (2020). Paleolimnological assessment of wildfire-derived atmospheric deposition of trace metal(loid)s and major ions to subarctic lakes (Northwest Territories, Canada). *Journal of Geophysical Research: Biogeosciences*, *125*, e2020JG005720. https://doi.org/10.1029/2020JG005720

Persico, L., & Meyer, G. (2009). Holocene beaver damming, fluvial geomorphology, and climate in Yellowstone National Park, Wyoming. *Quaternary Research*, *71*, 340–353. https://doi.org/10.1016/j.yqres.2008.09.007

Persico, L., & Meyer, G. (2013). Natural and historical variability in fluvial processes, beaver activity, and climate in the greater Yellowstone ecosystem. *Earth Surface Processes and Landforms*, *38*, 728–750. https://doi.org/10.1002/esp.3349

Peterson, B. J., Wollheim, W. M., Mulholland, P. J., Webster, J. R., Meyer, J. L., Tank, J. L., et al. (2001). Control of nitrogen export from watersheds by headwater streams. *Science*, *292*, 86–90. https://doi.org/10.1126/science.1056874

Pettit, N. E., & Naiman, R. J. (2005). Flood-deposited wood debris and its contribution to heterogeneity and regeneration in a semi-arid riparian landscape. *Oecologia*, *145*, 434–444. https://doi.org/10.1007/s00442-005-0143-z

Pettit, N. E., & Naiman, R. J. (2007). Fire in the riparian zone: characteristics and ecological consequences. *Ecosystems*, *10*, 673–687. https://doi.org/10.1007/s10021-007-9048-5

Pettit, N. E., Naiman, R. J., Rogers, K. H., & Little, J. E. (2005). Post-flooding distribution and characteristics of large woody debris piles along the semi-arid Sabie River, South Africa. *River Research and Applications*, *21*, 27–38. https://doi.org/10.1002/rra.812

Pierce, J., & Meyer, G. (2008). Long-term fire history from alluvial fan sediments: the role of drought and climate variability, and implications for management of Rocky Mountain forests. *International Journal of Wildland Fire*, *17*, 84–95. https://doi.org/10.1071/WF07027

Pierce, J. L., Meyer, G. A., & Jull, A. J. T. (2004). Fire-induced erosion and millennial-scale climate change in northern ponderosa pine forests. *Nature*, *432*, 87–90. https://doi.org/10.1038/nature03058

Pilotto, F., Harvey, G. L., Wharton, G., & Pusch, M. T. (2016). Simple large wood structures promote hydromorphological heterogeneity and benthic macroinvertebrate diversity in low-gradient rivers. *Aquatic Sciences*, *78*, 755–766. https://doi.org/10.1007/s00027-016-0467-2

Pinel-Alloul, B., Prepas, E., Planas, D., Steedman, R., & Charette, T. (2002). Watershed impacts of logging and wildfire: case studies in Canada. *Lake and Reservoir Management*, *18*, 307–318. https://doi.org/10.1080/07438140209353937

Pinto, P., Vaz, P., Robinson, C., & Morais, M. (2008). Wildfire impacts on aquatic ecosystems. In L. M. G. Duarte & P. Pinto (Eds.), *Sustainable development: Energy, environment and natural disasters* (pp. 25–34). Lisbon, Portugal: Fundacao Luis de Molina.

Polvi, L. E., & Wohl, E. (2012). The beaver meadow complex revisited: The role of beavers in post-glacial floodplain development. *Earth Surface Processes and Landforms*, *37*, 332–346. https://doi.org/10.1002/esp.2261

Pompeani, D. P., McLauchlan, K. K., Chileen, B. V., Calder, W. J., Shuman, B. N., & Higuera, P. E. (2020). The biogeochemical consequences of late Holocene wildfires in three subalpine lakes from northern Colorado. *Quaternary Science Reviews*, *236*, 106293. https://doi.org/10.1016/j.quascirev.2020.106293

Poulos, M. J., & Pierce, J. L. (2018). Alluvial fan depositional records from north and south-facing catchments in semi-arid montane terrain. *Quaternary Research*, *89*, 237–253. https://doi.org/10.1017/qua.2017.98

Power, M. J., Marlon, J., Ortiz, N., Bartlein, P. J., Harrison, S. P., Mayle, F. E., et al. (2008). Changes in fire regimes since the last glacial maximum: An assessment based on a global synthesis and analysis of charcoal data. *Climate Dynamics*, *30*, 887–907. https://doi.org/10.1007/s00382-007-0334-x

Praskievicz, S., & Sigdel, R. (2021). Loading of stream wood following the 2016 chimney tops 2 wildfire: Great Smoky Mountains National Park, Tennessee, USA. *River Research and Applications*, *37*, 475–483. https://doi.org/10.1002/rra.3765

Preston, D. L., Trujillo, J. L., Fairchild, M. P., Morrison, R. R., Fausch, K. D., & Kanno, Y. (2023). Short-term effects of wildfire on high elevation stream-riparian food webs. *Oikos*, *2023*, e09828. https://doi.org/10.1111/oik.09828

Pretty, T. J., Chanyi, C. M., Kuhn, C., & Gray, D. K. (2021). Factors influencing the structure of macroinvertebrate communities in subarctic lakes affected by wildfires. *Canadian Journal of Fisheries and Aquatic Sciences*, *78*, 218–231. https://doi.org/10.1139/cjfas-2020-0141

Railsback, S. F. (2022). What we don't know about the effects of temperature on salmonid growth. *Transactions of the American Fisheries Society*, *151*, 3–12. https://doi.org/10.1002/tafs.10338

Rathburn, S. L., Shahverdian, S. M., & Ryan, S. E. (2018). Post-disturbance sediment recovery: Implications for watershed resilience. *Geomorphology*, *305*, 61–75. https://doi.org/10.1016/j.geomorph.2017.08.039

Ré, A., Campos, I., Keizer, J. J., Goncalves, F. J. M., Pereira, J. L., & Abrantes, N. (2021). Effects of post-fire contamination in sediment-dwelling species of riverine systems. *Science of the Total Environment*, *771*, 144813. https://doi.org/10.1016/j.scitotenv.2020.144813

Reale, J. K., Van Horn, D. J., Condon, K. E., & Dahm, C. N. (2015). The effects of catastrophic wildfire on water quality along a river continuum. *Freshwater Science*, *34*, 1426–1442. https://doi.org/10.1086/684001

Reale, J. K., Archdeacon, T. P., Van Horn, D. J., Gonzales, E. J., Dudley, R. K., Turner, T. F., & Dahm, C. N. (2021). Differential effects of a catastrophic wildfire on downstream fish assemblages in an aridland river. *Aquatic Ecology*, *55*, 483–500. https://doi.org/10.1007/s10452-021-09839-4

Rhea, A. E., Covino, T. P., & Rhoades, C. C. (2021). Reduced N-limitation and increased in-stream productivity of autotrophic biofilms 5 and 15 years after severe wildfire. *Journal of Geophysical Research Biogeosciences, 126*, e2020JG006095. https://doi.org/10.1029/2020JG006095

Rhoades, C. C., Chow, A. T., Covino, T. P., Fegel, T. S., Pierson, D. N., & Rhea, A. E. (2019). The legacy of a severe wildfire on stream nitrogen and carbon in headwater catchments. *Ecosystems, 22*, 643–657. https://doi.org/10.1007/s10021-018-0293-6

Richardson, J. S., Bilby, R. E., & Bondar, C. A. (2005). Organic matter dynamics in small streams of the Pacific northwest. *Journal of the American Water Resources Association, 41*, 921–934. https://doi.org/10.1111/j.1752-1688.2005.tb03777.x

Richardson, J. S., & Danehy, R. J. (2007). A synthesis of the ecology of headwater streams and their riparian zones in temperate forests. *Forest Science, 53*, 131–147. https://doi.org/10.1093/forestscience/53.2.131

Rieman, B., Lee, D., Chandler, G., Myers, D. (1997). Does wildfire threaten extinction for salmonids? Responses of redband trout and bull trout following recent large fires on the Boise National Forest. *Proceedings – Fire Effects on Rare and Endangered Species and Habitats Conference* (47–57), Coeur d'Alene, Idaho: IAWF.

Rieman, B., Gresswell, R., & Rinne, J. (2012). Fire and fish: A synthesis of observation and experience. In C. Luce, P. Morgan, K. Dwire, D. Isaak, Z. Holden, & B. Rieman (Eds.), *Climate change, forests, fire, water, and fish: Building resilient landscapes, streams, and managers*. USDA Forest Service General Technical Report RMRS-GTR-290 (pp. 159–175). Fort Collins, CO: Rocky Mountain Research Station.

Riley, K., Pierce, J., & Meyer, G. A. (2015). Vegetation and climatic controls on Holocene wildfire and erosion recorded in alluvial fans of the middle fork Salmon River, Idaho. *The Holocene, 25*, 857–871. https://doi.org/10.1177/0959683615571423

Rinne, J. N. (1996). Management briefs: short-term effects of wildfire on fishes and aquatic macroinvertebrates in the South-Western United States. *North American Journal of Fisheries Management, 16*, 653–658. https://doi.org/10.1577/1548-8675(1996)016%3C0653:MBSTEO%3E2.3.CO;2

Rinne, J. N., & Carter, C. D. (2008). Short-term effects of wildfires on fishes in the southwestern United States, 2002: management implications. In M. G. Narog (Ed.), *Proceedings of the 2002 Fire Conference: Managing Fire and Fuels in The Remaining Wildlands and Open Spaces of the Southwestern United States. USDA Forest Service General Technical Report PSW-GTR-189*. Albany, California.

Roberts, J. H., & Hitt, N. P. (2010). Longitudinal structure in temperate stream fish communities: Testing conceptual models with temporal data. In K. Gido & D. Jackson (Eds.), *Community Ecology of North American Stream Fishes: Concepts, Approaches, and Techniques* (pp. 281–299). Bethesda, MD: American Fisheries Society.

Robinne, F. N., Hallema, D. W., Bladon, K. D., & Buttle, J. M. (2020). Wildfire impacts on hydrologic ecosystem services in North American high-latitude forests: A scoping review. *Journal of Hydrology*, *581*, 124360. https://doi.org/10.1016/j.jhydrol.2019.124360

Robinson, C. T., Uehlinger, U., & Minshall, G. W. (2005). Functional characteristics of wilderness streams twenty years following wildfire. *Western North American Naturalist*, *65*, 1–10. https://www.jstor.org/stable/41717417

Robson, B. J., Chester, E. T., Matthews, T. G., & Johnston, K. (2018). Post-wildfire recovery of invertebrate diversity in drought-affected headwater streams. *Aquatic Sciences*, *80*, 21. https://doi.org/10.1007/s00027-018-0570-7

Rodriguez-Cardona, B. M., Coble, A. A., Wymore, A. S., Kolosov, R., Podgorski, D. C., Zito, P., et al. (2020). Wildfires lead to decreased carbon and increased nitrogen concentrations in upland arctic streams. *Scientific Reports*, *10*, 8722. https://doi.org/10.1038/s41598-020-65520-0

Rodríguez-Lozano, P., Rieradevall, M., Rau, M. A., & Prat, N. (2015). Long-term consequences of a wildfire for leaf-litter breakdown in a Mediterranean stream. *Freshwater Science*, *34*, 1482–1493. https://doi.org/10.1086/683432

Roebuck, J. A., Bladon, K. D., Donahue, D., Graham, E. B., Grieger, S., Morgenstern, K., et al. (2022). Spatiotemporal controls on the delivery of dissolved organic matter to streams following a wildfire. *Geophysical Research Letters*, *49*, e2022GL099535. https://doi.org/10.1029/2022GL099535

Romanin, L. M., Hopf, F., Haberle, S. G., & Bowman, D. M. J. S. (2016). Fire regime and vegetation change in the transition from aboriginal to European land management in a Tasmanian eucalypt savanna. *Australian Journal of Botany*, *64*, 427–440. https://doi.org/10.1071/BT16032

Rosenberger, A. E., Dunham, J. B., Buffington, J. M., & Wipfli, M. S. (2011). Persistent effects of wildfire and debris flows on the invertebrate prey base of rainbow trout in Idaho streams. *Northwest Science*, *85*, 55–63. https://doi.org/10.3955/046.085.0105

Roth, H. K., Nelson, A. R., McKenna, A. M., Fegel, T. S., Young, R. B., Rhoades, C. C., et al. (2022). Impact of beaver ponds on biogeochemistry of organic carbon and nitrogen along a fire-impacted stream. *Environmental Science Processes and Impacts*, *24*, 1661. https://doi.org/10.1039/D2EM00184E

Rugenski, A. T., & Minshall, G. W. (2014). Climate-moderated responses to wildfire by macroinvertebrates and basal food resources in montane wilderness streams. *Ecosphere*, *5*, 25. https://doi.org/10.1890/ES13-00236.1

Rust, A. J., Randell, J., Todd, A. S., & Hogue, T. S. (2019). Wildfire impacts on water quality, macroinvertebrate, and trout populations in the Upper Rio Grande. *Forest Ecology and Management*, *453*, 117636. https://doi.org/10.1016/j.foreco.2019.117636

Ryan, R., Dosseto, A., Lemarchand, D., Dlapa, P., Thomas, Z., Simkovic, I., & Bradstock, R. (2023). Boron isotopes and FTIR spectroscopy to identify past high severity fires. *Catena*, *222*, 106887. https://doi.org/10.1016/j.catena.2022.106887

Ryan, S. E., & Dixon, M. K. (2008). Spatial and temporal variability in stream sediment loads using examples from the Gros Ventre Range, Wyoming, USA. In H. Habersack, H. Piegay, & M. Rinaldi (Eds.), *Gravel-bed rivers VI: From process understanding to river restoration* (pp. 387–407). Elsevier.

Ryan, S. E., Dwire, K. A., & Dixon, M. K. (2011). Impacts of wildfire on runoff and sediment loads at Little Granite Creek, western Wyoming. *Geomorphology*, *129*, 113–130. https://doi.org/10.1016/j.geomorph.2011.01.017

Sanborn, P., Geertsema, M., Jull, A. J. T., & Hawkes, B. (2006). Soil and sedimentary charcoal evidence for Holocene forest fires in an inland temperate rainforest, east-central British Columbia, Canada. *The Holocene*, *16*, 415–427. https://doi.org/10.1191/0959683606hl937rp

Schumm, S. A. (1973). Geomorphic thresholds and complex response of drainage systems. In M. Morisawa (Ed.), *Fluvial geomorphology* (pp. 299–310). SUNY Binghamton: Publications in Geomorphology.

Schumm, S. A., Harvey, M. D., & Watson, C. C. (1984). *Incised channels: Morphology, dynamics, and control*: Littleton, Colorado (p. 200). Littleton, Colorado: Water Resources Publications.

Scordo, F., Chandra, S., Suenaga, E., Kelson, S. J., Culpepper, J., Scaff, L., et al. (2021). Smoke from regional wildfires alters lake ecology. *Scientific Reports*, *11*, 10922. https://doi.org/10.1038/s41598-021-89926-6

Scott, D. F. (1993). The hydrological effects of fire in South African mountain catchments. *Journal of Hydrology*, *150*, 409–432. https://doi.org/10.1016/0022-1694(93)90119-T

Scrimgeour, G. J., Tonn, W. M., Paszkowski, C. A., & Goater, C. (2001). Benthic macroinvertebrate biomass and wildfires: evidence for enrichment of boreal subarctic lakes. *Freshwater Biology*, *46*, 367–378. https://doi.org/10.1046/j.1365-2427.2001.00682.x

Shakesby, R., & Doerr, S. (2006). Wildfire as a hydrological and geomorphological agent. *Earth-Science Reviews*, *74*, 269–307. https://doi.org/10.1016/j.earscirev.2005.10.006

Short, L. E., Gabet, E. J., & Hoffman, D. F. (2015). The role of large woody debris in modulating the dispersal of a post-fire sediment pulse. *Geomorphology*, *246*, 351–358. https://doi.org/10.1016/j.geomorph.2015.06.031

Shurin, J. B., Cottenie, K., & Hillebrand, H. (2009). Spatial autocorrelation and dispersal limitation in freshwater organisms. *Oecologia*, *159*, 151–159. https://doi.org/10.1007/s00442-008-1174-z

Silberstein, R. P., Dawes, W. R., Bastow, T. P., Byrne, J., & Smart, N. F. (2013). Evaluation of changes in post-fire recharge under native woodland using hydrological measurements, modelling and remote sensing. *Journal of Hydrology*, *489*, 1–15. https://doi.org/10.1016/j.jhydrol.2013.01.037

Silins, U., Bladon, K. D., Kelly, E. N., Esch, E., Spence, J. R., Stone, M., et al. (2014). Five-year legacy of wildfire and salvage logging impacts on nutrient runoff and aquatic plant, invertebrate, and fish diversity. *Ecohydrology*, *7*, 1508–1523. https://doi.org/10.1002/eco.1474

Silva, V., Abrantes, N., Costa, R., Keizer, J. J., Goncalves, F., & Pereira, J. L. (2016). Effects of ashloaded post-fire runoff on the freshwater clam *Corbicula fluminea*. *Ecological Engineering*, *90*, 180–189. https://doi.org/10.1016/j.ecoleng.2016.01.043

Smith, H. G., Sheridan, G. J., Lane, P. N. J., Nyman, P., & Haydon, S. (2011). Wildfire effects on water quality in forest catchments: A review with implications for water supply. *Journal of Hydrology*, *396*, 170–192. https://doi.org/10.1016/j.jhydrol.2010.10.043

Smokorowski, K. E., Pratt, T. C., Cole, W. G., McEachern, L. J., & Mallory, E. C. (2006). Effects on periphyton and macroinvertebrates from removal of submerged wood in three Ontario lakes. *Canadian Journal of Fisheries and Aquatic Sciences*, *63*, 2038–2049. https://doi.org/10.1139/f06-104

Stanford, J. A., Lorang, M. S., & Hauer, F. R. (2005). The shifting habitat mosaic of river ecosystems. *Verhandlungen der Internationalen Vereinigung für theoretische und angewandte Limnologie*, *29*, 123–136. https://doi.org/10.1080/03680770.2005.11901979

Stephan, K., Kavanagh, K. L., & Koyama, A. (2012). Effects of spring prescribed burning and wildfires on watershed nitrogen dynamics of central Idaho headwater areas. *Forest Ecology and Management*, *263*, 240–252. https://doi.org/10.1016/j.foreco.2011.09.013

Stephan, K., Kavanagh, K. L., & Koyama, A. (2015). Comparing the influence of wildfire and prescribed burns on watershed nitrogen biogeochemistry using 15N natural abundance in terrestrial and aquatic ecosystem components. *PLOS One*, *10*, e0119560. https://doi.org/10.1371/journal.pone.0119560

Stout, J. C., Rutherfurd, I. D., Grove, J., Webb, A. J., Kitchingman, A., Tonkin, Z., & Lyon, J. (2018). Passive recovery of wood loads in rivers. *Water Resources Research*, *54*, 8828–8846. https://doi.org/10.1029/2017WR021071

Strahler, A. N. (1952). Hypsometric (area-altitude) analysis of erosional topography. *Bulletin of the Geological Society of America*, *63*, 1117–1142. https://doi.org/10.1130/0016-7606(1952)63[1117,HAAOET]2.0.CO;2

Sutfin, N. A., & Wohl, E. (2019). Elevational differences in hydrogeomorphic disturbance regime influence sediment residence times within mountain river corridors. *Nature Communications*, *10*, 2221. https://doi.org/10.1038/s41467-019-09864-w

Townsend, C. R., & Hildrew, A. C. (1994). Species traits in relation to a habitat templet for river systems. *Freshwater Biology*, *31*, 265–275. https://doi.org/10.1111/j.1365-2427.1994.tb01740.x

Troendle, C. A., & Bevenger, G. S. (1996). Effect of fire on streamflow and sediment transport Shoshone National Forest, Wyoming. In J. Greenlee (Ed.), *The Ecological implications of fire in greater Yellowstone: Proceedings of the second biennial conference on the greater yellowstone ecosystem* (pp. 43–52). Fairfield, Washington: International Association of Wildland Fire.

Urmy, S. S., Williamson, C. E., Leach, T. H., Schladow, S. G., Overholt, E. P., & Warren, J. D. (2016). Vertical redistribution of zooplankton in an oligotrophic lake associated with reduction in ultraviolet radiation by wildfire smoke. *Geophysical Research Letters*, *43*, 3746–3753. https://doi.org/10.1002/2016GL068533

Vaz, P. G., Merten, E. C., Warren, D. R., Durscher, K., Tapp, M., Robinson, C. T., et al. (2015). Fire meets inland water via burned wood: and then what? *Freshwater Science*, *34*, 1468–1481. https://doi.org/10.1086/684095

Vaz, P. G., Warren, D. R., Pinto, P., Merten, E. C., Robinson, C. T., & Rego, F. C. (2011). Tree type and forest management effects on the structure of stream wood following wildfires. *Forest Ecology and Management*, *262*, 561–570. https://doi.org/10.1016/j.foreco.2011.04.026

Verkaik, I., Rieradevall, M., Cooper, S. D., Melack, J. M., Dudley, T. L., & Prat, N. (2013). Fire as a disturbance in Mediterranean climate streams. *Hydrobiologia*, *719*, 353–382. https://doi.org/10.1007/s10750-013-1463-3

Vieira, N. K., Clements, W. H., Guevara, L. S., & Jacobs, B. F. (2004). Resistance and resilience of stream insect communities to repeated hydrologic disturbances after a wildfire. *Freshwater Biology*, *49*, 1243–1259. https://doi.org/10.1111/j.1365-2427.2004.01261.x

Wagner, M. J., Bladon, K. D., Silins, U., Williams, C. H. S., Martens, A. M., Boon, S., et al. (2014). Catchment-scale stream temperature response to land disturbance by wildfire governed by surface-subsurface energy exchange and atmospheric controls. *Journal of Hydrology*, *517*, 328–338. https://doi.org/10.1016/j.jhydrol.2014.05.006

Wallace, J. B., & Benke, A. C. (1984). Quantification of wood habitat in subtropical coastal plain streams. *Canadian Journal of Fisheries and Aquatic Sciences*, *41*, 1643–1652. https://doi.org/10.1139/f84-203

Walling, D. E., & Fang, D. (2003). Recent trends in the suspended sediment loads of the world's rivers. *Global and Planetary Change*, *39*, 111–126. https://doi.org/10.1016/S0921-8181(03)00020-1

Warren, D. R., Roon, D. A., Swartz, A. G., & Bladon, K. D. (2022). Loss of riparian forests from wildfire led to increased stream temperatures in summer, yet salmonid fish persisted. *Ecosphere*, *13*, e4233. https://doi.org/10.1002/ecs2.4233

Warrick, J. A., Hatten, J. A., Pasternack, G. B., Gray, A. B., Goni, M. A., & Wheatcroft, R. A. (2012). The effects of wildfire on the sediment yield of a coastal California watershed. *Geological Society of America Bulletin*, *124*, 1130–1146. https://doi.org/10.1130/B30451.1

Webster, J. S., Gido, K. B., Hedden, S. C., Propst, D. L., & Whitney, J. E. (2022). Response of arid-land macroinvertebrate communities to extremes of drought, wildfire, and monsoonal flooding. *River Research and Applications*, *38*, 832–845. https://doi.org/10.1002/rra.3949

Weirich, J. J. (2021). *Beaver modified fire resistance in the North Cascades and potential for climate change adaptation*. MS thesis. Cheney, WA: Eastern Washington University.

Whipple, A. (2019). Riparian Resilience in the Face of Interacting Disturbances: Understanding Complex Interactions between Wildfire, Erosion, and Beaver (*Castor canadensis*) in Grazed Dryland Riparian Systems of Low Order Streams in North Central Washington State, USA (MS thesis). Cheney, WA: Eastern Washington University.

Whitney, J. E., Gido, K. B., Hedden, S. C., Macpherson, G. L., Pilger, T. J., Propst, D. L., & Turner, T. F. (2017). Identifying the source population of fish recolonizing an arid-land stream following wildfire-induced extirpation using otolith microchemistry. *Hydrobiologia*, *797*, 29–45. https://doi.org/10.1007/s10750-017-3143-1

Wilkinson, S. N., Wallbrink, P. J., Hancock, G. J., Blake, W. H., Shakesby, R. A., & Doerr, S. H. (2009). Fallout radionuclide tracers identify a switch in sediment sources and transport-limited sediment yield following wildfire in a eucalypt forest. *Geomorphology*, *110*, 140–151. https://doi.org/10.1016/j.geomorph.2009.04.001

Williams-Subiza, E. A., & Brand, C. (2021). Functional response of benthic macroinvertebrates to fire disturbance in Patagonian streams. *Hydrobiologia*, *848*, 1575–1591. https://doi.org/10.1007/s10750-021-04548-3

Williamson, C. E., Overholt, E. P., Brentrup, J. A., Pilla, R. M., Leach, T. H., Schladow, S. G., et al. (2016). Sentinel responses to droughts, wildfires, and floods: effects of UV radiation on lakes and their ecosystem services. *Frontiers in Ecology and the Environment*, *14*, 102–109. https://doi.org/10.1002/fee.1228

Wohl, E. (2015). Legacy effects on sediments in river corridors. *Earth-Science Reviews*, *147*, 30–53. https://doi.org/10.1016/j.earscirev.2015.05.001

Wohl, E. (2017a). Bridging the gaps: An overview of wood across time and space in diverse rivers. *Geomorphology*, *279*, 3–26. https://doi.org/10.1016/j.geomorph.2016.04.014

Wohl, E. (2017b). The significance of small streams. *Frontiers in Earth Science*, *11*, 447–456. https://doi.org/10.1007/s11707-017-0647-y

Wohl, E. (2021). An integrative conceptualization of floodplain storage. *Reviews of Geophysics*, *59*, e202RG000724. https://doi.org/10.1029/2020RG000724

Wohl, E. E., & Pearthree, P. A. (1991). Debris flows as geomorphic agents in the Huachuca Mountains of southeastern Arizona. *Geomorphology*, *4*, 273–292. https://doi.org/10.1016/0169-555X(91)90010-8

Wohl, E., Lininger, K. B., Rathburn, S. L., & Sutfin, N. A. (2020). How geomorphic context governs the influence of wildfire on floodplain organic carbon in fire-prone environments of the western United States. *Earth Surface Processes and Landforms*, *45*, 38–55. https://doi.org/10.1002/esp.4680

Wohl, E., Castro, J., Cluer, B., Merritts, D., Powers, P., Staab, B., & Thorne, C. (2021). Rediscovering, reevaluating, and restoring lost river-wetland corridors. *Frontiers in Earth Science*, *9*, 653623. https://doi.org/10.3389/feart.2021.653623

Wohl, E., Marshall, A. E., Scamardo, J., White, D., & Morrison, R. R. (2022). Biogeomorphic influences on river corridor resilience to wildfire disturbances in a mountain stream of the Southern Rockies, USA. *Science of the Total Environment*, *820*, 153321. https://doi.org/10.1016/j.scitotenv.2022.153321

Wohl, E., Marshall, A., Scamardo, J., & Rathburn, S. (2023). Biogeomorphic processes, spatial heterogeneity, and river corridor resilience to stand-killing wildfire. In J. L. Florsheim, A. P. O'Dowd, & A. Chin (Eds.), *Biogeomorphic responses to wildfire in fluvial ecosystems* (pp. 153–176). Boulder, CO: Geological Society of America Special Paper 562.

Wohl, E., Marshall, A. E., Triantafillou, S., Mobley, M., Means-Brous, M., & Morrison, R. R. (2024). Distribution of logjams in relation to lateral connectivity in the river corridor. *Geomorphology*, *451*. https://doi.org/10.1016/j.geomorph.2024.109100

Young, M. K. (1994). Movement and characteristics of stream-borne coarse woody debris in adjacent burned and undisturbed watersheds in Wyoming. *Canadian Journal of Forest Research*, *24*, 1933–1938. https://doi.org/10.1139/x94-248

Zeigler, J. M., & Whitledge, G. W. (2011). Otolith trace element and stable isotopic compositions differentiate fishes from the middle Mississippi River, its tributaries, and floodplain lakes. *Hydrobiologia*, *661*, 289–302. https://doi.org/10.1007/s10750-010-0538-7

Zelt, R. B., & Wohl, E. E. (2004). Channel and woody debris characteristics in adjacent burned and unburned watersheds a decade after wildfire, Park County, Wyoming. *Geomorphology*, *57*, 217–233. https://doi.org/10.1016/S0169-555X(03)00104-1

Zolkos, S., MacDonald, E., Hung, J. K. Y., Schade, J. D., Ludwig, S., Mann, P. J., et al. (2022). Physiographic controls and wildfire effects on aquatic biogeochemistry in tundra of the Yukon-Kukokwim Delta, Alaska. *Journal of Geophysical Research: Biogeosciences*, *127*, e2022JG006891. https://doi.org/10.1029/2022JG006891

5

FIRES AND HUMAN COMMUNITIES

Abstract

Wildfire hazard refers to the state of the fuel and is quantified with respect to likelihood and intensity of wildfire. Vulnerability describes exposure and susceptibility of assets and people. Risk includes the probability of an event, as well as values and expected losses, and is effectively the combination of hazard and vulnerability. Human attitudes toward wildfire have varied substantially through time. Many nomadic or migratory prehistoric people adapted fire as a tool for manipulating biotic communities. Settled communities are more likely to rigorously suppress fires to reduce risks. Contemporary human communities in fire-prone areas use a combination of fire suppression, fuels reduction, mitigation pre-and post-fire, and adaptation. All of these activities occur in a rapidly warming climate in which the changing characteristics of wildfires require adaptability from human communities close to fire-prone areas such as the wildland-urban interface (WUI) and at greater distances that can be affected by wildfire emissions. Economic limitations such as rising insurance costs in the WUI may help to slow human migration into areas with high fire risk, but the costs of preparing for fire, attempting to suppress ongoing fires, and dealing with post-fire hazards will likely continue to rise in future.

Landscapes on Fire: Impacts on Uplands, Rivers, and Communities, Special Publications 80,
First Edition. Ellen Wohl.

DOI: 10.1002/9781394235186.ch5

5.1. Introduction

To adapt an old saying, during the past century human communities in much of the United States and Europe decided that they could not live with fire, and then subsequently decided that they could not live without it. After a long history of human use of fire for resource management by indigenous societies (Bowman et al., 2011; Denevan, 1992), attitudes changed toward rigorous fire suppression during the 20th century (Ryan et al., 2013)—a change sometimes known as the war on fire (Tedim et al., 2020)—and then changed again toward attempting to manage fire with controlled burns during the late 20th and early 21st centuries (Steelman & Burke, 2007). This latest approach tries to balance the ecological necessity of wildfires and the impossibility of completely preventing them with the hazards that wildfires create for human communities (Tedim et al., 2020). In the European Union, for example, Integrated Fire Management recognizes traditional and continuing uses of fire as a land management tool, combines prevention and suppression strategies with ecological considerations that can maximize the net benefits of fires, and explicitly considers the four steps of emergency crisis management: mitigation, preparedness, response, and recovery (Fischer et al., 2016). Mitigation involves actions to reduce vulnerability such as reducing fuels or modifying human behavior via outreach or regulatory actions such as zoning or building codes (McCaffery et al., 2020). Preparedness involves activities undertaken to increase the ability to respond to a fire (McCaffery et al., 2020).

Stoof and Kettridge (2022) argue the need to change fire management from fire resistance to landscape resilience, a process described as "living with fire" by accepting fire's natural role in the landscape and emphasizing adaptation to fire over prevention and suppression of fire. Four axes of diversity in integrated fire management are important in this context: cross-geography, in which the knowledge and expertise developed in fire-prone countries are shared with regions likely to experience increased wildfire in the future; cross-risk, in which experience in managing floods is applied to managing fires; cross-section, which emphasizes stronger connections between science and practice; and social diversity, which emphasizes increasing the representation and participation of women and minorities in fire science and emergency management (Stoof & Kettridge, 2022).

Multiple investigators have called for greater incorporation of traditional and indigenous knowledge of fire management, but fundamental challenges remain for this strategy. These include increasing fire occurrence and growing human population; the largely static (rather than nomadic) structure of contemporary human communities and infrastructure; and environmental conditions that have moved outside the natural or historical range of variability (Murphy et al., 2018 and Bowman et al., 2020).

Wildfires continue to create risk for humans in the 21st century because fires involve so many variables that cannot be fully controlled, including weather, fuels, and natural ignition sources such as lightning and accidents, criminal arson, and intentional attacks on national security (Busenberg, 2004). Consequently, fire remains an enormous challenge to most sedentary human societies, with the most immediate effects at what is referred to as either the WUI (wildland-urban interface) or the WHI (wildland-human interface).

Fire is challenging because of the hazard, vulnerability, and risk during the fire and for many years after fire. Hazard, vulnerability, and risk are terms used in many contexts. In the context of wildfire, hazard typically refers to the state of the fuel and is quantified with respect to likelihood and intensity of wildfire (McCaffery et al., 2020; Scott et al., 2013). As with other potentially hazardous natural processes, a wildfire is perceived as a hazard when it negatively affects something that humans value, whether that is air quality, infrastructure, a scenic view, or an endangered species (McCaffery et al., 2020).

Vulnerability typically describes exposure and susceptibility, which can be differentiated as assets vulnerability and human vulnerability. In the context of technical risk assessments, risk includes the probability of an event, as well as values and expected losses (Calkin et al., 2015): in other words, the probability of an event with harmful consequences, which can be thought of as the combination of hazard and vulnerability. In the context of fire, however, risk may refer only to the probability of natural or human-caused ignition (Hardy, 2005), or risk may be used to refer to hazard and the susceptibility of values-at-risk (Finney, 2005).

For simplicity, I will use hazard in this chapter very broadly to refer to a potential negative consequence of wildfire for human communities. Hazards used in this context can describe those that occur during the wildfire, which typically have a duration of weeks to months, and those that occur following the wildfire, which can persist for many years.

5.2. Hazards During Fire

The primary hazards to humans and infrastructure during fire are smoke and death or injury to humans and damage to infrastructure because of combustion, smoke stains, and ash deposition. Wildfire smoke can temporarily reduce air quality and visibility and these effects can extend hundreds of kilometers beyond the area actively burning

Figure 5.1 Wildfire smoke. At left, rising from the 2012 High Park Fire in Colorado, USA, as seen from the author's backyard on 17 June 2012, approximately 50 km east of the burning area. At right, rising from the 2020 Cameron Peak Fire in Colorado as seen from approximately 100 km southeast (upper) and 40 km southeast (lower). Upper right courtesy of Katherine B. Lininger and lower right courtesy of Anna E. Marshall.

(Busenberg, 2004) to cause health effects in more densely populated regions (O'Dell et al., 2021).

Wildfire smoke (Figure 5.1), as noted earlier, may contain particulate matter that is more toxic than equal doses of ambient particulate matter in the air (Aguilera et al., 2021). Fine particulate matter known as $PM_{2.5}$ gets the most attention in this regard (Liu et al., 2015; Reid et al., 2005). These particles, which have an aerodynamic diameter $\leq 2.5\,\mu m$, can be inhaled into the deepest recesses of the lungs and can enter the bloodstream, impairing vital organs including the lungs (Pope & Dockery, 2006; Xing et al., 2016).

The composition of wildfire smoke is complex and dynamic. The chemical species found in smoke reflect factors unique to the burn site, including the type of vegetation burned and weather conditions (Hayden et al., 2022; Pokhrel et al., 2021; Urbanski, 2013). Efforts to study smoke effects are limited by the availability of air-quality measures and inconsistent air quality reporting (Black et al., 2017). However, $PM_{2.5}$ particles in wildfire smoke are particularly dangerous because they are composed mostly of carbon compounds with more oxidative potential than ambient urban particulates (Adetona et al., 2016). These compounds generate more free radicals, which are unstable molecules that can damage other molecules such as DNA, lipids, and proteins. Carbon compounds in wildfire smoke also have a greater potential to cause inflammation and oxidative stress in the lungs (Williams et al., 2013). Volatile organic compounds such as the carcinogen benzene can also be present in wildfire smoke (Dickinson et al., 2022). Multiple studies document increased hospitalizations related to respiratory mortality during wildfire (e.g., Aguilera et al., 2021; Bowman & Johnston, 2005; Nawaz & Henze, 2020). Respiratory mortality includes asthma, chronic obstructive pulmonary disease, bronchitis, and pneumonia (Cascio, 2018).

Health hazards from wildfire smoke are not limited to small areas. Global mortality from wildfire smoke has been estimated at an astonishing level of 339,000 people per year (Johnston et al., 2012) and investigators estimate that each year more than 200 million people are exposed to wildfire smoke (Black et al., 2017). This may partly reflect the fact that wildfires contribute substantially to the average annual $PM_{2.5}$. The projected trends of increased global wildfire activity with warming climate also imply that human health hazards associated with smoke will increase (Ford et al., 2018; Xu et al., 2020). These issues became highly visible during the summer of 2023 when massive wildfires in Canada created hazardous smoke that drifted southward to densely populated areas of the northeastern United States, including New York City (Shakoor et al., 2023). Prolonged exposure to wildfire smoke can create psychological distress (Burkhardt et al., 2020). Lowe and Garfin (2023) note that crisis-related media exposure, such as the apocalyptic photos of world-famous landmarks barely visible through fire-induced haze, also carries mental health risks.

People in the burned area can experience solastalgia (Eisenman et al., 2015), or psychological distress from losing a valued landscape to a wildfire (Sullivan et al., 2022). People in burned areas are also likely to experience other sources of disruption and stress including disruptions of the economy and loss of employment, damage to transportation and other infrastructure, and evacuation from homes and associated anxieties as to whether domestic animals, pets, and homes will survive.

Smoke shading can also create surface cooling. This cooling can reduce regional wind speeds associated with circulation patterns such as a sea breeze, creating regional stagnation and associated impaired air quality, as well as reduced solar and wind energy production (Lassman et al., 2023).

Although wildfire smoke exposure can affect broad swaths of the population, the ability to shelter within indoor spaces that have filtered air varies between individuals and groups within a population (D'Evelyn et al., 2022). Smoke exposure can be reduced using public warnings to stay indoors and avoid strenuous exercise (Figure 5.2; Text Box 5.1) and by using portable air filters or other in-home interventions to improve the quality of indoor air during wildfires (Cascio, 2018). Numerical weather prediction

AQI	Air Pollution Level	Health Implications	Cautionary Statement (for PM2.5)
0–50	Good	Air quality is considered satisfactory, and air pollution poses little or no risk	None
51–100	Moderate	Air quality is acceptable; however, for some pollutants there may be a moderate health concern for a very small number of people who are unusually sensitive to air pollution.	Active children and adults, and people with respiratory disease, such as asthma, should limit prolonged outdoor exertion.
101–150	Unhealthy for Sensitive Groups	Members of sensitive groups may experience health effects. The general public is not likely to be affected.	Active children and adults, and people with respiratory disease, such as asthma, should limit prolonged outdoor exertion.
151–200	Unhealthy	Everyone may begin to experience health effects; members of sensitive groups may experience more serious health effects	Active children and adults, and people with respiratory disease, such as asthma, should avoid prolonged outdoor exertion; everyone else, especially children, should limit prolonged outdoor exertion
201–300	Very Unhealthy	Health warnings of emergency conditions. The entire population is more likely to be affected.	Active children and adults, and people with respiratory disease, such as asthma, should avoid all outdoor exertion; everyone else, especially children, should limit outdoor exertion.
300+	Hazardous	Health alert: everyone may experience more serious health effects	Everyone should avoid all outdoor exertion

Figure 5.2 Examples of air-quality ratings posted online and updated every 15 minutes. These ratings come from Larimer County, Colorado, USA and are used continuously (rather than only during wildfires). I find the upper right face to be particularly effective at encouraging me to remain indoors.

Text Box 5.1 How Do We Know? Air Quality Ratings and Activity Recommendations

National directives on air quality require local and national governments to monitor and report on air quality so that the public can be informed on ambient pollution levels (Plaia & Ruggieri, 2011). Because the concentration of individual pollutants can be confusing, concentration measurements are typically transformed into an air quality index that facilitates comparisons through time and space based on a single number. Air quality is commonly assessed using monitoring networks that continuously record air pollution concentrations at multiple points within a city or other geographic area. Air quality is complex because of potential additive, synergistic, or antagonistic effects between different chemicals, and multiple versions of an air quality index exist, but these complications are typically distilled into a rating system that varies from good to bad quality (Plaia & Ruggieri, 2011). The World Health Organization has issued air quality guidelines targeting the common aim of protecting public health and individual national agencies (e.g., in the United States, the Environmental Protection Agency) also have specific guidelines. Among the commonly measured compounds are CO (carbon monoxide), NO_2 (nitrogen dioxide), O_3 (ozone), particulate matter (solid particles and liquid droplets), and SO_2 (sulfur dioxide). The value of the air quality index is typically defined by the value of the pollutant with the highest concentration relative to its standard on a given day; in other words, the index typically does not account for the combined effects of multiple pollutants (Kyrkilis et al., 2007). The health effects of a pollutant are estimated by toxicologists focusing on the link between personal exposure and health effects, commonly using laboratory tests on animals, and by epidemiologists focusing on correlations between pollutant levels and health in human communities (McClellan, 2002).

can be coupled with online atmospheric chemistry conditions to simulate smoke surface concentrations and this approach can be used to forecast wildfire behavior and smoke dispersal (Lassman et al., 2023), enhancing the predictive capability of health advisories issued during a fire.

A small proportion of human deaths each year is attributable to direct and immediate wildfires (Figure 5.3). The World Health Organization attributes 2,400 deaths worldwide to wildfires and volcanic activities during the period 1998–2017, for example, whereas climate-related and

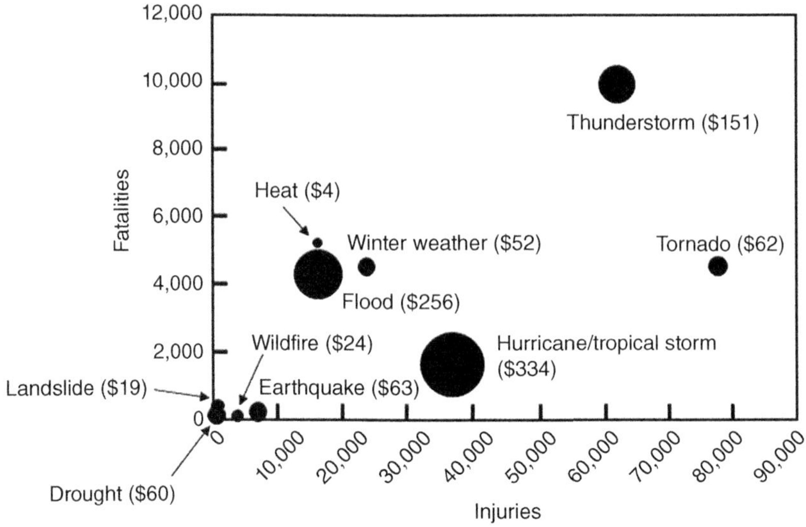

Figure 5.3 Historical U.S. hazard losses from the top 10 loss-causing hazards. The position of each hazard represents cumulative injuries (x-axis) and fatalities (y-axis) over the period 1960–2017. Dot size is proportional to monetary losses adjusted to 2017 U.S. dollars from each hazard type. Numbers in parentheses represent cumulative dollar losses in billions for each hazard type. Thunderstorm losses combine data for hail, thunderstorm wind, severe thunderstorm, and lightning hazards. (Schumann et al., 2020 / with permission of ELSEVIER.)

geophysical disasters killed 1.3 million people during this period (CRED, 2018). However, wildfires account for significant damages to infrastructure.

As might be expected, communities in the WUI where forested land borders and overlaps with infrastructure and development are at highest risk for human casualties, damage to civil infrastructure, and destruction of personal property (Radeloff et al., 2018). In the United States, WUI communities are differentiated as interface where the population density is ≥96 per square kilometer intermix where the population density is 11 to 96 per square kilometer; and occluded where structures border a contained area of wildland fuels such as a city park (USDA and USDI, 2001). An estimated one-third of all U.S. residents live in a WUI and about 17.4 million people in the United States are affected by the annual threat of wildfire hazards (Thomas & Butry, 2014).

Globally, the WUI is expanding rapidly as urban populations grow. These trends are particularly strong in southern Europe, western North America, and southern Australia, which have the greatest concentrations of wildfire fatalities and infrastructure damage (Bowman et al., 2017; Moritz et al., 2014). Sometimes the individual choice to live in such an area reflects a desire to live in a more natural environment with lower population density, but sometimes it reflects the economic necessity of more affordable housing in less intensely developed areas.

Economic damages to infrastructure also threaten insurance companies and governmental authorities. These costs have risen significantly (Lüthi et al., 2021; Sullivan et al., 2022), causing insurance companies to significantly raise insurance premiums or decline to provide coverage in some areas, such as in California, USA. Wildfire accounted for <2% of the total insured losses globally during 1985–2015, for example, but 12.4% during 2016–2020 (Swiss Re, 2021). Wildfires cause direct, tangible damage to structures but also damage to operations such as disruption of traffic, the electrical grid, or cellular communications (Anderson et al., 2020; Habermann & Hedel, 2018). Disruption of electrical systems and utilities in turn affects the ability of facilities such as hospitals to receive and treat people injured in the fire (e.g., Schulze et al., 2020). Secondary economic costs follow during the weeks and months after the fire associated with lost business economic activity, unemployment insurance, disaster relief (Diaz, 2012), and the post-fire hazards of floods, debris flows, and impaired water quality.

Over longer time periods and larger areas, the emission of carbon to the atmosphere during fire also constitutes a hazard (e.g., Jerrett et al., 2022). Estimates of annual global carbon emissions by wildfire range from 1410 to 3139 Tg of carbon per year (Schultz et al., 2008), with an average of 2078–2460 Tg per year (Lavoué et al., 2000; Schultz et al., 2008; Van der Werf et al., 2006) (Text Box 5.2). This is 26–31% of current global CO_2 emissions from fossil fuels and industrial processes (Raupach et al., 2007). Australian wildfires during the continent's Black Summer of 2019–2020 released an estimated 963 million tons of carbon dioxide into the atmosphere, which offset ~50% of the global reduction in carbon dioxide emissions associated with the Covid-19 pandemic in 2020 (Hong et al., 2023). Wildfires also release about 40% of global black carbon emissions (Bond et al., 2004). Black carbon is the second strongest contributor to global warming after CO_2 (IPCC, 2007). It is important to note, however, that unlike fossil-fuel emissions, at least some portion of wildfire-released carbon emissions can be sequestered again by regrowing vegetation. Wildfires associated with peatlands and with deforestation are most likely to be net emitters of carbon (van der Werf et al., 2006). Burned areas and total fire emissions are also largely decoupled: total carbon emissions track burning in forested areas, whereas burned area largely reflects savanna fires (van der Werf et al., 2006).

Text Box 5.2 How Do We Know? Carbon Emissions from Wildfire

Many papers use the Seiler-Crutzen ratio (Seiler & Crutzen, 1980) to estimate carbon emissions from wildfire. This ratio is based on

$$C = A \times B \times CE \times \beta$$

where A is area burned by fire (m^2), B is biomass density or pre-fire fuel load (kg/m^2), CE is the carbon fraction of the biomass (g/kg), and β is the fraction of biomass consumed during wildfire. Forest type characteristics, pre-fire fuel loads, and β can be summarized from field or remote sensing data (e.g., Ponomarev et al., 2021). The increasing spatial extent, frequency, information content, and availability of remote sensing data have substantially improved ability to characterize the spatial extent of fire, fuel characteristics (fuel type, fuel load, plant physiological and moisture conditions), pre- and post-fire site characteristics, and environmental conditions during fire, which influence fire severity and intensity (French et al., 2011). Comparisons of different methods for estimating carbon emissions from wildfires indicate that different methods and models generally produce similar results (French et al., 2011).

At smaller spatial scales, on-site measurements in areas affected by wildfire plumes of smoke and ash can be compared to analogous but unaffected areas. This approach indicated surface-water concentrations of dissolved black carbon up to 13% in the Santa Barbara Channel off California, USA during the fire (Wagner et al., 2021). Lidar overflights of smoke plumes can estimate CO_2 in atmospheric columns above the fire based on surface reflected sunlight (Mao et al., 2021).

The great hazard in wildfire carbon emissions is a positive feedback in which warmer, drier climate creates conditions conducive to more fire, which increases carbon emissions and enhances global warming (Lavorel et al., 2006). Feedbacks appear to be occurring in the Arctic, where increasing concentrations of black carbon have contributed substantially to rapid Arctic warming during recent decades (Shindell & Faluvegi, 2009). Black carbon within soot deposited over snow and ice significantly increases solar absorption and melting (Ramanathan & Carmichael, 2008), for example, likely contributing to retreat of Arctic sea ice (Flanner et al., 2007) and accelerated melting of mountain snowpacks and glaciers (e.g., Kaspari et al., 2015; Koshkin et al., 2022; You & Xu, 2022).

Fires also have a more indirect effect on global carbon cycles by altering carbon storage in soils and carbon uptake and release via plant communities. Terrestrial ecosystems account for more than half of the CO_2 naturally exchanged between the biosphere and atmosphere (120 of >200 Pg carbon per year; IPCC, 2007). Terrestrial ecosystems store as much as 3,000 Pg organic carbon, which is more than four times the atmospheric CO_2 pool, and constitute the largest sources of methane (CH_4) and nitrous oxide (N_2O) to the atmosphere (Grimm et al., 2013). Consequently, changes in ecosystem function associated with wildfire have the potential to alter the climate via either positive or negative feedbacks (Grimm et al., 2013). In California's Sierra Nevada, for example, projected decreases in tree species richness associated with warming climate and increased wildfire are likely to cause more of the forested area to change to a carbon source, rather than sink, thus reducing the landscape-level carbon sequestration potential (Liang et al., 2017). In the tropics, woody plants store ~230 Pg C in their aboveground living biomass, but Brando et al. (2019) suggest that rates of carbon sequestration have decreased in recent decades as a result of droughts and wildfires. As the intensity and frequency of droughts and fires increase, the severe and repeated disturbances may exceed the ability of tropical forests to recover, causing declining productivity and carbon stocks (Brando et al., 2019).

Wildfires can also enhance dust emissions (e.g., Sankey et al., 2009; Hahnenberger and Nicoll, 2014). Changes in land cover/land use and warming climate have together created increased global fluxes of dust (Tegen et al., 2004; Webb & Pierre, 2018). Dust in this context refers to silt and clay-sized particles transported by wind. Increasing dust deposition is of concern because of multiple secondary effects. Dust darkens the surface of glaciers, icefields, and seasonal snow accumulations, allowing the ice and snow to absorb more solar energy and melt faster (Oerlemans et al., 2009; Skiles and Painter, 2017; Sarangi et al., 2020). Dust can include wind-borne fungal spores that cause infection and death in humans (Zhang et al., 2016). Dust can include wind-borne fungal spores that cause infection and death in humans (Zhang et al., 2016). Dust can increase nutrient and carbon inputs to soils, rivers, lakes, and the oceans, affecting soil and water chemistry, microbial populations, and higher levels of the food web (Goudie, 2009; Lequy et al., 2012; Psenner, 1999; Reche et al., 2009).

Large wildfires reduce vegetation cover and soil moisture, creating at least a temporarily increased potential for silt and clay to be eroded from the ground surface by the wind. Satellite measurements of active fires, aerosol abundance, vegetation cover, and soil moisture from 2003 to 2020 demonstrate that enhanced dust emission followed 54% of the examined ~150,000 global large wildfires (Yu & Ginoux, 2022). The occurrence and duration of post-fire dust emissions primarily reflect the extent of wildfire and resultant vegetation changes, as well as pre-fire drought conditions (Yu & Ginoux, 2022).

5.2.1. Prediction and Mitigation of the Hazard of Fire

The primary approaches to attempting to prevent uncontrollable wildfires from occurring are the fuel treatments of thinning and prescribed burning. During mechanical thinning, fuels are physically removed from fire-prone ecosystems by machine or by hand. Prescribed burns are deliberately set, low-intensity fires.

By the end of the 20th century, the effects of more than 50 years of fire suppression in the United States produced massive fuel accumulations and increased wildfire risks across large scales (Busenberg, 2004; Carle, 2002). Learning and adaptation during this period focused on wildland firefighting, including the use of airplanes and helicopters to detect fires, deliver wildland firefighters into remote areas, and drop water and fire-retardant chemicals on fires (Busenberg, 2004). Road networks in national forests expanded to facilitate ground access by firefighters and approaches such as the Incident Command System facilitated interagency coordination of firefighting resources and personnel (Busenberg, 2004). Wildland fuel reduction did not keep pace, partly because the U.S. Forest Service was resistant to controlled burns due to the risk of these fires escaping control and causing unintended damages, even though more than 99% of prescribed fires are successfully held within planned perimeters (Dether & Black, 2006). As Ryan et al. (2013) note, prescribed burns that go well get little public attention, but escaped burns can create enormous consequences for future prescribed burning.

During the period 1968–1978, the U.S. National Park Service and the U.S. Forest Service enacted new policies to allow the widespread use of prescribed fire for fuel reduction (Pyne, 1982; van Wagtendonk, 1991). These new policies have not yet adequately counteracted the hazardous accumulations of fuel in American wildlands, however, partly because. Negative public and political reactions to prescribed burns that got out of control or air-quality concerns associated with smoke have led both to repeated suspensions of federal prescribed burning programs and reluctance to use prescribed burns during periods of high temperatures, high winds or drought. (Busenberg, 2004; Ryan et al., 2013).

Mechanical thinning has fewer publicly perceived negative side effects but requires enormous levels of personnel and equipment to implement widely. Mechanical thinning also generates substantial loads of organic materials that must somehow be disposed of lest they increase fuel loads and the risk of future fires (Busenberg, 2004).

In a review and meta-analysis of the effects of forest thinning and burning treatments in western United States pine forests, Fulé et al. (2012) note that decades of fire suppression have altered some forests to the point where the surface fires used in prescribed burns are insufficient to reduce dense stands (Miller & Urban, 2000). Fulé et al. (2012) find that combined thinning and burning most effectively reduce surface fuels and stand density relative to either just thinning or just burning.

Reviewing prescribed burns, Ryan et al. (2013) highlight some of the differences between frequent, low-severity natural fires, and prescribed burns. Natural fires burn in a patchy mosaic, leaving unburned refugia for fire-sensitive plants and small animals with limited mobility. Increased forest density and accumulation of litter and duff in areas with a history of fire suppression create a more continuous, uniformly flammable fuel-bed (Knapp & Keeley, 2006), so that prescribed burns leave few refugia. Prescribed fires are commonly ignited in linear strips or at multiple points along regular grids and this uniform ignition also produces more uniform burns with fewer residual unburned patches. Prescribed burns are likely to be conducted in different seasons and under higher moisture conditions than natural fires. This limits their fuel consumption and ability to kill encroaching plant communities not adapted to fire, leading to shifts in species distribution. Ryan et al. (2013) also discuss how prescribed burns are conducted within the framework of multiple environmental laws, such as the Clean Air Act that may limit smoke emissions and the Endangered Species Act, which requires managers to analyze immediate, short-term risks associated with prescribed burning but not the long-term risks associated with lack of fire. The end result is that prescribed burns typically do not create the same type of environmental disturbance as "natural" fires to which native organisms may be adapted and resilient.

Ager et al. (2015) describe how fuel reduction activities may be conducted at mismatched scales because of the different spatial and temporal scales at which biophysical and social processes operate (Guerrero et al., 2013; Kondolf & Podolak, 2014). Individuals and communities plan at local scales and act to reduce ignition risk nearby (Jakes et al., 2011), whereas wildfires may ignite in a location distant from the community and then burn across large areas into the community (Ager et al., 2012; Williams, 2013). Ager et al. (2015) instead propose planning mitigation activities at the scale of a 'fireshed' that includes ignition locations that have or are reasonably capable of transmitting fire to a community.

5.3. Post-fire Hazards

Hazards present after a wildfire include the effects of accelerated fluxes from uplands and through river networks—floods, debris flows, and degraded water quality. These hazards can be addressed through prediction and mitigation. Both prediction and mitigation can target the hazardous processes, the characteristics of infrastructure, or human perceptions and behavior.

5.3.1. Prediction and Mitigation of Post-fire Floods, Debris Flow, and Degraded Water Quality

The accelerated fluxes from uplands and through river networks in the days to years following wildfire create post-fire hazards of hillslope mass movements (Nyman et al., 2011), flooding, channel movement, and degraded water quality and water storage (Hallema et al., 2018; Murphy et al., 2018; Belongia et al., 2023). These hazards can affect irreplaceable archeological sites (e.g., Ortega-Becerril et al., 2024), as well as contemporary infrastructure.

Assessment of potential hazards from accelerated fluxes typically starts with a post-fire assessment of burn spatial extent and severity. In the United States, a formal protocol developed by the U.S. Forest Service and known as the Burn Area Emergency Response (BAER) is used when fires occur on public lands (Hudak et al., 2007). BAER deploys teams of experts to make an emergency assessment of any burned area covering more than 120 ha and to recommend stabilization or mitigation measures.

The process starts by identifying, mapping, and field-verifying soil burn severity based on the loss of soil and aboveground organic matter (Keeley, 2009). This can be done using aerial reconnaissance or satellite imagery, with satellite imagery typically used to produce a Burn Area Reflectance Classification (BARC) (Bernkopf et al., 2021). BARC is derived from an image transformation algorithm known as the normalized burn ratio, which uses near-infrared and short-wave infrared bands from Landsat satellite sensors (Clark, 2013; Clark & Bobbe, 2007). The BARC data layer is field-validated or adjusted to match ground conditions and then used with topography, soil type, and pre-fire vegetation data in subsequent modeling of the post-fire hydrologic response of the watershed (Clark, 2013). In the United States, these resources are posted on a public site (https://mtbs.gov), which makes it much easier to obtain burn severity and wildfire perimeter data. Those involved in post-fire work have developed simplified user interfaces for widely used models such as the Water Erosion Prediction

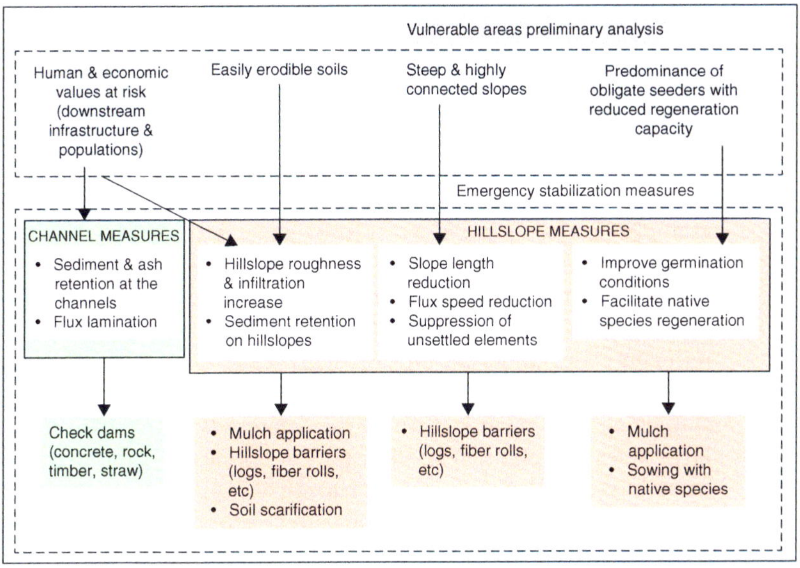

Figure 5.4 Schematic illustration of relations between emergency stabilization measures, risks, and objectives, as well as types of channel and hillslope mitigation measures. (Lucas Borja, 2021 / with permission of ELSEVIER.)

Project (WEPP; Flanagan & Livingston, 1995) and limited needed user inputs to a few essential variables (Robichaud & Ashmun, 2013).

At the upstream end of the post-fire sediment cascade, municipalities and government agencies commonly attempt to minimize sediment movement from uplands and damage associated with fluxes through river networks using one or more approaches (Figure 5.4) that include:

- aerial spreading of mulch, which can be straw, wood chips, bark strands, and other materials (Fernandez & Vega, 2014; Groen & Woods, 2008; Robichaud et al., 2014);
- aerial spreading of seeds to accelerate regrowth of upland vegetation (Peppin et al., 2011; Pyke et al., 2013);
- felling trees so that they lie along slope contours and can trap sediment moving downslope, increase infiltration, and favor plant regeneration (Lucas-Borja, 2021; Robichaud et al., 2008; Adams et al., 2023);
- installing other structures along slope contours, such as straw wattles and fiber rolls (Groen & Woods, 2008; Wohlgemuth et al., 2007), or trenching along contours (Neary, 2009);

- installing barriers to mitigate rockfall hazard above high-value infrastructure such as roads and homes (DeGraff & Gallegos, 2007);
- scarification and ripping of hillslope surfaces to promote infiltration (Neary, 2009);
- installing detention basins or sediment traps using straw bales, rock, or log check dams along small ephemeral channels (Kastridis et al., 2022; Lucas-Borja, 2021; Robichaud et al., 2014);
- protecting transportation corridors by enlarging or armoring culverts and ditches and installing trash racks and debris fences (Clark, 2013; Neary, 2009);
- reducing road and trail erosion with water bars and enhanced drainage (Wohlgemuth et al., 2007); and
- stabilizing stream banks with various materials from partly buried large wood or planted woody riparian species to gabions (wire-mesh baskets filled with rocks) (Haring et al., 2021).

These emergency stabilization measures are cumulatively very expensive and come with diverse secondary costs. Following the 2000 and 2001 fire seasons, for example, the U.S. Department of the Interior and the U.S. Forest Service spent US$118 million and US$182 million, respectively, for post-fire rehabilitation and restoration (Groen & Woods, 2008).

Mulching via helicopter drops, using blowers, or by hand can protect the soil by providing cover that reduces raindrop impact, prevents soil sealing, promotes infiltration, and slows runoff. Mulching can be effective in reducing post-fire soil erosion (e.g., Robichaud et al., 2013; Wagenbrenner et al., 2006) and can promote vegetation establishment by retaining soil moisture in dry sites (Peterson et al., 2009). At some sites, however, mulch can inhibit native seed germination (Robichaud, 2005) and vegetation regrowth (Dodson & Peterson, 2010) and introduce exotic species (Kruse et al., 2004). Straw mulch is easily transported by wind, leaving bare areas, but can act more quickly to reduce erosion during the first year than aerial seeding. Although most state and federal agencies in the United States require certified weed-free straw for post-fire mulching, I have observed such straw giving rise to extensive stands of the invasive species cheatgrass (*Bromus tectorum*) in the Southern Rockies of Colorado. Plants such as cheatgrass, buffelgrass (*Pennisetum ciliare*), and red brome (*Bromus rubens*), which may be present as contaminants in weed-free straw mulch, respond rapidly to fire, out-compete native grasses, and then alter the fire regime toward more frequent, hotter fires (Neary, 2009). This promotes desertification in burned areas (Neary, 2009).

Aerial seeding is relatively low cost and easy to apply, but its effectiveness depends on the seed density and the amount and timing of rainfall after seed application (Wagenbrenner et al., 2006). Too much rain can wash seeds off the slopes and too little rain can limit germination (Robichaud et al., 2000). Consequently, many grass-seeding treatments are ineffective, particularly in the first year after fire when burned slopes are especially vulnerable (Groen & Woods, 2008; Peppin et al., 2011; Pyke et al., 2013).

Seeding can also inhibit natural vegetation recovery because non-native plant seeds have been widely used (Neary, 2009; Peppin et al., 2011). Keeley (2004) describes a burn area in California, USA, which was seeded with a non-persistent variety of wheat. Although this enhanced ground cover in the first growing season, wheat was the dominant species and corresponded to a decrease in species richness relative to unseeded areas and suppression of post-fire endemic species whose lifecycle is restricted to immediate post-fire environments. The wheat seeding also caused an order-of-magnitude decline in native *Pinus ponderosa* seedling recruitment and was likely to continue inhibiting seedling recruitment for years because it produced thick layers of thatch (Keeley, 2004). (Thatch is a tightly intermingled layer of living and dead stems, leaves, and roots, which accumulates between soil and actively growing plants.)

In the five or so years following a wildfire, upland rehabilitation and restoration activities tend to focus on afforestation with native plant species and fuel management. These can be undertaken in part to counter post-fire changes in vegetation distribution and community composition caused by emergency stabilization measures that create water and sediment disconnectivity in uplands (Lucas-Borja, 2021).

Within river corridors, efforts to increase water and sediment retention and to stabilize channel banks include the use of diverse materials and structures, including human-built analogs of beaver dams and engineered logjams that are anchored or unanchored. These structures can be very effective at attenuating downstream post-fire fluxes of materials, especially if the structures are in portions of the river corridor with adjustment space (e.g., the potential for the active channel to migrate laterally around a retention structure once it fills with sediment) and built in series with multiple structures along a length of channel or floodplain (Figure 5.5). The structures may be less effective if they are in portions of a river network dominated by transport (e.g., steep, laterally confined river segments), unless the primary purpose of the structure is to dissipate flow energy. Structures should be sized and spaced based on reasonable expectations of post-fire sediment yields (Robichaud et al., 2019) and must also be designed so that they do not create sediment depletion in downstream reaches or limit movement of aquatic organisms (Willardson, 2020).

(a)

(b)

Figure 5.5 Examples of instream retention structures installed after fire. (a) Image from Little Beaver Creek, Colorado, USA, taken during snowmelt peak flow, includes a single, post-assisted log structure designed to enhance overbank flow and sediment deposition. Channel is approximately 20m wide. (b) Image from Black Hollow Creek, Colorado, taken during intermediate flow levels, includes multiple rock check dams designed to retain sediment in the lower, laterally confined portion of the channel. These check dams, which began with 4-m deep pools, filled within days of installation as seasonal snowmelt peak flows began to rise during the first year after the wildfire. Arrows indicate flow direction.

At the downstream end of the post-fire sediment cascade, sedimentation in reservoirs can damage critical infrastructure for water resources by reducing reservoir storage capacity, causing intake valves to shut down (Warziniack & Thompson, 2013), and can supply excess nutrients that cause harmful algal blooms, unpleasant taste in drinking-water supplies, and toxic disinfection byproducts (Hohner et al., 2016; Meixner & Wohlgemuth, 2004; Smith et al., 2011). Dredging sediment from reservoirs, treating water, and undertaking capital improvements to pipes, pumps, and water filtration to maintain water supplies represent major costs post-fire (Kinoshita et al., 2016), so most mitigation related to reservoirs and water quality focuses on keeping sediment from reaching reservoirs and water intakes by stabilizing uplands and small channels.

5.3.2. Prediction and Mitigation of Human Perceptions and Behavior Related to Hazards

To state the obvious, fire-related hazards occur because people and their infrastructure are in the path of the fire or the post-fire fluxes of material. Consequently, management of humans and their infrastructure is a key component of reducing fire-related hazards.

Integrated fire hazard assessments may combine GIS-based analyses of fire ignition and spread with risk response models that evaluate how easily people respond to and potentially suppress fire. Human responses are evaluated based on factors such as location of fire with respect to settlements and resources such as firefighters, communications, and transport infrastructure available in the area (e.g., Asori et al., 2020; Oliveira et al., 2021).

Although physical processes and the functioning of infrastructure may include unexpected components that are difficult to predict, human perceptions and behavior are the proverbial "wild card" in such analyses. As Kumagai et al. (2004) express it, "... people tend to be unduly optimistic in the face of environmental hazards such as fire and ... the characteristics of the affected community are at least as important as those of the fire in understanding social impacts" (p. 28).

Kumagai et al. (2004) explain how people may accept recommendations on preparing for potential hazards but may also ignore recommendations that they perceive to be encroaching on their freedom. People repeatedly told to evacuate areas at risk during each storm following a wildfire may also get warning fatigue and stop evacuating if no debris flows or flash floods have occurred following past evacuation warnings (LeClerc and Joslyn, 2015; McLennan et al., 2019). Individuals may not be as personally concerned if they perceive risk as being handled appropriately by government agencies but may then be angry and feel betrayed when authorities do not successfully ward off hazards. Experience with past natural hazards such as fire can also create a false sense of security if individuals expect the return interval of the natural hazard to be beyond their lifetime or if the hazard did not directly affect the individual (Kumagai et al., 2004).

Individuals or communities with a different history of interaction with the surrounding environment may also perceive and respond to wildfires differently (Paveglio & Edgeley, 2017). Examining uneven levels of preparation among landowners in at-risk fire areas in southeastern Australia, Eriksen and Prior (2011) note that the diversity of people migrating into the WUI leads to a community with widely varying experiences, beliefs, attitudes, and values relating to fire. Drawing on Kolb's experiential learning theory (Kolb, 1984), in which knowledge is a process rather than a product, Eriksen and Prior (2011) emphasize the importance of engagement with communities rather than traditional, unidirectional risk communication. Successful community outreach programs build on existing local knowledge of diverse community members while simultaneously adapting to better reflect changing social, cultural, environmental, and economic needs (Eriksen & Prior, 2011).

Of perhaps greatest importance is the phenomenon demonstrated by human reaction to many hazards, from cigarette smoking to Covid vaccines to climate change: people are not simply passive receptors of risk information but instead play a critical interactive role in the process of risk communication (Fessenden-Raden et al., 1987). This could be paraphrased as "Just because you say it doesn't mean I believe it." Social scientists find that people are more likely to believe risk information that is specific, consistent, certain, and repeatedly disseminated, although the source of the information is also important (Kumagai et al., 2004). Examining community response to two catastrophic wildfires in Colorado, USA, Koebele et al. (2015) highlight the importance of "citizen entrepreneurs"—highly motivated community members who can help resource-constrained wildfire agencies encourage mitigation on private property by directly engaging with residents in the WUI.

Roncoli et al. (2012) provide an example of how improving the effectiveness of wildfire risk forecast evolved iteratively based on consultations with residents in the region. Scientists developed a seasonal climate forecast that integrates the mean monthly values of a drought index and communicated this information in a visual representation that can be easily understood. The representation takes the form of a single interactive map on a website with small, color-coded circles that indicate probabilities of abnormally dry conditions (fire weather) at multiple sites across a region.

The map is updated regularly and includes categories of abnormally, moderately, severely, and extremely dry.

Wildland fires are in some sense a hybrid natural hazard in that they can be viewed as a natural disaster—which tends to bring communities together to face a shared threat—or as a human-caused disaster—which is more likely to lead to divisive assignment of blame—because of the influence on the fire of resource management actions such as fire suppression, controlled burns, and "fire-safeing" of residential properties (Kumagai et al., 2004). Schumann et al. (2020) propose that the period following a destructive wildfire may be a "hot moment" for community adaptation and notes that changes during postfire recovery will shape a community's vulnerability to future fires.

Regardless of how individuals or communities perceive wildfire, the costs associated with wildfire damage and management continue to increase substantially in the United States. (Comparable data from Australia and from Europe, including fire-prone Mediterranean countries, are incomplete or missing; Mattioli et al., 2022.) This partly reflects climate change toward warmer and drier conditions in much of the fire-prone region of the western United States, along with the continuing effects of past fire suppression and the trend toward increasing human population in the exurban areas known as the WUI or the red zone (Calkin et al., 2015). The WUI in the western United States expanded over 60% between 1970 and 2007 (Theobald & Romme, 2007), for example, and about 70% of this land is in private ownership (Schoennagel et al., 2009). Wildfire-related insured losses in the United States during 2002–2011 totaled US$7.9 billion, which represents a $6.2 billion increase over the previous decade (Haldane, 2013). During 2013-2022, these losses rose to $98.87 billion globally and $62.92 billion insured losses globally (Swiss Re Group, 2023). Considering that climate change, population shifts, and wildfire suppression all continue, wildfire-related costs are likely to also continue rising into the future. The absence of integrated frameworks for addressing wildfire with respect to ecosystem processes and risk management for human communities creates similar trends of increasingly damaging fires in parts of the European Union and Australia (Moritz et al., 2014).

5.4. Resilience

Resilience is now a word widely used in many different contexts. Wildfire is certainly one of those contexts, both because fires are increasing in frequency, extent, and severity in forested environments (but not grasslands and savannas; Andela et al., 2017) and because—despite years of investments to support ecological and socioeconomic resilience to wildfires—impacts to people, property, and ecosystems continue to increase (McWethy et al., 2019). Between 2003 and 2012, for example, over 34,000 homes were destroyed by wildfires in the United States and suppression costs ranged from US$1 to 2 billion per year (Ager et al., 2015).

The concept of resilience has broadened from ecological resilience (Holling, 1973) to individual, societal, and other forms of resilience. Basic resilience describes return to a state or conditions similar to those before the disturbance (Pelling, 2012). Adaptive resilience describes how human communities or social systems adapt to new or dynamic conditions by changing fundamental characteristics of the system, such as zoning land use to limit fire hazards (Olsson et al., 2015; Walker et al., 2004). Transformative resilience describes the creation of fundamentally new systems (McDonald et al., 2016; McWethy et al., 2019). With respect to wildfire, communities can emphasize fire-resistant structures via building codes, requirements for building materials, and creation of buffer zones between forest and infrastructure, and then combine these with resilience measures such as establishing safe zones, identifying evacuation routes, and reducing fuel loads in adjacent forested areas (Tedim et al., 2020).

Resilience planning can be expressed as three stages of action. Before fire outbreak, resilience measures include fuel reduction, building codes, zoning, emergency planning, and regulations such as campfire or fireworks bans during specified periods. During the fire, resilience measures include crisis communication such as evacuation warnings, firefighting activities, and addressing community needs (e.g., safe zones, evacuation routes). After the fire, resilience measures include addressing needs for individual and community recovery, hazard mitigation, reconstruction of infrastructure, and ecosystem recovery (Tedim et al., 2020).

5.5. Summary

Human attitudes toward wildfire have varied substantially through time as people keep trying to find some achievable balance with fire. Many nomadic or migratory prehistoric people adapted fire as a tool for manipulating biotic communities. Settled communities that developed subsequently were more likely to employ rigorous fire suppression to reduce wildfire risks. Contemporary human communities in fire-prone areas use a varying combination of fire suppression, fuels reduction via thinning and prescribed burns, mitigation pre- and post-fire, and adaptation. All of these activities occur in a rapidly warming climate in which the changing characteristics of wildfires require adaptability from human communities.

Human communities affected by wildfires can also be hundreds or even thousands of kilometers away from the actual burn zone because of emissions of wildfire smoke that includes fine particulate organic matter, black carbon, and CO_2.

Economic limitations such as rising insurance costs in the WUI may help to slow human migration into areas with high fire risk. Nonetheless, the costs of preparing for fire, attempting to suppress ongoing fires, and dealing with post-fire hazards will likely continue to rise in future as global human population increases and as people continue to choose the many lifestyle-amenities of living in the WUI.

References

Adetona, O., Reinhardt, T. E., Domitrovich, J., Broyles, G., Adetona, A. M., Kleinman, M. T., et al. (2016). Review of the health effects of wildland fire smoke on wildland firefighters and the public. *Inhalation Toxicology, 28*, 95–139.

Ager, A. A., Kline, J. D., & Fischer, A. P. (2015). Coupling the biophysical and social dimensions of wildfire risk to improve wildfire mitigation planning. *Risk Analysis, 35*, 1393–1406.

Ager, A. A., Vaillant, N. M., Finney, M. A., & Preisler, H. K. (2012). Analyzing wildfire exposure and source–sink relationships on a fire prone forest landscape. *Forest Ecology and Management, 267*, 271–283.

Aguilera, R., Corringham, T., Gershunov, A., & Benmarhnia, T. (2021). Wildfire smoke impacts respiratory health more than fine particles from other sources: Observational evidence from southern California. *Nature Communications, 12*, 1493.

Anderson, S., Barford, C., & Barford, P. (2020). *Five alarms: Assessing the vulnerability of US cellular communication infrastructure to wildfires* (pp. 162–175). IMC'20, Association for Computing Machinery.

Asori, M., Emmanuel, D., & Dumedah, G. (2020). Wildfire hazard and risk modelling in the northern regions of Ghana using GIS-based multi-criteria decision making analysis. *Journal of Environment and Earth Science, 10*, 11–28.

Bernkopf, R., Kuwayama, Y., Gibson, R., Blakely, J., Mabee, B., Clifford, T. J., et al. (2021). Monetising the savings of remotely sensed data and information in burn area emergency response (BAER) wildfire assessment. *International Journal of Wildland Fire, 30*, 18–29.

Black, C., Tesfaigzi, Y., Bassein, J. A., & Miller, L. A. (2017). Wildfire smoke exposure and human health: Significant gaps in research for a growing public health issue. *Environmental Toxicology and Pharmacology, 55*, 185–195.

Bond, T. C., Streets, D. G., Yarber, K. F., Nelson, S. M., Woo, J.-H., & Klimont, Z. (2004). A technology-based global inventory of black and organic carbon emissions from combustion. *Journal of Geophysical Research, 109*, D14203. http://dx.doi.org/10.1029/2003JD003697

Bowman, D. M. J. S., & Johnston, F. H. (2005). Wildfire smoke, fire management, and human health. *EcoHealth*, *2*, 76–80.

Bowman, D. M. J. S., Balch, J., Artaxo, P., Bond, W. J., Cochrane, M. A., D'Antonio, C. M., et al. (2011). The human dimension of fire regimes on Earth. *Journal of Biogeography*, *38*, 2223–2236.

Bowman, D. M. J. S., Kolden, C. A., Abatzoglou, J. T., Johnstons, F. H., van der Werf, G. R., & Flannigan, M. (2020). Vegetation fires in the Anthropocene. *Nature Reviews Earth and Environment*, *1*, 500–515.

Bowman, D. M. J. S, Williamson, G. J. et al. (2017). Human exposure and sensitivity to globally extreme wildfire events. *Nature Ecology and Evolution*, *1*, 0058.

Brando, P. M., Paolucci, L., Ummenhofer, C. C., Ordway, E. M., Hartmann, H., Cattau, M. E., et al. (2019). Droughts, wildfires, and forest carbon cycling: A pantropical synthesis. *Annual Review of Earth and Planetary Sciences*, *47*, 555–581.

Burkhardt, J., Bayham, J., Wilson, A., Berman, J. D., O'Dell, K., Ford, B., et al. (2020). The relationship between monthly air pollution and violent crime across the United States. *Journal of Environmental Economics and Policy*, *9*, 188–205. http://dx.doi.org/10.1080/21606544.2019.1630014

Busenberg, G. J. (2004). Adaptive policy design for the management of wildfire hazards. *American Behavioral Scientist*, *48*, 314–326.

Calkin, D. E., Thompson, M. P., & Finney, M. A. (2015). Negative consequences of positive feedbacks in US wildfire management. *Forest Ecosystems*, *2*, 9.

Carle, D. (2002). *Burning questions: America's fight with nature's fire*. NY: Praeger.

Cascio, W. E. (2018). Wildland fire smoke and human health. *Science of the Total Environment*, *624*, 586–595.

Clark, J. (2013). Remote sensing and geospatial support to burned area emergency response (BAER) teams in assessing wildfire effects to hillslopes. In C. Margottini, et al. (Eds.), *Landslide science and practice* (Vol. 4, pp. 211–215). Berlin: Springer-Verlag.

Clark, J., & Bobbe, T. (2007). Using remote sensing to map and monitor fire damage in forest ecosystems. In M. A. Wulder & S. E. Franklin (Eds.), *Understanding forest disturbance and spatial patterns: Remote sensing and GIS approaches* (p. 246). London: Taylor & Francis.

CRED (Centre for Research on the Epidemiology of Disaster). (2018). *Economic losses, poverty and disasters 1998–2017* (p. 31). Louvain, Belgium: United National Office for Disaster Risk Reduction.

DeGraff, J. V., & Gallegos, A. J. (2007). The challenge of improving identification of rockfall hazard after wildfires. *Environmental and Engineering Geoscience*, *18*, 389–397.

Denevan, W. M. (1992). The pristine myth: The landscape of the Americas in 1492. *Annals of the Association of American Geographers*, *82*, 369–385.

Dether, D., & Black, A. (2006). Learning from escaped prescribed fires – Lessons for high reliability. *Fire Management Today*, *66*, 50–56.

D'Evelyn, S. M., Jung, J., Alvarado, E., Baumgartner, J., Caligiuri, P., Hagmann, R. K., et al. (2022). Wildfire, smoke exposure, human health, and environmental justice need to be integrated into forest restoration and management. *Current Environmental Health Reports*, *9*, 366–385.

Diaz, J. M. (2012). Economic impacts of wildfire. *Southern Fire Exchange, SFE Fact Sheet 2012-7*. Joint Fire Science Program.

Dickinson, G. N., Miller, D. D., Bajracharya, A., Bruchard, W., Durbin, T. A., McGarry, J. K. P., et al. (2022). Health risk implications of volatile organic compounds in wildfire smoke during the 2019 FIREX-AQ campaign and beyond. *GeoHealth*, *6*, e2021GH000546. http://dx.doi.org/10.1029/2021GH000546

Dodson, E. K., & Peterson, D. W. (2010). Mulching effects on vegetation recovery following high severity wildfire in north-central Washington state, USA. *Forest Ecology and Management*, *260*, 1816–1823.

Eisenman, D., McCaffrey, S., Donatello, I., & Marshall, G. (2015). An ecosystems and vulnerable populations perspective on solastalgia and psychological distress after a wildfire. *EcoHealth*, *12*, 602–610. http://dx.doi.org/10.1007/s10393-015-1052-1

Eriksen, C., & Prior, T. (2011). The art of learning: Wildfire, amenity migration and local environmental knowledge. *International Journal of Wildland Fire*, *20*, 612–624.

Fernandez, C., & Vega, J. A. (2014). Efficacy of bark strands and straw mulching after wildfire in NW Spain: Effects on erosion control and vegetation recovery. *Ecological Engineering*, *63*, 50–57.

Fessenden-Raden, J., Fitchen, J. M., & Heath, J. S. (1987). Providing risk information in communities: Factors influencing what is heard and accepted. *Science, Technology, and Human Values*, *12*, 94–101.

Finney, M. A. (2005). The challenge of quantitative risk assessment for wildland fire. *Forest Ecology and Management*, *211*, 97–108.

Fischer, A. P., Spies, T. A., Steelman, T. A., Moseley, C., Johnston, B. R., Bailey, J. D., et al. (2016). Wildfire risk as a socioecological pathology. *Frontiers in Ecology and the Environment*, *14*, 276–284.

Flanagan, D. C., Livingston, S. J. (eds.) (1995). WEPP user summary. US Department of Agriculture, Agricultural Research Service, National Soil Erosion Research Laboratory, NSERL Report No.11. West Lafayette, IN.

Flanner, M. G., Zender, C. S., Randerson, J. T., & Rasch, P. J. (2007). Present-day forcing and response from black carbon in snow. *Journal of Geophysical Research*, *112*, D11202. http://dx.doi.org/10.1029/2006JD008003

Ford, B., Martin, M. V., Zelasky, S. E., Fischer, E. V., Anenberg, S. C., Heald, C. L., & Pierce, J. R. (2018). Future fire impacts on smoke concentrations, visibility, and health in the contiguous United States. *GeoHealth*, *2*, 229–247.

French, N. H. F., de Groot, W. J., Jenkins, L. K., Rogers, B. M., Alvarado, E., Amiro, B., et al. (2011). Model comparisons for estimating carbon emissions from North American wildland fire. *Journal of Geophysical Research Biogeosciences*, *116*, G00K05. http://dx.doi.org/10.1029/2010JG001469

Fulé, P. Z., Crouse, J. E., Roccaforte, J. P., & Kalies, E. L. (2012). Do thinning and/or burning treatments in western USA ponderosa or Jeffrey pine-dominated forests help restore natural fire behavior? *Forest Ecology and Management*, *269*, 68–81.

Goudie, A. S. (2009). Dust storms: recent developments. *Journal of Environmental Management, 90*, 89–94. http://dx.doi.org/10.1016/j.jenvman.2008.07.007

Grimm, N. B., Chapin, F. S., Bierwagen, B., Gonzalez, P., Groffman, P. M., Luo, Y., et al. (2013). The impacts of climate change on ecosystem structure and function. *Frontiers in Ecology and the Environment, 11*, 474–482.

Groen, A. H., & Woods, S. W. (2008). Effectiveness of aerial seeding and straw mulch for reducing post-wildfire erosion, north-western Montana, USA. *International Journal of Wildland Fire, 17*, 559–571.

Guerrero, A. M., McAllister, R. R. J., Corcoran, J., & Wilson, K. A. (2013). Scale mismatches, conservation planning, and the value of social network analyses. *Conservation Biology, 27*, 34–44.

Habermann, N., & Hedel, R. (2018). Damage functions for transport infrastructure. *International Journal of Disaster Resilience in the Built Environment, 9*, 420–434.

Haldane, M. (2013). Insurers, government grapple with costs of growth in wildland urban interface. *Insurance Journal*. Available at: http://www.insurancejournal.com/news/national/2013/08/15/301833.htm.

Hardy, C. C. (2005). Wildland fire hazard and risk: Problems, definitions, and context. *Forest Ecology and Management, 211*, 73–82.

Haring, C. P., Altmann, G. L., Suedel, B. C., & Brown, S. W. (2021). Using engineering with nature (EWN) principles to manage erosion of watersheds damaged by large-scale wildfires. *Integrated Environmental Assessment and Management, 17*, 1194–1202.

Hayden, K. L., Li, S. M., Liggio, J., Wheeler, M. J., Wentzell, J. J. B., Leithead, A., et al. (2022). Reconciling the total carbon budget for boreal forest wildfire emissions using airborne observations. *Atmospheric Chemistry and Physics, 22*, 12493–12523. http://dx.doi.org/10.5194/acp-22-12493-2022

Hohner, A. K., Cawley, K., Oropeza, J., Summers, R. S., & Rosario-Ortiz, F. L. (2016). Drinking water treatment response following a Colorado wildfire. *Water Research, 105*, 187–198.

Holling, C. S. (1973). Resilience and stability of ecological systems. *Annual Review of Ecology and Evolutionary Systems, 4*, 1–23.

Hudak, A. T., Morgan, P., Bobbitt, M. J., Smith, A. M. S., Lewis, S. A., Lentile, L. B., et al. (2007). The relationship of multispectral satellite imagery to immediate fire effects. *Fire Ecology, 3*, 64–90.

IPCC. (2007). Climate Change 2007: Synthesis Report. In R. K. Pachauri & A. Reisinger (Eds.), *Contribution of Working Groups I, II and III to the Fourth Assessment Report of the Intergovernmental Panel on Climate Change*. Geneva, Switzerland.

Jakes, P. J., Nelson, K. C., Enzler, S. A., Burns, S., Cheng, A. S., Sturtevant, V., et al. (2011). Community wildfire protection planning: is the Healthy Forests Restoration Act's vagueness genius? *International Journal of Wildland Fire, 20*, 350–363. https://doi.org/10.1071/WF10038

Johnston, F. H., Henderson, S. B., Chen, Y., Randerson, J. T., Marlier, M., Defries, R. S., et al. (2012). Estimated global mortality attributable to smoke from landscape fires. *Environmental Health Perspectives, 120*, 695–701.

Kaspari, S., Skiles, S. M., Delaney, I., Dixon, D., & Painter, T. H. (2015). Accelerated glacier melt on snow dome, Mount Olympus, Washington, USA, due to deposition of black carbon and mineral dust from wildfire. *Journal of Geophysical Research: Atmospheres,120*,2793–2807.http://dx.doi.org/10.1002/2014JD022676

Kastridis, A., Margiorou, S., & Sapountzis, M. (2022). Check-dams and silt fences: Cost-effective methods to monitor soil erosion under various disturbances in forest ecosystems. *Land*, *11*, 2129.

Keeley, J. E. (2004). Ecological impacts of wheat seeding after a Sierra Nevada wildfire. *International Journal of Wildland Fire*, *13*, 73–78.

Keeley, J. (2009). Fire intensity, fire severity and burn severity: A brief review and suggested usage. *International Journal of Wildland Fire*, *18*, 116–126. http://dx.doi.org/10.1071/WF07049

Kinoshita, A. M., Chin, A., Simon, G. L., Briles, C., Hogue, T. S., O'Dowd, A. P., et al. (2016). Wildfire, water, and society: Toward integrative research in the "Anthropocene". *Anthropocene*, *16*, 16–27.

Knapp, E. E., & Keeley, J. E. (2006). Heterogeneity in fire severity with early season and late season prescribed burns in a mixed conifer forest. *International Journal of Wildland Fire*, *15*, 37–45.

Koebele, E., Crow, D. A., Lawhon, L. A., Kroepsch, A., Schild, R., & Clifford, K. (2015). Wildfire outreach and citizen entrepreneurs in the wildland-urban interface: A cross-case analysis in Colorado. *Society & Natural Resources*, *28*, 918–923. http://dx.doi.org/10.1080/08941920.2015.1054975

Kolb, D. A. (1984). *Experiential learning: Experience as the source of learning and development*. Englewood Cliffs, NJ: Prentice-Hall.

Kondolf, G. M., & Podolak, K. (2014). Space and time scales in human-landscape systems. *Environmental Management*, *53*, 76–87.

Koshkin, A. L., Hatchett, B. J., & Nolin, A. W. (2022). Wildfire impacts on western United States snowpacks. *Frontiers in Water*, *4*, 971271. http://dx.doi.org/10.3389/frwa.2022.971271

Kruse, R., Bend, E., & Bierzychudek, P. (2004). Native plant regeneration and introduction of non-natives following post-fire rehabilitation with straw mulch and barley seeding. *Forest Ecology and Management*, *196*, 299–310.

Kumagai, Y., Carroll, M. S., & Cohn, P. (2004). Coping with interface wildfire as a human event: Lessons from the disaster/hazards literature. *Journal of Forestry*, *102*, 28–32.

Kyrkilis, G., Chaloulakou, A., & Kassomenos, P. A. (2007). Development of an aggregate air quality index for urban Mediterranean agglomeration: Relation to potential health effects. *Environment International*, *33*, 670–676. http://dx.doi.org/10.1016/j.envint.2007.01.010

Lassman, W., Mirocha, J. D., Arthur, R. S., Kochanski, A. K., Caus, A. F., Bagley, A. M., et al. (2023). Using satellite-derived fire arrival times for coupled wildfire-air quality simulations at regional scales of the 2020 California wildfire season. *Journal of Geophysical Research: Atmospheres*, *128*, e2022JD037062. http://dx.doi.org/10.1029/2022JD037062

Lavorel, S., Flannigan, M. D., Lambin, E. F., & Scholes, M. C. (2006). Vulnerability of land systems to fire: Interactions among humans, climate, the atmosphere, and ecosystems. *Mitigation and Adaptation Strategies for Global Change*, *12*, 33–53. http://dx.doi.org/10.1007/S11027-006-9046-5

Lavoué, D., Liousse, C., Cachier, H., Stocks, B. J., & Goldammer, J. G. (2000). Modeling of carbonaceous particles emitted by boreal and temperate wildfires at northern latitudes. *Journal of Geophysical Research*, *105*, 26.871–26.890. http://dx.doi.org/10.1029/2000JD900180

Lequy, E., Conil, S., & Turpault, M. P. (2012). Impacts of aeolian dust deposition on European forest sustainability: A review. *Forest Ecology and Management*, *267*, 240–252. http://dx.doi.org/10.1016/j.foreco.2011.12.005

Liang, S., Hurteau, M. D., & Westerling, A. L. (2017). Response of Sierra Nevada forests to projected climate-wildfire interactions. *Global Change Biology*, *23*, 2016–2030. http://dx.doi.org/10.1111/gcb.13544

Liu, J. C., Pereira, G., Uhl, S. A., Bravo, M. A., & Bell, M. L. (2015). A systematic review of the physical health impacts from non-occupational exposure to wildfire smoke. *Environmental Research*, *136*, 120–132.

Lowe, S. R., & Garfin, D. R. (2023). Crisis in the air: The mental health implications of the 2023 Canadian wildfires. *The Lancet: Planetary Health*, *7*, e732–e733. http://dx.doi.org/10.1016/S2542-5196(23)00188-2

Lucas-Borja, M. E. (2021). Efficiency of postfire hillslope management strategies: Gaps of knowledge. *Current Opinion in Environmental Science and Health*, *21*, 100247.

Lüthi, S., Aznar-Siguan, G., Fairless, C., & Bresch, D. N. (2021). Globally consistent assessment of economic impacts of wildfires in CLIMADA v2.2. *Geoscientific Model Development*, *14*, 7175–7187.

Mao, J., Abshire, J. B., Kawa, S. R., Riris, H., Sun, X., Andela, N., & Kolbeck, P. T. (2021). Measuring atmospheric CO_2 enhancements from the 2017 British Columbia wildfires using a lidar. *Geophysical Research Letters*, *48*, e2021GL093805. http://dx.doi.org/10.1029/2021GL093805

Mattioli, W., Ferrara, C., Lombardo, E., Barbati, A., Salvati, L., & Tomao, A. (2022). Estimating wildfire suppression costs: A systematic review. *International Forestry Review*, *24*, 15–29. http://dx.doi.org/10.1505/146554822835224801

McCaffery, S., McGee, T. K., Coughlan, M., & Tedim, F. (2020). Understanding wildfire mitigation and preparedness in the context of extreme wildfires and disasters. In F. Tedim, V. Leone, & T. K. McGee (Eds.), *Extreme wildfire events and disasters* (pp. 155–174). Elsevier.

McClellan, R. (2002). Setting ambient air quality standards for particulate matter. *Toxicology*, *181–182*, 329–347. http://dx.doi.org/10.1016/S0300-483X(02)00459-6

McDonald, T., Gann, G. D., Jonson, J., & Dixon, K. W. (2016). *International standards for the practice of ecological restoration – Including principles and key concepts*. Society for Ecological Restoration.

McWethy, D. B., Schoennagel, T., Higuera, P. E., Krawchuk, M., Harvey, B. J., Metcalf, E. C., et al. (2019). Rethinking resilience to wildfire. *Nature Sustainability*, *2*, 797–804.

Meixner, T., & Wohlgemuth, P. (2004). Wildfire impacts on water quality. *Journal of Wildland Fire*, *13*, 27–35.

Miller, C., & Urban, D. L. (2000). Modeling the effects of fire management alternatives on Sierra Nevada mixed-conifer forests. *Ecological Applications*, *10*, 85–94.

Moritz, M. A., Batllori, E., Bradstock, R. A., Gill, A. M., Handmer, J., Hessburg, P. F., et al. (2014). Learning to coexist with wildfire. *Nature*, *515*, 58–66.

Nawaz, M. O., & Henze, D. K. (2020). Premature deaths in Brazil associated with long-term term exposure to $PM_{2.5}$ from Amazon fires between 2016 and 2019. *GeoHealth*, *4*, e2020GH000268. http://dx.doi.org/10.1029/2020GH000268

Neary, D. G. (2009). Post-wildland fire desertification: Can rehabilitation treatments make a difference? *Fire Ecology*, *5*, 129–144.

Nyman, P., Sheridan, G. J., Smith, H. G., & Lane, P. N. (2011). Evidence of debris flow occurrence after wildfire in upland catchments of south-east Australia. *Geomorphology*, *125*, 383–401.

O'Dell, K., Bilsback, K., Ford, B., Martenies, S. E., Magzamen, S., Fischer, E. V., & Pierce, J. R. (2021). Estimating mortality and morbidity attributable to smoke plumes in the United States: Not just a western US problem. *GeoHealth*, *5*, e2021GH000457. http://dx.doi.org/10.1029/2021GH000457

Oerlemans, J., Giesen, R. H., & Van Den Broeke, M. R. (2009). Retreating alpine glaciers: Increased melt rates due to accumulation of dust (Vadret da Morteratsch, Switzerland). *Journal of Glaciology*, *55*, 729–736. http://dx.doi.org/10.3189/002214309789470969

Oliveira, S., Rocha, J., & Sá, A. (2021). Wildfire risk modeling. *Current Opinion in Environmental Science and Health*, *23*, 100274.

Olsson, L., Jerneck, A., Thoren, H., Persson, J., & O'Byrne, D. (2015). Why resilience is unappealing to social science: Theoretical and empirical investigations of the scientific use of resilience. *Science Advances*, *1*, e1400217.

Paveglio, T., & Edgeley, C. (2017). Community diversity and hazard events: Understanding the evolution of local approaches to wildfire. *Natural Hazards*, *87*, 1083–1108.

Pelling, M. (2012). *The vulnerability of cities: Natural disasters and social resilience*. Routledge.

Peppin, D., Fule, P. Z., Sieg, C. H., Beyers, J. L., & Hunter, M. E. (2010). Post-wildfire seeding in forests of the western United States: An evidence-based review. *Forest Ecology and Management*, *260*, 573–586.

Peppin, D.L., Fule, P. Z., Sieg, C. H., Beyers, J. L., Hunter, M. E., & Robichaud, P.R. (2011). Recent trends in post-wildfire seeding in western US forests: *costs and seed mixes. International Journal of Wildland Fire*, *20*, 702–708. https://doi.org/10.1071/WF10044

Peterson, D. W., Dodson, E. K., & Harrod, R. J. (2009). Fertilization and seeding effects on vegetative cover after wildfire in north-central Washington state. *Forest Science*, *55*, 494–502.

Plaia, A., & Ruggieri, M. (2011). Air quality indices: A review. *Reviews in Environmental Science and Bio/Technology*, *10*, 165–179. http://dx.doi.org/10.1007/s11157-010-9227-2

Pokhrel, R. P., Gordon, J., Fiddler, M. N., & Bililign, S. (2021). Determination of emission factors of pollutants from biomass burning of African fuels in laboratory measurements. *Journal of Geophysical Research: Atmospheres*, *126*, e2021JD034731. http://dx.doi.org/10.1029/2021JD034731

Ponomarev, E., Yakimov, N., Ponomareva, T., Yakubailik, O., & Conard, S. G. (2021). Current trend of carbon emissions from wildfires in Siberia. *Atmosphere*, *12*, 559. http://dx.doi.org/10.3390/atmos12050559

Pope, C. A., & Dockery, D. W. (2006). Health effects of fine particulate air pollution: Lines that connect. *Journal of Air and Waste Management Association*, *56*, 709–742.

Psenner, R. (1999). Living in a dusty world: Airborne dust as a key factor for alpine lakes. *Water, Air, and Soil Pollution*, *112*, 217–227. http://dx.doi.org/10.1023/A:1005082832499

Pyke, D. A., Wirth, T. A., & Beyers, J. L. (2013). Does seeding after wildfires in rangelands reduce erosion or invasive species? *Restoration Ecology*, *21*, 415–421.

Pyne, S. J. (1982). *Fire in America: A cultural history of wildland and rural fire*. Princeton, NJ: Princeton University Press.

Radeloff, V. C., Helmers, D. P., Kramer, H. A., Mockrin, M. H., Alexandre, P. M., Bar-Massada, A., et al. (2018). Rapid growth of the US wildland-urban interface raises wildfire risk. *Proceedings of the National Academy of Sciences*, *115*, 3314–3319. http://dx.doi.org/10.1073/pnas.17188

Ramanathan, V., & Carmichael, G. (2008). Global and regional climate changes due to black carbon. *Nature Geoscience*, *1*, 221–227. http://dx.doi.org/10.1038/NGEO156

Raupach, M. R., Marland, G., Ciais, P., Le Quéré, C., Canadell, J. G., Klepper, G., & Field, C. B. (2007). Global and regional drivers of accelerating CO2 emissions. *Proceedings of the National Academy of Sciences of the United States of America*, *104*, 10.288–10.293. http://dx.doi.org/10.1073/PNAS.0700609104

Reche, I., Ortega-Retuerta, E., Romera, O., Villena, E. P., Baquero, R. M., & Casamayor, E. O. (2009). Effect of Saharan dust inputs on bacterial activity and community composition in Mediterranean lakes and reservoirs. *Limnology and Oceanography*, *54*, 869–879. http://dx.doi.org/10.4319/lo.2009.54.3.0869

Reid, J. S., Koppmann, R., Eck, T. F., & Eleuterio, D. P. (2005). A review of biomass burning emissions part II: Intensive physical properties of biomass burning particles. *Atmospheric Chemistry and Physics*, *5*, 799–825.

Robichaud, P. R. (2005). Measurement of post-fire hillslope erosion to evaluate and model rehabilitation treatment effectiveness and recovery. *International Journal of Wildland Fire*, *14*, 475–485.

Robichaud, P. R., & Ashmun, L. E. (2013). Tools to aid post-wildfire assessment and erosion-mitigation treatment decisions. *International Journal of Wildland Fire*, *22*, 95–105.

Robichaud, P. R., Beyers, J. L., & Neary, D. G. (2000). Evaluating the effectiveness of post-fire rehabilitation treatments. USDA Forest Service, Rocky Mountain Research Station, *General Technical Report RMRS-GTR-63*. Fort Collins, CO.

Robichaud, P. R., Lewis, S. A., Wagenbrenner, J. W., Ashmun, L. E., & Brown, R. E. (2013). Post-fire mulching for runoff and erosion mitigation. Part I: Effectiveness at reducing hillslope erosion rates. *Catena*, *105*, 75–92.

Robichaud, P. R., Rhee, H., & Lewis, S. A. (2014). A synthesis of post-fire burned area reports from 1972 to 2009 for western US Forest Service lands: Trends in wildfire characteristics and post-fire stabilization treatments and expenditures. *International Journal of Wildland Fire*, *23*, 929–944

Robichaud, P. R., Rhee, H., & Lewis, S. A. (2014). A synthesis of post-fire burned area reports from 1972 to 2009 for western US Forest Service lands: Trends in wildfire characteristics and post-fire stabilization treatments and expenditures. *International Journal of Wildland Fire*, *23*, 929–944.

Robichaud, P. R., Storrar, K. A., & Wagenbrenner, J. W. (2019). Effectiveness of straw bale check dams at reducing post-fire sediment yields from steep ephemeral channels. *Science of the Total Environment*, *676*, 721–731. http://dx.doi.org/10.1016/j.scitotenv.2019.04.246

Robichaud, P. R., Wagenbrenner, J. W., Brown, R. E., Wohlgemuth, P. M., & Beyers, J. L. (2008). Evaluating the effectiveness of contour-felled log erosion barriers as a post-fire runoff and erosion mitigation treatment in the western United States. *International Journal of Wildland Fire*, *17*, 255–273.

Roncoli, C., Breuer, N., Zierden, D., Fraisse, C., Broad, K., & Hoogenboom, G. (2012). The art of science: Climate forecasts for wildfire management in the southeastern United States. *Climatic Change*, *113*, 1113–1121. http://dx.doi.org/10.1007/s10584-012-0526-1

Ryan, K. C., Knapp, E. E., & Varner, J. M. (2013). Prescribed fire in North American forests and woodlands: History, current practice, and challenges. *Frontiers in Ecology and the Environment*, *11*, e15–e24.

Sarangi, C., Qian, Y., Rittger, K., Leung, L. R., Chand, D., Bormann, K. J., & Painter, T. H. (2020). Dust dominates high-altitude snow darkening and melt over high-mountain Asia. *Nature Climate Change*, *10*, 1045–1051. http://dx.doi.org/10.1038/s41558-020-00909-3

Schoennagel, T., Nelson, C. R., Theobald, D. M., Carnwath, G. C., & Chapman, T. B. (2009). Implementation of National Fire Plan treatments near the wildland – Urban interface in the western United States. *Proceedings of the National Academy of Sciences*, *106*, 10706–10711.

Schultz, M. G., Heil, A., Hoelzemann, J. J., Spessa, A., Thonicke, K., Goldammer, J. G., et al. (2008). Global wildland fire emissions from 1960 to 2000. *Global Biogeochemical Cycles*, *22*, GB2002. http://dx.doi.org/10.1029/2007GB003031

Schulze, S. S., Fischer, E. C., Hamideh, S., & Mahmoud, H. (2020). Wildfire impacts on schools and hospitals following the 2018 California camp fire. *Natural Hazards*, *104*, 901–925.

Schumann, R. L., Mockrin, M., Syphard, A. D., Whittaker, J., Price, O., Gaither, C. J., et al. (2020). Wildfire recovery as a "hot moment" for creating fire-adapted communities. *International Journal of Disaster Risk Reduction*, *42*, 101354.

Scott, J. H., Thompson, M. P., & Calkin, D. E. (2013). A wildfire risk assessment framework for land and resource management. *USDA Forest Service General Technical Report RMRS-GTR-315*. Fort Collins, CO.

Seiler, W., & Crutzen, P. J. (1980). Estimates of gross and net fluxes of carbon between the biosphere and the atmosphere from biomass burning. *Climatic Change*, *2*, 207–247. http://dx.doi.org/10.1007/BF00137988

Shakoor, A., Farooq, T. H., Arif, M. S., & Shahzad, S. M. (2023). Unprecedented wildfires in Canada and transboundary effects of carbon monoxide pollution. *Natural Hazards*, *118*, 2711–2713. http://dx.doi.org/10.1007/s11069-023-06117-4

Shindell, D., & Faluvegi, G. (2009). Climate response to regional radiative forcing during the twentieth century. *Nature Geoscience*, *2*, 294–300. http://dx.doi.org/10.1038/NGEO473

Smith, H. G., Sheridan, G. J., Lane, P. N., Nyman, P., & Haydon, S. (2011). Wildfire effects on water quality in forest catchments: A review with implications for water supply. *Journal of Hydrology*, *396*, 170–192.

Steelman, T. A., & Burke, C. A. (2007). Is wildfire policy in the United States suitable? *Journal of Forestry*, *105*, 67–72.

Stoof, C. R., & Kettridge, N. (2022). Living with fire and the need for diversity. *Earth's Futures*, *10*, e2021EF002528. http://dx.doi.org/10.1029/2021EF002528

Sullivan, A., Baker, E., & Kurvits, T. (Eds.) (2022). *Spreading like wildfire: The rising threat of extraordinary landscape fires*. United Nations Environment Programme.

Swiss Re. (2021). sigma 1/2021. Natural catastrophes in 2020. Tech. Rep., 1, Swiss Re, Zurich, available at: https://www.swissre.com/institute/research/sigma-research/sigma-2021-01.html, last access: 9 December 2022.

Tedim, F., McCaffrey, S., Leone, V., Delogu, G. M., Castelnou, M., McGee, T. K., & Aranha, J. (2020). What can we do differently about the extreme wildfire problem: An overview. In F. Tedim, V. Leone, & T. K. McGee (Eds.), *Extreme wildfire events and disasters* (pp. 233–263). Elsevier.

Tegen, I., Werner, M., Harrison, S. P., & Kohfeld, K. E. (2004). Relative importance of climate and land use in determining present and future global soil dust emission. *Geophysical Research Letters*, *31*, L05105. http://dx.doi.org/10.1029/2003GL019216

Theobald, D. M., & Romme, W. H. (2007). Expansion of the US wildland-urban interface. *Landscape and Urban Planning*, *83*, 340–354.

Thomas, D. S., & Butry, D. T. (2014). Areas of the US wildland-urban interface threatened by wildfire during the 2001–2010 decade. *Natural Hazards*, *71*, 1561–1585. https://doi.org/10.1007/s1106%209-013-0965-7

Urbanski, S. P. (2013). Combustion efficiency and emission factors for wildfire-season fires in mixed conifer forests of the northern Rocky Mountains, US. *Atmospheric Chemistry and Physics*, *13*, 7241–7262.

USDA and USDI. (2001). Urban wildland interface communities within vicinity of federal lands that are at high risk from wildfire. *Federal Register*, *66*, 751–777.

Van der Werf, G. R., Randerson, J. T., Giglio, L., Collatz, G. J., Kasibhatla, P. S., & Arellano, A. F. (2006). Interannual variability in global biomass burning emissions from 1997 to 2004. *Atmospheric Chemistry and Physics*, *6*, 3423–3441.

Van Wagtendonk, J. W. (1991). The evolution of National Park Service fire policy. *Fire Management Notes*, *52*, 10–15.

Wagenbrenner, J. W., MacDonald, L. H., & Rough, D. (2006). Effectiveness of three postfire rehabilitation treatments in the Colorado front range. *Hydrological Processes*, *20*, 2989–3006.

Wagner, S., Harvey, E., Baetge, N., McNair, H., Arrington, E., & Stubbins, A. (2021). Investigating atmospheric inputs of dissolved black carbon to the Santa Barbara Channel during the Thomas fire (California, USA). *Journal of Geophysical Research: Biogeosciences*, *126*, e2021JG006442. http://dx.doi.org/10.1029/2021JG006442

Walker, B., Holling, C. S., Carpenter, S. R., & Kinzig, A. (2004). Resilience, adaptability and transformability in social–ecological systems. *Ecology and Society*, *9*, 5.

Warziniack, T., & Thompson, M. (2013). Wildfire risk and optimal investments in watershed protection. *Western Economics Forum*, *12*, 19–28.

Webb, N. P., & Pierre, C. (2018). Quantifying anthropogenic dust emissions. *Earth's Futures*, *6*, 286–295. http://dx.doi.org/10.1002/2017EF000766

Willardson, B. (2020). Sustainable debris basins for post-fire protection. In *2020 Intermountain engineering, technology and computing* (pp. 1–6). Orem, UT. http://dx.doi.org/10.1109/IETC47856.2020.9249179

Williams, J. (2013). Exploring the onset of high-impact mega-fires through a forest land management prism. *Forest Ecology and Management*, *294*, 4–10.

Williams, K. M., Franzi, L. M., & Last, J. A. (2013). Cell-specific oxidative stress and cytotoxicity after wildfire coarse particulate matter installation into mouse lung. *Toxicology and Applied Pharmacology*, *266*, 48–55.

Wohlgemuth, P. M., Hubbert, K. R., Beyers, J. L., & Weise, D. R. (2007). Evaluating the effectiveness of burned area emergency response (BAER) efforts after the 2003 wildfires, southern California. In *Watershed management on the edge: Scarcity, quality and distribution* (pp. 129–137). San Diego, CA: University of California Water Resources Center Report No. 109.

Xing, Y. F., Xu, Y. H., Shi, M. H., & Lian, Y. X. (2016). The impact of $PM_{2.5}$ on the human respiratory system. *Journal of Thoracic Disease*, *8*, E69.

Xu, R., Yu, P., Abramson, M. J., Johnston, R. H., Samet, J. M., Bell, M. L., et al. (2020). Wildfires, global climate change, and human health. *The New England Journal of Medicine*, *383*, 2173–2181.

You, C., & Xu, C. (2022). Himalayan glaciers threatened by frequent wildfires. *Nature Geoscience*, *15*, 956–957. http://dx.doi.org/10.1038/s41561-022-01076-0

Yu, Y., & Ginoux, P. (2022). Enhanced dust emission following large wildfires due to vegetation disturbance. *Nature Geoscience*, *15*, 878–884. http://dx.doi.org/10.1038/s41561-022-01046-6

Zhang, X., Zhao, L., Tong, D. Q., Wu, G., Dan, M., & Teng, B. (2016). A systematic review of global desert dust and associated human health effects. *Atmosphere*, *7*, 158. http://dx.doi.org/10.3390/atmos7120158

6

FIRES IN THE FUTURE

Abstract

Increases in the frequency and extremity of fire weather have been globally pervasive during 1979–2019, largely due to human-driven climate change. Consequently, landscapes across the world are primed to burn more frequently, with notable increases in burned area recorded for extra-tropical and high-latitude forests. Fire-prone areas are generally projected to have more frequent, severe, and spatially extensive fires, and the areas adjacent to fire-prone regions will become more fire-prone. More frequent, severe, or extensive wildfires, and wildfires coupled with changes in other disturbances, are creating perceptible changes in the geographic ranges of individual species, ecosystems, and biomes. Among the key future research directions related to wildfires are models and measurements of fire; prediction and mitigation of post-fire water and sediment hazards; species, community, and ecosystem responses to fire; and approaches that can reduce fire-related hazards to human communities at the wildland-urban interface as well as communities far distant from fire that are affected by smoke. Mitigation and adjustment to wildfire can focus on the fire regime, the ecosystem effects of wildfires, restoring landscape resilience, and the hazards to human communities during and after fire.

Landscapes on Fire: Impacts on Uplands, Rivers, and Communities, Special Publications 80,
First Edition. Ellen Wohl.

DOI: 10.1002/9781394235186.ch6

Change is the only constant. (paraphrased from Heraclitus)

"May you live in interesting times" is an English expression that is supposed to be translated from a Chinese curse. We live in interesting times in many respects, not least with respect to wildfires. As noted in the first chapter, wildfires are having a moment, except that this moment will continue for decades into the future. As Southwell et al. (2022) describe the situation after the 2019–2020 fires in Australia: "We must anticipate wildfires rather than respond to them reactively as they will occur more frequently due to climate change." (p. 559).

The regional and global trends revealed in fire weather indices from the recent past and forecast for the near future are sobering. Examples from the recent past:

- In Mediterranean France, anthropogenic climate change was responsible for nearly half of the long-term increases in fire weather and fire-related drought during 1958–2017 and has significantly elevated the likelihood of summers with extreme fire danger (Barbero et al., 2020).
- Proxies for fuel aridity, including fire weather indices, indicate drying trends across much of the globe over the past four decades (Jolly et al., 2015).
- Observed fire weather in Australia during 1973–2010 increased significantly at a majority of stations (Clarke & Evans, 2019; Clarke et al., 2013).
- Catastrophic autumn wildfires in the western United States co-occurred with record or near-record autumn fire weather indices associated with extreme fuel dryness and strong offshore dry winds, conditions which have become more frequent (Hawkins et al., 2022).
- Observed global trends in annual extreme (95th percentile) values of the fire weather index from 1979 to 2020 indicate significant increases over a quarter to almost half of the global burnable land mass (Jain et al., 2022). These trends are driven by decreasing relative humidity and increasing temperature.
- An average increase of 18.7% in the length of the fire-weather season has occurred across global burnable lands from 1979 to 2013 (McRae et al., 2015) and increases have continued and are projected to continue into the 21st century (Senande-Rivera et al., 2022).
- Either direct (land cover change) or indirect (climate change) human influences have substantially increased fire globally (Figure 6.1) (Sullivan et al., 2022).

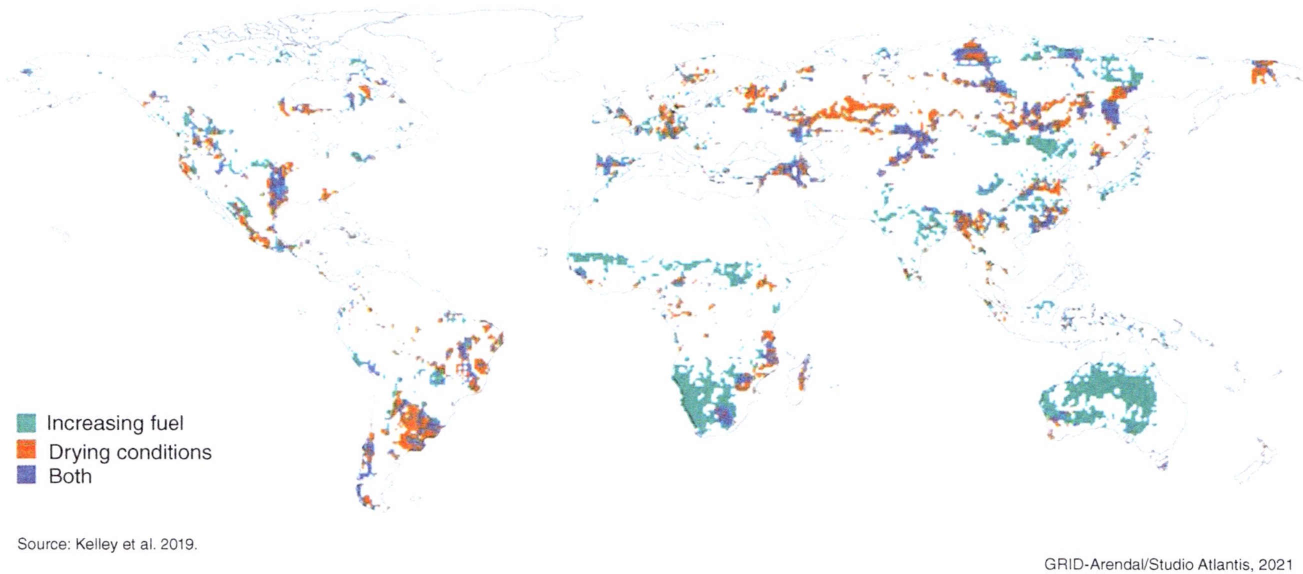

Figure 6.1 Increases in burnt area due to changing fuel and/or moisture from 2001 to 2014. (From GRID-Arendal/Studio Atlantis, https://www.grida.no/resources/15560.)

Fundamentally, increases in the frequency and extremity of fire weather have been globally pervasive during 1979–2019, largely due to human-driven climate change (e.g., Zhuang et al., 2021). This means that landscapes across the world are primed to burn more frequently (Jones et al., 2022). Increases in burned area of 50% or higher have occurred in some extra-tropical forest ecoregions and high-latitude forests during 2001–2019, although other factors can override the relationship between burned area and fire weather (Jones et al., 2022). Burned area in savannas, for example, relates to patterns of fuel production or the fragmentation of fire-prone landscapes by agriculture. Burned area in tropical forests relates more strongly to deforestation rates and forest degradation than to fire weather (Jones et al., 2022).

Wildfire-related trends predicted for the near future include:

- Modeling suggests a significant increase in future fire weather risk in parts of the Iberian Peninsula (Calheiros et al., 2021). Central Europe will tend toward more years with wet, warm winters and dry, hot summers, as well as fuel accumulation, increasing fire risk (Trnka et al., 2021). Areas previously limited by flammability are becoming less so as climate warms.
- Wildfire can be considered an emergent risk even in regions with relatively minor past extreme fire history, such as the United Kingdom, due to climate change (Perry et al., 2022). In such regions, changes in flammability may be an important factor. Projections for wildfire in the western United States for 2030 to 2099, for example, indicate massive increases in burned area in currently flammability-limited systems versus substantial decreases in fuel-limited non-forested systems (Littell et al., 2018).
- Anthropogenic climate change has increased the likelihood of extreme fire weather indices by 40% in parts of northern California and Oregon, USA, where recent extreme fires have occurred in autumn (Hawkins et al., 2022).
- Anthropogenic increases in extreme fire weather index occur for 22% of burnable land area globally by 2019, including much of the Mediterranean and the Amazon. This proportion increases to 33–62% of burnable lands by the mid-21st century and becomes more widespread as the rate of global temperature change increases (Abatzoglou et al., 2019).
- Climate model ensembles suggest that increases in extreme fire weather driven by greenhouse gas emissions were balanced by aerosol-driven cooling throughout the 20th century (Touma et al., 2021). Modeling projects that this compensation will disappear due to future reductions in aerosol emissions, causing "unprecedented increases in extreme fire weather risk in the 21st century" (Touma et al., 2021, p. 1).

- Projected changes through 2100 AD in extreme measures of fire danger indices suggest more dangerous near-surface fire weather conditions for Australia, and increased risk of pyroconvection and extreme fires (Dowdy et al., 2019).
- A 41% model-projected global increase in the long-continuing-current lightning flashes that are an important source of wildfire ignitions (Pérez-Invernón et al., 2023).

As Jones et al. (2022) summarize the situation, climate change increases the frequency and intensity of fire weather, and this pressure will escalate with each increment of global warming.

Environmental historian Stephen J. Pyne has a distinctive view of fires. He notes that anthropogenic fire history tracks the burning of living landscapes versus the burning of fossil landscapes (fossil fuels) and suggests that both forms of burning are causing the relics of the Pleistocene (ice sheets, glaciers, permafrost) to disappear with warming climate. He suggests that the Anthropocene could be defined by humanity's power source as the Pyrocene, "a time when humanity's firepower could interact in ways that promoted conditions that favored further fire" (Pyne, 2020, p. 1).

As we move into a more fire-prone future in much of the world, artists can contribute to our growing understanding of wildfires and their consequences. The website Following Fire (https://www.followingfire.com/) provides photographs and text examining forest resilience after the 2020 Holiday Farm Fire in Oregon, USA, for example, and the website Fireforest (https://www.fireforestphoto.com/) visually explores wildfire and restoration in northern Colorado, USA.

6.1. Changing Fire Regimes

The general projection is that fire-prone areas will have more frequent, severe, and spatially extensive fires and the areas adjacent to fire-prone regions will themselves become more fire-prone. In the well-studied western United States, the largest increases in rates of burned area during 1984–2017 occurred at elevations above 2,500 m (Alizadeh et al., 2021). The upslope advance of wildfires is consistent with observed warming and warm season vapor pressure deficits as limitations on flammability decrease.

On the other hand, fires associated with deforestation and vegetation degradation are modeled to decrease during the next century because of the combined effects of population and economic growth and increasing precipitation (Park et al., 2021). These changes are projected to be most pronounced in Sub-Saharan Africa, the tropics of southern Asia and Mexico, and the temperate latitudes of the Northern Hemisphere.

Deforestation and degradation fires are projected to increase in the Amazon basin and in much of Australia (Park et al., 2021).

Fires are burning larger areas. The positive exponential function in annual forest area burned in the western United States during recent decades is largely driven by exponential growth rates of individual large fires that have more potential for growth because of more extensive peripheries (Juang et al., 2022). The portion of the year in which fires occur is also growing longer in this region (McKenzie et al., 2004). This is exemplified by trends in California, where the winter rainy season has become progressively delayed, increasing the temporal overlap between extremely dry vegetation conditions and fire-promoting downslope winds in late autumn (Luković et al., 2021; Swain, 2021). In deserts of the western United States, invasive annual grasses are well adapted to fire. These grasses are also doing well in a warming climate that reduces the length of the freeze-free season and brings winter moisture but also an earlier onset of the fire season, all of which favor the fire-invasive grass feedback loop (Abatzoglou & Kolden, 2011).

In the Mediterranean, projections of increased burnt area vary from 5% to 50% per decade, depending on climate trajectories, fuel availability, and human actions (Dupuy et al., 2020). Human actions can reduce fires through thinning and prescribed burns. Human actions can also increase fires and fire risk through growing population in the wildland-urban interface; tree plantations of flammable species and local communities that target these plantations in arson attacks, as in Chile (Úbeda & Sarricolea 2016); and increasing recreation and associated accidental human-caused fire ignitions in wildland areas (Ganteaume et al., 2013; Preisler et al., 2016). Overall, fire seasons are lengthening for temperate and boreal regions (Flannigan et al., 2009; Jain et al., 2017; Senande-Rivera et al., 2022).

Despite these general trends, which are already observable and likely to continue into the future, the details of changing fire regimes are not straightforward or easy to predict. The availability of fuels, ignition agents, human actions, and climate conditions conducive to combustion jointly determine fire regimes, but regional increases in fire probabilities in some portions of the globe will likely be counter-balanced by decreases in other regions (Krawchuk et al., 2009). A net balance, however, can still equate to substantial changes in wildfire regime across large portions of the globe.

6.2. Ecosystem Changes and Species Ranges

Many environmental variables that influence plant distributions are changing. Projections suggest that by 2100 these changes may alter biome composition across 5–20% of land area in the United States (Grimm et al., 2013). Wildfire is likely to create the main constraint on plant distributions in fire-prone regions because the fire regime is changing more rapidly than vegetation can respond (Krawchuk et al., 2009; McKenzie et al., 2004). Changes in tree species range associated with changing climate may be delayed by the long lifespan of trees, their capacity to withstand long periods of physiological stress, and limited dispersal. Wildfire could disrupt this inertia by killing adult trees that could otherwise persist in conditions no longer suitable for seedling establishment and survival (Campbell & Shinneman, 2017; Davis et al., 2019). Fire may be more important in eliminating species along trailing edges of their geographic distribution than facilitating establishment along leading edges (Campbell & Shinneman, 2017).

More frequent, severe, or extensive wildfires, and wildfires coupled with changes in other disturbances, are already leading to perceptible changes in the geographic ranges of individual species, ecosystems, and biomes. Models also suggest more substantial changes in future (Text Box 6.1), as illustrated by conifer forests in the western United States.

The fire regime in western U.S. conifer forests is changing from more frequent but smaller, moderate intensity fires to high-severity fires that prevent conifer regeneration, in part by creating large, severely burned patches with long distances to seed sources (Keeley & Syphard, 2019). Conifer forests are transitioning toward dominance by fire-adapted hardwoods, shrubs, meadows, and grasslands (Coop et al., 2020; Jager et al., 2021). The process is starting with dry forests at the edge of their climatic tolerance (Hessburg et al., 2019; Stevens-Rumann & Morgan, 2019; Stevens-Rumann et al., 2018), but the trajectory of change will likely vary with latitude and elevation (Jager et al., 2021). High-elevation forests and alpine

Text Box 6.1 How Do We Know? Predicting Biotic Changes

Most modeled projections of changes in plant and animal species start with modeled projections of changes in temperature, precipitation, and other climate variables. The climate projections are then combined with forest landscape simulation models, stream temperature models, or other simulations of physical variables and individual species or communities, as appropriate to the scenario being investigated. The uncertainty of detailed future climate trends introduces uncertainties in these simulations, but the response of biota can to some extent be verified by comparison to changes observed during the accelerated warming and wildfires of the latter 20th century (e.g., Bloom et al., 2018; Stevens-Rumann et al., 2018).

plant communities are especially vulnerable as snowpack decreases (Bloom et al., 2018; Triepke et al., 2019). Low-elevation forests are particularly vulnerable because they may be at the temperature- and moisture-limits for seedling establishment and survival or tree growth (Davis et al., 2019; Halofsky et al., 2018). Mid-elevation forests may be the most stable under changing climate and fire regimes (Halofsky et al., 2018).

The transition may be gradual at some sites. Hardwood species with lightweight seeds that disperse long distances by wind and can resprout, such as aspen (*Populus tremuloides*), are favored by changing fire regimes in some coniferous forests (Whitman et al., 2019), whereas loss of snowpack and moisture stress may favor replacement of conifer forest by grasslands in other areas (Barber et al., 2018). At other sites, a threshold response seems to characterize changes in vegetation communities such that fire-killed stands do not regenerate as forest (Davis et al., 2019).

The areal extent of these changes is likely to be substantial in some regions. Models suggest that one- to two-thirds of upland mixed and conifer forest in Alberta, Canada will be replaced by grassland-dominated systems by the end of the 21st century, for example (Stralberg et al., 2018). Projections also suggest that the synergistic effects of climate change and wildfire are likely to encourage invasive species in the western United States (McKenzie et al., 2004). Even in areas that remain forested, models suggest a change toward more drought-tolerant species and decreased tree species richness (Liang et al., 2017). In the Blue Mountains of Oregon, USA, for example, subalpine tree species are likely to decline whereas lower elevation species may increase, homogenizing the remaining forest across elevational gradients (Cassell et al., 2019).

Warming, drought, and wildfire are tightly coupled, but they also interact with other forms of forest disturbance such as insect infestations and blowdowns caused by winds that in diverse environments vary from microbursts (Wohl, 2013b) to hurricanes (Phillips and Park, 2009). Stand-replacing fire and drought-induced mortality both contribute to species changes and non-native plant invasions (Littell et al., 2016) (Figure 6.2).

One of the key considerations in understanding biome-level response to changing fire characteristics is that forest community distributions may not change as intact units but may instead disaggregate as the least-fire-tolerant species disappear from a site (Campbell & Shinneman, 2017; Liang et al., 2017). Plant macrofossils that record past migration patterns and rates in response to changes in climate and fire regime indicate differences in rates among individual species (Payette et al., 2022). Foster et al. (2019) provide an example of how to address these issues by using an individual tree-based vegetation model to estimate current and future forest conditions in the boreal regions of interior Alaska underlain by

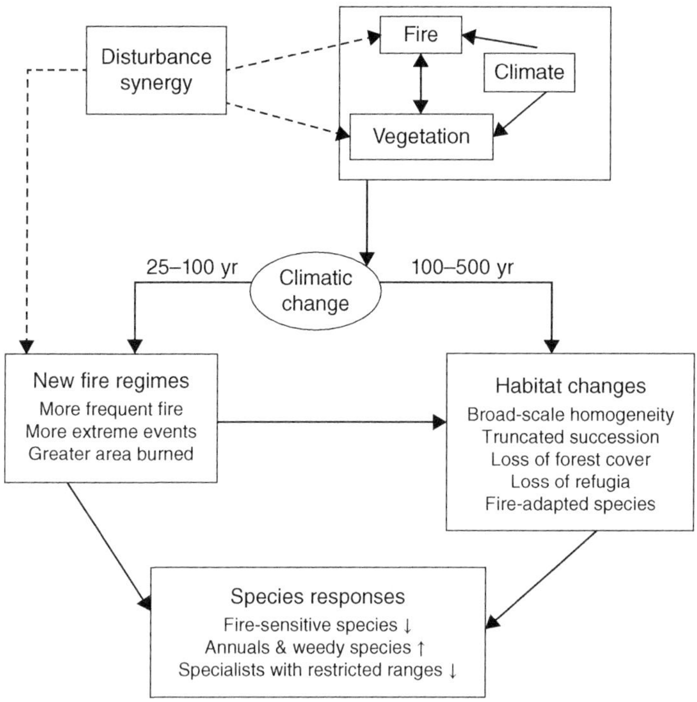

Figure 6.2 Schematic illustration of interactions between climate, vegetation, and fire and how these will shift with climate change. Fire will provide the main constraint on vegetation in the western United States because fire regimes will change more rapidly than vegetation can respond to climate alone (numbers are approximate). Species responses will vary, but the synergistic effects of climate change and fire are expected to favor invasive species. (From McKenzie et al., 2004, Figure 2.)

permafrost. Modeling results suggest an overall decrease in biomass despite increases in some species such as aspen, with increasing biomass at cold, wet locations and high elevations and decreasing biomass in dry locations. Fire-adapted species, such as pyrophilous fungi, may expand their species ranges and increase local biodiversity in some areas (Hughes et al., 2020).

Analogously, wildlife communities will not change evenly. Habitat-specialists and those with limited dispersal abilities are more likely to be at risk than habitat-generalists and those with greater dispersal abilities. At the extreme, some species may be trapped in sites that are population sinks by returning to suboptimal breeding habitat dramatically altered by disturbance. O'Neil et al. (2020) illustrate this with a study of the greater sage-grouse (*Centrocercus urophasianus*). Strong site fidelity causes female sage-grouse to continue nesting in fire-degraded habitats, despite reduced survival of nests in these areas.

Changes in plant communities will affect all the other organisms that interact with the plants, from soil microbial communities and mycorrhizal fungi through animals. Aquatic communities will also be affected by changes in riparian shading and inputs of coarse particulate organic matter. As with forest communities, these interactions are difficult to predict. Coldwater fishes in mountainous regions, for example, are vulnerable to thermal changes both because of their physiology and because their movements are constrained to linear networks that are easily fragmented (Isaak et al., 2010). Downscaling climatic conditions to local habitats reveals increases in basin-average mean and maximum stream temperatures in central Idaho, USA, during 1993–2006, primarily due to 30–50-year trends in air temperatures and stream flows. Within burned areas, however, stream temperature increases were 2–3 times greater than basin averages because of radiation gains (Isaak et al., 2010).

Faced with changing climate and fire regimes, natural resources managers are now thinking about responses in the context of a RAD framework based on strategies to resist, accept, or direct change (Kocik et al., 2022; Rahel & Lynch, 2022). Ecosystem transformations can be resisted by promoting the persistence of current or historical ecosystem composition, structure, and processes (Thompson et al., 2021). Alternatively, transformations can be accepted because they cannot be feasibly stopped, they are not sufficiently impactful to warrant a response, they are considered acceptable, or there is a lack of will to resist change. Alternatively, transformations can be directed toward a specific alternative ecosystem configuration (Thompson et al., 2021). Choosing a strategy involves the level of scientific understanding of the ecosystem, the factors driving change, and the trajectory of change, as well as the ability (financial, legal, technical intervention capability, stakeholder desire) to adopt specific management strategies.

6.3. Synergistic Effects

One of the great challenges to qualitatively or quantitatively predicting ongoing interactions among climate, wildfire, and biotic and human communities is that details matter—greatly. A few examples can be used to illustrate this very general statement.

1. One of the many concerns with climate warming is accelerated melting of glaciers and ice sheets and the associated rise in sea level and changes in atmospheric and oceanic circulation patterns

and water supplies (Biemans et al., 2019; Defrance et al., 2020). Glaciers melt faster as air temperatures warm, but wildfires can influence these factors. On the Athabasca Glacier in the Canadian Rocky Mountains, upwind wildfires can decrease surface albedo on the glacier via deposition of soot, which accelerates melting. There is a partial compensating effect, however, in that wildfires can also change atmospheric conditions above the glacier via dense smoke that reduces the solar energy reaching the ice (Aubry-Wake et al., 2022). Smoke decreases glacier melt rate, which increased by only 6.8% in years with dense smoke, compared to an increase greater than 10% in years. with fire effects from only soot but not active smoke (Aubry-Wake et al., 2022). The bottom line is that wildfires increase glacier melting, but the amount of increase reflects the details of smoke and soot.

2. Ice cores extracted from an ice field in southeastern Alaska, USA provide a continuous, high-resolution record of soot and ammonium generated by wildfires in the region between 1933 and 2001. Prior to 1984, years with high soot/ammonium/fire activity occurred during periods of below-normal precipitation. After 1984, these fire years coincided with years of normal and low precipitation but elevated temperatures. These results suggest that the primary influence on fires shifted from drought to air temperature (Sierra-Hernández et al., 2022), a trend that has continued into the 21st century.
3. "When it rains it pours" may describe human misfortunes or meteorological scenarios, but "when there's fire there's a firestorm" may describe more of the world's future. Multiple fires occurring across a region limit the availability of resources for fire suppression as well as creating compounded hazards from smoke and limits to evacuation routes or shelter areas. Abatzoglou et al. (2021) document a 25-day increase in the annual number of days with synchronous fire danger in forests of the western United States during 1979–2020 and project a doubling of such days by 2051–2080.
4. The Australian platypus (*Ornithorhynchus anatinus*) is a zoological curiosity—an egg-laying mammal—and an iconic freshwater species that faces numerous threats in its home range. Platypus distribution coincides with wildfire-prone areas. Although there is some evidence that the animals are resilient to fire, they need riparian vegetation to provide structural bank stability for their burrows and to provide shading and inputs of organic material for

the stream macroinvertebrates on which platypus feed. Platypus are mainly aquatic and are more susceptible to predation and heat stress when out of the water. Increased frequency and severity of fires, combined with drought, is leading to reduced water availability and insect prey, and platypus numbers in burned sites are significantly lower than those in unburnt sites (Bino et al., 2021)

5. Wildfires increasingly occur as a result of human activities (e.g., Balch et al., 2017). Post-fire changes in runoff and sediment are in many environments occurring in a landscape modified by humans in diverse ways (e.g., Backer et al., 2004; Sosa-Pérez and MacDonald, 2017) and response trajectories of biotic communities following fire are commonly influenced by pre- and post-fire human alterations of biotic communities (e.g., changes in connectivity and dispersal for organisms, as well as changes in species assemblages) (e.g., Luce et al., 2012). Consequently, the direct and indirect effects of human alterations of fire weather, fire ignition, landscapes, and ecosystems constitute a key synergistic effect for wildfires. A population of fishes may survive in a severely burned catchment if they can migrate to refugia, for example, but they may be extirpated where a structure such as a culvert under a road creates a migration barrier (e.g., Howell, 2006).

6.4. Research Directions

A plethora of interesting and important research directions are available to those interested in understanding more about wildfire and its numerous effects. This section briefly summarizes four key areas: models and measurements of fire; prediction and mitigation of post-fire water and sediment hazards; species, community, and ecosystem responses to fire; and approaches that can reduce fire-related hazards to human communities at the wildland-urban interface as well as communities far distant from fire that are affected by smoke.

6.4.1. Models and Measurements of Fire

Numerical simulations of individual wildfires can help to predict fire spread and flame length (Bakhshaii & Johnson, 2019). Initial approaches relied on easily measured variables related to fuel and surface weather and were largely empirical. Subsequent iterations included topographic and wind characteristics, with a focus on fire behavior rather than mechanisms of combustion and heat transfer. The most recent simulations use mechanistic combustion models and large-eddy simulation to define the flaming combustion and the mechanism of rate of spread. The wildfire model is then coupled to a mesoscale weather model. Deep Learning and Artificial Intelligence approaches are also being used to improve data-driven anticipation of wildfires (Kondylatos et al., 2022). Continuing development of models that couple the numerous factors that influence fire behavior, such as lightning density (Di Giuseppe, 2022), is a major research direction (Bakhshaii & Johnson, 2019).

Other techniques are being applied to understanding regional- to continental-scale patterns of wildfires and these efforts are likely to grow in the future. Wang et al. (2021), for example, describe a machine learning model that incorporates local meteorology, land-surface characteristics, and socioeconomic variables to predict monthly burned area at grid cells of about 750 km^2 resolution across the contiguous United States. In addition to potential predictive capabilities of such a model, the results reveal the role of longitude and latitude in delineating fire regimes with different temporal patterns of burned area.

Weather radar can be used as a fire-perimeter tracking tool to provide more timely and accurate updates on the location and rate of spread of large wildfires (Lareau et al., 2022). Instruments that can be deployed on the ground and from towers to measure fire properties in real-time include sonic anemometers to measure turbulent velocity and air temperatures, thermocouples to measure fuel-level air and flame temperatures, and thermal image velocimetry and image segmentation to measure flaming zone dynamics (Katurji et al., 2021). Increasing resolution of spatial, spectral, and temporal data from satellite imagery facilitates characterization of fires, as do deployment of lidar from various platforms, and digital aerial photogrammetry using unmanned aircraft (drones) (Rego et al., 2021).

Wireless sensor networks are being used to expand existing sampling capabilities by enlarging data collection across large areas with high temporal frequency (Rego et al., 2021). These sensors can collect physical parameters such as temperature and humidity, chemical data such as carbon dioxide, and images. Continued development and deployment of these and other techniques will facilitate more accurate characterization and modeling of fire behavior.

6.4.2. Prediction and Mitigation of Post-fire Water and Sediment Hazards

Studies analyzing large data sets also continue to illuminate patterns across diverse wildfires. Analyzing more than 1.8 million individual wildfires from 1992 to 2015 in the United States, for example, Abatzoglou et al. (2018) found that large human-caused wildfires tend to coincide with higher wind speeds than either small human-caused or large lightning-caused wildfires. Human-caused fires in the United States also tend to occur under conditions of lower lightning frequency and higher fuel moisture than lightning-caused fires (Balch et al., 2017). Understanding these patterns becomes important in the context of models of fire risk, many of which do not explicitly incorporate winds or differentiate the co-occurrence of winds and human-ignition agents such as roads and powerlines (Abatzoglou et al., 2018).

Prediction of post-fire hazards such as debris flows continues to be a major effort, especially in the western United States. An increasing source of concern related to this is how post-fire sedimentation can threaten water supplies via reduced water quality (Murphy et al., 2023) and reservoir sedimentation and reduced water storage in drylands (Hallema et al., 2018; Murphy et al., 2018; Belongia et al., 2023). Refining prediction of conditions under which debris flows occur, developing model output at relevant spatial and temporal scales for hazard planning (Oakley, 2021; Rengers and McGuire, 2021), and developing ensembles of linked models for different processes (Sankey et al., 2017; Ebel et al., 2023) will continue to constitute substantial research directions.

As catchment-scale models of post-fire runoff and upland and riverine fluxes continue to be developed (e.g., Canfield et al., 2005; Wang et al., 2020), research is likely to focus on (i) more effective parameterization that includes factors such as spatial differentiation in processes and nonlinear effects of area on flux magnitudes (Kampf et al., 2020) or the effect of temporal sequencing and frequency of fire on hydrologic dynamics (Wine & Cadol, 2016), (ii) integrative models that link multiple processes, such as fire effects, water availability, vegetative growth, and carbon cycling (Bart et al., 2020), or water and sediment fluxes with water quality parameters (Basso et al., 2021), and (iii) applications of these models to management to facilitate decisions such as where in a catchment to prioritize protection and restoration of natural features that can mitigate post-fire hazards associated with water and sediment.

6.4.3. Species, Community, and Ecosystem Responses to Fire

Greater understanding of the effects on and responses of individual species and biodiversity to wildfire requires reconnaissance surveys followed by monitoring in burned areas (Southwell et al., 2022). Ideally, surveys (i) occur immediately after fire to ground-truth fire severity mapping and provide a benchmark for assessing recovery and (ii) include burnt and unburnt sites in regions with historical data so that relevant variables can be compared with baseline estimates (Southwell et al., 2022). Given the problems that some post-fire mitigation such as mulching or aerial seeding can create for native species recovery and biodiversity, research focusing on mitigation strategies that reduce post-fire hazards associated with fluxes downslope and downstream, and simultaneously minimize negative impacts on native biota, will continue to be a priority.

Limited information on how species will respond to continued climate change, repeated disturbances, and other site factors following wildfires constrains ability to determine future trends in forest regeneration (Stevens-Rumann & Morgan, 2019). Stevens-Rumann and Morgan (2019) provide a decision tree to assist managers in prioritizing post-fire reforestation based on climate and forest type, distance to living trees of the target species, aspect, elevation, hillslope gradient, competing vegetation, soils, microsite conditions, and seedling species and genotypes. Research on species-, ecosystem-, and biome-level responses to the combined effects of changing climate and fire regimes will be a major emphasis in future, along with applications of resulting understanding such as the decision tree described above.

The potential for novel ecosystems to form following wildfire, rather than gradual re-formation of pre-fire species associations and communities, complicates the ability to predict the geographic distribution of species and the abundance of individuals within a species (McKelvey et al., 2021; Yu et al., 2016). This creates challenges for conservation biology and the attempt to preserve sufficient habitat and connectivity to maintain species (Hobbs et al., 2014). Uncertainty about future biotic communities also creates challenges for predictions of future fire hazard if the fuel loads and fire susceptibility of novel ecosystems differ from contemporary ecosystems (Metz et al., 2013).

6.4.4. Fire-Related Hazards to Human Communities

Bowman et al. (2020) note the need for data collection to improve quantification and modeling of interactions between fire activity and human populations in a manner that includes individual and community socioeconomic status and historical, cultural, and political legacies. As summarized in the preceding chapter, multiple published case studies from diverse regions and countries illustrate the effectiveness of engaging local communities in creating and communicating understanding of wildfire hazards and mitigation.

Social scientists, artists, and individuals skilled in scientific communication can facilitate knowledge transfer between scientists and communities affected by fire, but transdisciplinary research remains hampered by intellectual and institutional barriers, including limited interactions between the humanities and physical sciences (Bowman et al., 2020; Castree, 2017). Faced with a world largely modified by humans and in which humans and wildfire increasingly interact, it makes sense to for scientists interested in any aspect of wildfire to be proactive about working with human communities and with scholars from the humanities who can potentially facilitate such interactions.

Colavito et al. (2020) describe the importance of boundary organizations as a means of bridging barriers. Boundary organizations develop and deliver actionable scientific information and facilitate ongoing communication among different communities of knowledge (Guston, 2001). The U.S. Joint Fire Science Program Fire Science Exchange Network is an example of a boundary organization that works in fire and forest management by facilitating knowledge exchange about wildfire among scientists, managers, policymakers, and other stakeholders. Colavito et al. (2020) describe *Fires of Change*, a collaborative, multimedia art exhibit designed to enhance understanding of the shifting fire regimes of the southwestern United States. Two boundary organizations convened scientists, managers, and artists in the co-production of science-based artwork. Surveys of exhibit visitors indicate that the exhibits increased visitors' understanding of the effects of changing climate on fire regimes and increased visitors' support for management actions to address these effects. Development of the exhibit also created new relationships and networks among the participants and increased appreciation for collaborations among scientists, managers, and artists.

Swanson et al. (2012) summarize other collaborative programs hosted by the Joint Fire Science Program in the United States, including creative residencies; conferences, workshops, and symposia; art-science collaborative field studies; and community programs and partnerships. This summary notes that collaborative humanities programs may help in a cultural reframing of the wildfire dilemma—the situation in which the public is either unconcerned with fire or views it solely as a destructive force—which results in part from the fact that wildfire has been framed solely in terms of a technical, political, and scientific matter. Swanson et al. (2012) describe several examples of successful collaborative programs centered on wildfire and note that "Artist's explorations of notions of resilience, reconciliation, harmony, chaos, synergy, sustainability, and place overlap tremendously with issues of wildland fire." (p. 7)

6.5. Summary: Mitigation and Adjustment

In the context of wildfires, mitigation and adjustment can focus on the fire regime, the ecosystem effects of wildfires, restoring landscape resilience, and the hazards to human communities during and after fire.

6.5.1. The Fire Regime

We are likely powerless in the immediate term (decades, perhaps to centuries) to significantly control ongoing climate warming and its secondary effects such as drought.

Human-caused wildfire ignitions will continue to occur as long as individuals do foolish or malicious things to start fires, but communities and power companies can reduce human-caused ignitions by modifying infrastructure such as putting powerlines underground (Keeley and Syphard, 2018). The combination of thinning and prescribed burning appears to be the best approach to reducing fuels that have accumulated in areas subject to past fire suppression (Fulé et al., 2012). These activities alter ecosystems in sometimes unpredictable ways because they do not fully mimic natural fire regimes to which species have adapted (Ryan et al., 2013), but ecosystems and biomes are already changing rapidly due to global warming and its many secondary effects. Environmental management now commonly involves a triage approach of trying to identify the actions that will create the least harm. Thinning and prescribed burns may fall into the category of lesser harm than catastrophic wildfires.

Questions of how and when to thin and burn remain. Given the rapid changes in fire regimes, it is reasonable to ask whether historical fire regimes remain relevant to managing fires today. Historical fire regimes provide models for prescribed fires based on the assumptions that (i) historical fire regimes can be accurately inferred, (ii) fire-suppressed communities are ecologically degraded, and (iii) reinstating historical fire regimes is desirable despite changing conditions (Freeman et al., 2017). Each of these assumptions can break down in some conditions. Ecological outcomes of fire suppression are complex and may not result in degradation, for example, and habitat fragmentation and invasive species can interact with prescribed fire to create negative ecological outcomes (Freeman et al., 2017). Understanding of reference conditions remains relevant, however, because they encompass the evolutionary history of organisms and can provide insights into past forest adaptations and migrations under varying climates (Fulé, 2008). A key component of reference conditions in some environments can be indigenous traditional uses of fire, which can be effective at thinning the forest understory with low-severity fire (Marks-Marino, et 2021). The best advice may be to proceed with caution and seek to understand contemporary fire-biota interactions across trophic levels, functional groups, spatial and temporal scales, and management contexts (Freeman et al., 2017).

6.5.2. Ecosystem Effects of Wildfires

Mitigation and adjustment may involve assisted colonization by organisms or human-assisted expansion of existing geographic ranges, although this strategy remains under debate by conservation biologists and ecologists (Hunter, 2007; Leech et al., 2011; McLachlan et al., 2007). Humans already have a long history of such manipulations with accidental or deliberate introductions of every kind of living creature. Conservation biologists recognize that human fragmentation of ecosystems and biomes has made it challenging or impossible for many species to disperse to potentially suitable new habitat without direct human assistance.

6.5.3. Restoring Landscape Resilience

A recently growing emphasis within catchment management is restoring natural resilience that has been reduced through past human activities. The first written descriptions of river corridors across the United States, for example, clearly indicate that extensive portions of river networks were river-wetland corridors (Wohl et al., 2021) with multiple channels that branched and rejoined downstream around extensive floodplain wetlands. Channels were hard to navigate even in small boats because of the numerous blockages—beaver (*Castor canadensis*) dams in smaller channels and logjams or wood rafts (Triska, 1984) in even the largest rivers (Wohl, 2014). Place names across Europe and the United Kingdom also indicate the past abundance of slow-moving channels surrounded by abundant wetlands (Brown et al., 2018). These river corridors had substantial lateral connectivity between the channel and floodplain and vertical connectivity between the surface, hyporheic zone, and groundwater, but limited downstream connectivity (Burchsted et al., 2010; Wohl et al., 2021). They were perfect natural fire breaks because of their high moisture levels in saturated soils, wet vegetation, and standing water at the surface (Fairfax & Whittle, 2020).

River-wetland corridors also significantly attenuated post-fire fluxes. Following a fire in South Africa, for example, riverine and hillslope seepage wetlands had significantly lower soil water repellency than adjacent soils (Grenfell et al., 2022). The emergence of new vegetation within the wetlands, likely from intact rhizome networks of predominant sprouting vegetation, enhanced post-fire infiltration due to macropore flow along rhizome networks. The wetlands reduced early wet season runoff and erosion due to rainfall-intercepting leaf and stem structures. The wetlands also exhibited suspended sediment trap efficiencies of 87–99% (Grenfell et al., 2022).

In diverse river networks, the densely growing vegetation and limited downstream conveyance of river-wetland corridors reduce peak flows and increase infiltration (e.g., Curran & Hession, 2013; Naiman & Decamps, 1997), promoting denitrification (e.g., Fisher & Acreman, 2004). River-wetland corridors store sediment and anything moving with that sediment, such as phosphate (Wiener et al., 2022). They serve as fire refugia from which organisms can repopulate adjacent areas (McLaughlin et al., 2017; Selwood & Zimmer, 2020) and they host greater biomass and biodiversity than simpler portions of the river network (Bellmore & Baxter, 2014). Consequently, the abundance, spatial distribution, and characteristics of river-wetland corridors influence the spread and severity of wildfires, as well as post-fire resilience.

River-wetland corridors can extend continuously for many kilometers along a river network, but they are typically interspersed downstream with narrower, steeper segments of valley floor. The wider, lower gradient segments that have more spatially extensive floodplains are sometimes referred to as beads (Stanford et al., 1996; Wohl et al., 2018) and the narrower intervening portions as strings (Figure 6.3). Downstream alternations between beads and strings occur in low-relied watersheds, but are especially pronounced in high-relief terrain, where the spatial pattern of beads reflects differences in bedrock lithology (Wohl et al., 1994); the geometry of bedrock fractures that strongly influence weathering and erosion (Ehlen & Wohl, 2002); river response to changes in base level (Duvall et al., 2004); or glaciation (Hauer et al., 2016; Wohl, 2013a).

Figure 6.3 An example of downstream alternations between beads and strings, here along North St. Vrain Creek, Colorado, USA. White dashed line indicates approximate boundary of river corridor. Arrows indicate flow direction. Photo at upper left shows a string; photo at upper right indicates a bead.

Centuries of human modifications designed to increase downstream conveyance in channels have greatly simplified and homogenized river corridors in much of the temperate latitudes (Peipoch et al., 2014; Poff et al., 2007; Wohl et al., 2017). Channelization (straightening, dredging), bank stabilization, artificial levees, floodplain drainage and deforestation, flow regulation, removal of large wood, and killing of beavers have cumulatively resulted in river corridors that are literally streamlined (Wohl, 2016). Everything moves downstream faster and more efficiently, especially after wildfires. Now human communities are determining whether all the king's horses and all the king's men can put some portions of Humpty together again. In other words, people are trying to restore spatial heterogeneity and diversity to river corridors with the objectives of restoring lost functions including biodiversity, nutrient uptake, and resilience to disturbances such as wildfire (Pugh et al., 2022) (Figure 6.4).

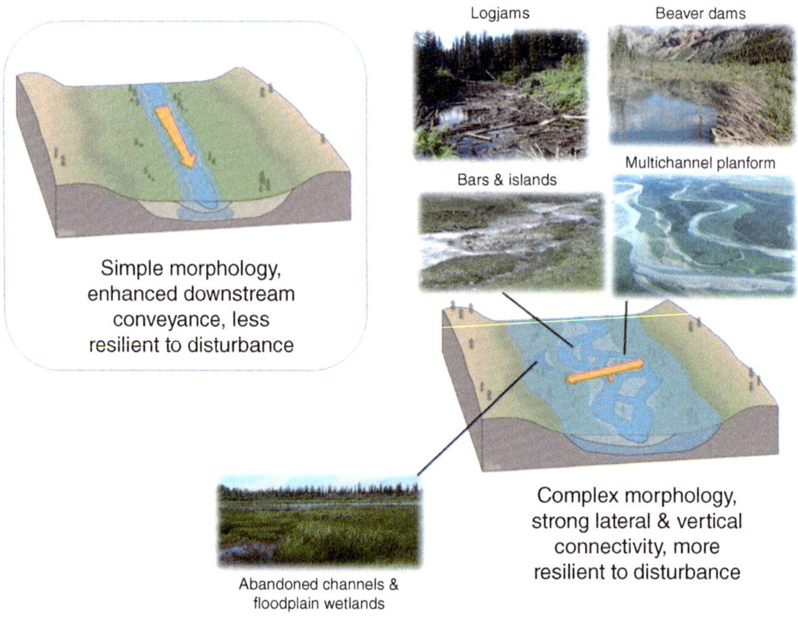

Figure 6.4 Schematic illustration of a stream corridor anthropogenically modified to maximize downstream (upper left) conveyance versus a natural stream corridor in which various forms of spatial heterogeneity enhance lateral and vertical connectivity and create resilience to disturbances including wildfire. Blue overlay indicates extent of flooding during peak flows and orange arrows indicate relative dimensions of connectivity. Inset photos illustrate sources of spatial heterogeneity. Block diagrams courtesy of Julianne Scamardo.

This type of restoration is vitally important but raises questions of scale: How many beads need to be restored? Where do restored, fully functional river beads most effectively create fire breaks or promote post-fire resilience—in headwaters or farther downstream? If a bead is partly restored, promoting lateral connectivity with only a portion of the natural floodplain, how much effect does that have? River scientists have broadly advocated for allowing space for rivers to adjust and respond and for maintaining and restoring enhanced structural and functional complexity within river corridors for decades (e.g., Biron et al., 2014; Downs & Gregory, 2004; Downs et al., 2002; Rathburn et al., 2018; Wohl et al., 2018, 2022). As resource managers maintain this emphasis going forward, we also need more careful monitoring of the effectiveness of restoration techniques being employed in this context and continuing development of integrated predictive models that can be used to prioritize management efforts.

Fundamentally, restoring landscape resilience involves managing the wildfire disturbance cascade (Figure 6.5) by working with natural physical processes and landforms in a manner that enhances feedbacks capable of attenuating the cascade. This approach is referred to as nature-based storage or natural infrastructure (Norman et al., 2022; Robotham et al., 2023), re-wilding (Harvey & Henshaw, 2023; Rideout et al., 2021), and natural flood management (Dadson et al., 2017).

6.5.4. Hazards to Human Communities During and After Fire

It would be most effective to limit population growth in the wildland-urban interface, but at least some democratic societies, such as the United States, are challenged to exert sufficient control on their citizens to prevent such growth. Actions such as putting powerlines underground (Keeley and Syphard, 2018), prohibiting campfires during fire-prone conditions, and requiring proper storage of flammable liquids in households can help reduce human-caused fires, but wildfire-related hazards will likely continue to increase in regions with increasing human presence. In some cases, these hazards are exacerbated by other climate-related changes. Short-duration extreme rainfalls after wildfire are very effective triggers for debris flows and flash floods, for example, and observations, theory, and modeling studies indicate that both short- and long-duration rainfall extremes are intensifying with global warming (Fowler et al., 2021).

Consequently, mitigating hazards to human communities focuses on other approaches beyond limiting the locations of people and infrastructure. Enhancing identification of hazard-prone areas in burned landscapes

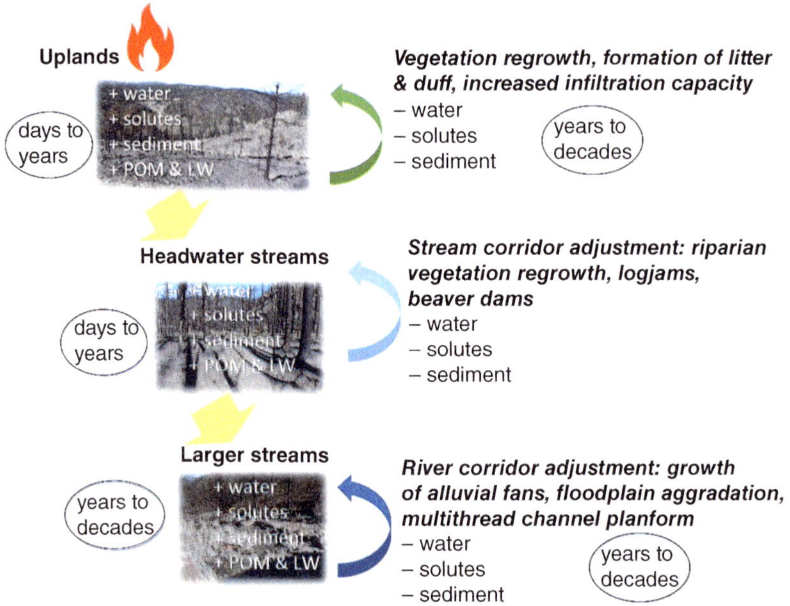

Figure 6.5 The wildfire disturbance cascade, which starts with significantly increased connectivity for all materials from uplands to headwater streams and down river networks immediately after fire and continuing for years (left side of figure). Feedbacks (indicated by curved arrows) such as regrowth of upland vegetation and channel adjustment subsequently decrease downslope and downstream connectivity over periods of years to decades (right side of figure). POM is particulate organic matter; LW is large wood.

(e.g., debris flow sources) can help to limit infrastructure within these specific areas. Enhancing catchment resilience by restoring natural features that retain or attenuate post-fire fluxes from uplands and through river networks (e.g., via beaver reintroduction in the western United States) can mitigate excess water and sediment yields to downstream human communities. Proactively working with communities at the wildland-urban interface on wildfire preparedness and emergency response is also an important component of mitigating and adjusting to changing wildfire regimes. Recent views and perspectives note the need to shift from observation and modeling of diverse components that influence wildfire and hazards toward more integrative and predictive approaches that can identify means to mitigate and adapt in an increasingly flammable world (Shuman et al., 2022).

Change is occurring. The test for humanity will be how we react.

References

Abatzoglou, J. T., Juang, C. S., Williams, A. P., Kolden, C. A., & Westerling, A. L. (2021). Increasing synchronous fire danger in forests of the western United States. *Geophysical Research Letters*, *48*, e2020GL091377. http://dx.doi.org/10.1029/2020GL091377

Abatzoglou, J. T., & Kolden, C. A. (2011). Climate change in western US deserts: Potential for increased wildfire and invasive annual grasses. *Rangeland Ecology & Management*, *64*, 471–478.

Abatzoglou, J. T., Williams, A. P., & Barbero, R. (2019). Global emergence of anthropogenic climate change in fire weather indices. *Geophysical Research Letters*, *46*, 326–336. http://dx.doi.org/10.1029/2018GL080959

Alizadeh, M. R., Abatzoglou, J. T., Luce, C. H., Adamowski, J. F., Farid, A., & Sadegh, M. (2021). Warming enabled upslope advance in western US forest fires. *Proceedings of the National Academy of Sciences*, *118*, e2009717118.

Aubry-Wake, C., Bertoncini, A., & Pomeroy, J. W. (2022). Fire and ice: The impact of wildfire-affected albedo and irradiance on glacier melt. *Earth's Future*, *10*, e2022EF002685. http://dx.doi.org/10.1029/2022EF002685

Bakhshaii, A., & Johnson, E. A. (2019). A review of a new generation of wildfire-atmosphere modeling. *Canadian Journal of Forest Research*, *49*, 565–574.

Barber, Q. E., Parisien, M. A., Whitman, E., Stralberg, D., Johnson, C. J., St-Laurant, M. H., et al. (2018). Potential impacts of climate change on the habitat of boreal woodland caribou. *Ecosphere*, *9*, e02472. http://dx.doi.org/10.1002/ecs2.2472

Barbero, R., Abatzoglou, J. T., Pimont, F., Ruffault, J., & Curt, T. (2020). Attributing increases in fire weather to anthropogenic climate change over France. *Frontiers in Earth Science*, *8*, 104. http://dx.doi.org/10.3389/feart.2020.00104

Bart, R. R., Kennedy, M. C., Tague, C. L., & McKenzie, D. (2020). Integrating fire effects on vegetation carbon cycling within an ecohydrologic model. *Ecological Modelling*, *416*, 108880.

Basso, M., Mateus, M., Ramos, T. B., & Vieira, D. C. S. (2021). Potential post-fire impacts on a water supply reservoir: An integrated watershed-reservoir approach. *Frontiers in Environmental Science*, *9*, 684703.

Bellmore, J. R., & Baxter, C. V. (2014). Effects of geomorphic process domains on river ecosystems: A comparison of floodplain and confined valley segments. *River Research and Applications*, *30*, 617–630.

Biemans, H., Siderius, C., Lutz, A. F., Nepal, S., Ahmad, B., Hassan, T., et al. (2019). Importance of snow and glacier meltwater for agriculture on the Indo-Gangetic Plain. *Nature Sustainability*, *2*, 594–601. http://dx.doi.org/10.1038/s41893-019-0305-3

Bino, G., Hawke, T., & Kingsford, R. T. (2021). Synergistic effects of a severe drought and fire on platypuses. *Science of the Total Environment*, *777*, 146137. http://dx.doi.org/10.1016/j.scitotenv.2021.146137

Biron, P., Buffin-Belanger, T., Larocque, M., Chone, G., Cloutier, C., Ouellet, M., et al. (2014). Freedom space for rivers: A sustainable management approach to enhance river resilience. *Environmental Management*, *54*, 1056–1073.

Bloom, T. D. S., Flower, A., Medler, M., & DeChaine, E. G. (2018). The compounding consequences of wildfire and climate change for a high-elevation wildflower (*Saxifraga austromontana*). *Journal of Biogeography*, *45*, 2755–2765.
Bowman, D. M. J. S., Kolden, C. A., Abatzoglou, J. T., Johnstons, F. H., van der Werf, G. R., & Flannigan, M. (2020). Vegetation fires in the Anthropocene. *Nature Reviews Earth and Environment*, *1*, 500–515. https://doi.org/10.1038/s43017-020-0085-3
Brown, A. G., Lespez, L., Sear, D. A., Macaire, J. J., Houben, P., Klimek, K., et al. (2018). Natural vs anthropogenic streams in Europe: History, ecology and implications for restoration, river-rewilding and riverine ecosystem services. *Earth-Science Reviews*, *180*, 185–205.
Burchsted, D., Daniels, M., Thorson, R., & Vokoun, J. (2010). The river discontinuum: Applying beaver modifications to baseline conditions for restoration of forested headwaters. *BioScience*, *60*, 908–922.
Calheiros, T., Pereira, M. G., & Nunes, J. P. (2021). Assessing impacts of future climate change on extreme fire weather and pyro-regions in Iberian Peninsula. *Science of the Total Environment*, *754*, 142233. https://doi.org/10.1016/j.scitotenv.2020.142233
Campbell, J. L., & Shinneman, D. J. (2017). Potential influence of wildfire in modulating climate-induced forest redistribution in a central Rocky Mountain landscape. *Ecological Processes*, *6*, 7. http://dx.doi.org/10.1186/s13717-017-0073-9
Canfield, H. E., Goodrich, D. C., & Burns, I. S. (2005). Selection of parameter values to model post-fire runoff and sediment transport at the watershed scale in southwestern forests. In *Managing Watersheds for Human and Natural Impacts: Engineering, Ecological, and Economic Challenges*. Williamsburg, VA: ASCE Watershed Management Conference.
Cassell, B. A., Scheller, R. M., Lucash, M. S., Hurteau, M. D., & Loudermilk, E. L. (2019). Widespread severe wildfires under climate change lead to increased forest homogeneity in dry mixed-conifer forests. *Ecosphere*, *10*, e02934.
Castree, N. (2017). Speaking for the 'people disciplines': global change science and its human dimensions. *Anthropocene Review*, *4*, 160–182.
Clarke, H., & Evans, J. P. (2019). Exploring the future change space for fire weather in southeast Australia. *Theoretical and Applied Climatology*, *136*, 513–527. http://dx.doi.org/10.1007/s00704-018-2507-4
Clarke, H., Lucas, C., & Smith, P. (2013). Changes in Australian fire weather between 1973 and 2010. *International Journal of Climatology*, *33*, 931–944. http://dx.doi.org/10.1002/joc.3480
Colavito, M., Wolfson, B. S., Thode, A. E., Haffey, C., & Kimball, C. (2020). Integrating art and science to communicate the social and ecological complexities of wildfire and climate change in Arizona, USA. *Fire Ecology*, *16*, 19. https://doi.org/10.1186/s42408-020-00078-w
Coop, J. D., Parks, S. A., Stevens-Rumann, C. S., Crausbay, S. D., Higuera, P. E., Hurteau, M. D., et al. (2020). Wildfire-driven forest conversion in western North American landscapes. *BioScience*, *70*, 659–673. https://doi.org/10.1093/biosci/biaa061

Curran, J. C., & Hession, W. C. (2013). Vegetative impacts on hydraulics and sediment processes across the fluvial system. *Journal of Hydrology, 505*, 364–376.

Dadson, S. J., Hall, J. W., Murgatroyd, A., Acreman, M., Bates, P., Beven, K., et al. (2017). A restatement of the natural science evidence concerning catchment-based 'natural' flood management in the UK. *Proceedings of the Royal Society A, 473*, 20160706. https://doi.org/10.1098/rspa.2016.0706

Davis, K. T., Dobrowski, S. Z., Higuera, P. E., Holden, Z. A., Veblen, T. T., Rother, M. T., et al. (2019). Wildfires and climate change push low-elevation forests across a critical threshold for tree regeneration. *Proceedings of the National Academy of Science, 116*, 6193–6198.

Defrance, D., Catry, T., Rajaud, A., Dessay, N., & Sultan, B. (2020). Impacts of Greenland and Antarctic ice sheet melt on future Köppen climate zone changes simulated by an atmospheric and oceanic general circulation model. *Applied Geography, 119*, 102216. http://dx.doi.org/10.1016/j.apgeog.2020.102216

Di Giuseppe, F. (2022). The value of probabilistic prediction for lightning ignited fires. *Geophysical Research Letters, 49*, e2022LF099669. http://dx.doi.org/10.1029/2022GL099669

Downs, P. W., & Gregory, K. J. (2004). *River Channel Management: Towards Sustainable Catchment Hydrosystems*. London, UK: Hodder Arnold.

Downs, P. W., Skinner, K. S., & Kondolf, G. M. (2002). Rivers and streams. In M. R. Perrow & A. J. Davy (Eds.), *Handbook of Ecological Restoration, v 2: Restoration in Practice* (pp. 267–296). Cambridge, UK: Cambridge University Press.

Dupuy, J. L., Fargeon, H., Martin-St Paul, N., Pimont, F., Ruffault, J., Guijarro, M., et al. (2020). Climate change impact on future wildfire danger and activity in southern Europe: A review. *Annals of Forest Science, 77*, 35.

Duvall, A., Kirby, E., & Burbank, D. (2004). Tectonic and lithologic controls on bedrock channel profiles and processes in coastal California. *Journal of Geophysical Research - Earth Surface, 109*, F03002. http://dx.doi.org/10.1029/2003JF000086

Ehlen, J., & Wohl, E. (2002). Joints and landform evolution in bedrock canyons. *Transactions. Japan Geomorphological Union, 23*, 237–255.

Fairfax, E., & Whittle, A. (2020). Smokey the beaver: beaver-dammed riparian corridors stay green during wildfire throughout the western United States. *Ecological Applications, 30*, e02225. Fireforest website: https://www.fireforestphoto.com

Fisher, J., & Acreman, M. C. (2004). Wetland nutrient removal: A review of the evidence. *Hydrology and Earth System Sciences, 8*, 673–685.

Flannigan, M. D., Krawchuk, M. A., de Groot, W. J., Wotton, B. M., & Gowman, L. M. (2009). Implications of changing climate for global wildland fire. *International Journal of Wildland Fire, 18*, 483–507. Following Fire website: https://www.followingfire.com

Foster, A. C., Armstrong, A. H., Shuman, J. K., Shugart, H. H., Rogers, B. M., Mack, M. C., et al. (2019). Importance of tree- and species-level interactions with wildfire, climate, and soils in interior Alaska: Implications for forest change under a warming climate. *Ecological Modelling, 409*, 108765.

Fowler, H. J., Lenderink, G., Prein, A. F., Westra, S., Allan, R. P., Ban, N., et al. (2021). Anthropogenic intensification of short-duration rainfall extremes. *Nature Reviews Earth and Environment*, *2*, 107–122.

Freeman, J., Kobziar, L., Rose, E. W., & Cropper, W. (2017). A critique of the historical-fire-regime concept in conservation. *Conservation Biology*, *31*, 976–985.

Fulé, P. Z. (2008). Does it make sense to restore wildland fire in changing climate? *Restoration Ecology*, *16*, 526–531.

Fulé, P. Z., Crouse, J. E., Roccaforte, J. P., & Kalies, E. L. (2012). Do thinning and/or burning treatments in western USA ponderosa or Jeffrey pine-dominated forests help restore natural fire behavior? *Forest Ecology and Management*, *269*, 68–81.

Ganteaume, A., Camia, A., Jappiot, M., San-Miguel-Ayanz, J., Long-Fournel, M., & Lampin, C. (2013). A review of the main driving factors of forest fire ignition over Europe. *Environmental Management*, *51*, 651–662.

Grenfell, M. C., Abrahams, E., & Fisher, R. M. (2022). Analysis and conceptual geospatial modelling of the intermediary role of wetlands in drylands in post-fire material flux dynamics, Silvermine River catchment, Cape Town. *Wetlands Ecology and Management*, *30*, 623–645.

Grimm, N. B., Chapin, F. S., Bierwagen, B., Gonzalez, P., Groffman, P. M., Luo, Y., et al. (2013). The impacts of climate change on ecosystem structure and function. *Frontiers in Ecology and the Environment*, *11*, 474–482. http://dx.doi.org/10.1890/120282

Guston, D. H. (2001). Boundary organizations in environmental policy and science: An introduction. *Science, Technology & Human Values*, *26*, 399–408. https://doi.org/10.1177/016224390102600401

Halofsky, J. S., Conklin, D. R., Donato, D. C., Halofsky, J. E., & Kim, J. B. (2018). Climate change, wildfire, and vegetation shifts in a high-inertia forest landscape: Western Washington, USA. *PLoS One*, *13*, e0209490.

Harvey, G. L., & Henshaw, A. J. (2023). Rewilding and the water cycle. *WIREs Water*, *10*, e1686. http://dx.doi.org/10.1002/wat2.1686

Hauer, F. R., Locke, H., Dreitz, V. J., Hebblewhite, M., Lowe, W. H., Muhlfeld, C. C., et al. (2016). Gravel-bed river floodplains are the ecological nexus of glaciated mountain landscapes. *Science Advances*, *2*, e1600026.

Hawkins, L. R., Abatzoglou, J. T., Li, S., & Rupp, D. E. (2022). Anthropogenic influence on recent severe autumn fire weather in the west coast of the United States. *Geophysical Research Letters*, *49*, e2021GL095496. http://dx.doi.org/10.1029/2021GL095496

Hessburg, P. F., Miller, C. L., Parks, S. A., Povak, N. A., Taylor, A. H., Higuera, P. E., et al. (2019). Climate, environment, and disturbance history govern resilience of western North American forests. *Frontiers in Ecology and Evolution*, *7*. http://dx.doi.org/10.3389/fevo.2019.00239

Hobbs, R. J., Higgs, E., Hall, C. M., Bridgewater, P., Chapin, F. S., Ellis, E. C., et al. (2014). Managing the whole landscape: Historical, hybrid, and novel ecosystems. *Frontiers in Ecology and the Environment*, *12*, 557–564. http://dx.doi.org/10.1890/130300

Hughes, K. W., Matheny, P. B., Miller, A. N., Petersen, R. H., Iturriaga, T. M., Johnson, K. D., et al. (2020). Pyrophilous fungi detected after wildfires in the Great Smoky Mountains National Park expand known species ranges and biodiversity estimates. *Mycologia*, *112*, 677–698.

Hunter, M. L. (2007). Climate change and moving species: Furthering the debate on assisted colonization. *Conservation Biology*, *21*, 1356–1358.

Isaak, D. J., Luce, C. H., Rieman, B. E., Nagel, D. E., Peterson, E. E., Horan, D. L., et al. (2010). Effects of climate change and wildfire on stream temperatures and salmonid thermal habitat in a mountain river network. *Ecological Applications*, *20*, 1350–1371.

Jager, H. I., Long, J. W., Malison, R. L., Murphy, B. P., Rust, A., Silva, L. G. M., et al. (2021). Resilience of terrestrial and aquatic fauna to historical and future wildfire regimes in western North America. *Ecology and Evolution*, *11*, 12259–12284.

Jain, P., Castellanos-Acuna, D., Coogan, S. C. P., Abatzoglou, J. T., & Flannigan, M. D. (2022). Observed increases in extreme fire weather driven by atmospheric humidity and temperature. *Nature Climate Change*, *12*, 63–70. http://dx.doi.org/10.1038/s41558-021-01224-1

Jolly, W. M., Cochrane, M. A., Freeborn, P. H., Holden, Z. A., Brown, T. J., Williamson, G. J., et al. (2015). Climate-induced variations in global wildfire danger from 1979 to 2013. *Nature Communications*, *6*, 7537. https://doi.org/10.1038%2Fncomms8537

Jones, M. W., Abatzoglou, J. T., Ververbeke, S., Andela, N., Lasslop, G., Forkel, M., et al. (2022). Global and regional trends under climate change. *Reviews of Geophysics*, *60*, e2020RG000726. https://doi.org/10.1029/2020RG000726

Juang, C. S., Williams, A. P., Abatzoglou, J. T., Balch, J. K., Hurteau, M. D., & Moritz, M. A. (2022). Rapid growth of large forest fires drives the exponential response of annual forest-fire area to aridity in the western United States. *Geophysical Research Letters*, *49*, e2021GL097131. http://dx.doi.org/10.1029/2021GL097131

Kampf, S. K., Gannon, B. M., Wilson, C., Saavedra, F., Miller, M. E., Heldmyer, A., et al. (2020). PEMIP: post-fire erosion model inter-comparison project. *Journal of Environmental Management*, *268*, 110704.

Katurji, M., Zhang, J., Satinsky, A., McNair, H., Schumacher, B., Strand, T., et al. (2021). Turbulent thermal image velocimetry at the immediate fire and atmospheric interface. *Journal of Geophysical Research: Atmospheres*, *126*, e2021JD035393. http://dx.doi.org/10.1029/2021JD035393

Keeley, J. E., & Syphard, A. D. (2019). Twenty-first century California, USA, wildfires: Fuel-dominated vs. wind-dominated fires. *Fire Ecology*, *15*, 24. https://fireecology.springeropen.com/articles/10.1186/s42 408-019-0041-0

Kocik, J. F., Hayes, S. A., Carlson, S. M., & Cluer, B. (2022). A resist-accept-direct (RAD) future for salmon in Maine and California: Salmon at the southern edge. *Fisheries Management and Ecology*, *29*, 456–474.

Kondylatos, S., Prapas, I., Ronco, M., Papoutsis, I., Camps-Valls, G., Piles, M., et al. (2022). Wildfire dangers prediction and understanding with deep learning. *Geophysical Research Letters*, *49*, e2022GL099368. http://dx.doi.org/10.1029/2022GL099368

Krawchuk, M. A., Moritz, M. A., Parisien, M. A., Van Dorn, J., & Hayhoe, K. (2009). Global pyrogeography: the current and future distribution of wildfire. *PLoS One*, *4*, e5102.

Lareau, N. P., Donohue, A., Roberts, M., & Ebrahimian, H. (2022). Tracking wildfires with weather radars. *Journal of Geophysical Research: Atmospheres*, *127*, e2021JD036158. http://dx.doi.org/10.1029/2021JD036158

Leech, S. M., Almuedo, P. L., & O'Neill, G. (2011). Assisted migration: Adapting forest management to a changing climate. *BC Journal of Ecosystems and Management*, *12*, 18–34.

Liang, S., Hurteau, M. D., & Westerling, A. L. (2017). Response of Sierra Nevada forests to projected climate-wildfire interactions. *Global Change Biology*, *23*, 2016–2030. http://dx.doi.org/10.1111/gcb.13544

Littell, J. S., Peterson, D. L., Riley, K. L., Liu, Y., & Luce, C. H. (2016). A review of the relationships between drought and forest fire in the United States. *Global Change Biology*, *22*, 2353–2369.

Luković, J., Chiang, J. C. H., Blagojević, D., & Sekulić, A. (2021). A later onset of the rainy season in California. *Geophysical Research Letters*, *48*, e2020GL09350. http://dx.doi.org/10.1029/2020GL090350

McKelvey, K. S., Block, W. M., Jain, T. B., Luce, C. H., Page-Dumroese, D. S., Richardson, B. A., et al. (2021). Adapting research, management, and governance to confront socioecological uncertainties in novel ecosystems. *Frontiers in Forests and Global Change*, *4*, 644696. http://dx.doi.org/10.3389/ffgc.2021.644696

McKenzie, D., Gedalof, Z., Peterson, D. L., & Mote, P. (2004). Climatic change, wildfire, and conservation. *Conservation Biology*, *18*, 890–902.

McLachlan, J. S., Hellmann, J. J., & Schwartz, M. W. (2007). A framework for debate of assisted migration in an era of climate change. *Conservation Biology*, *21*, 297–302.

McLaughlin, B. C., Ackerly, D. D., Klos, P. Z., Natali, J., Dawson, T. E., & Thompson, S. E. (2017). Hydrologic refugia, plants, and climate change. *Global Change Biology*, *23*, 2941–2961.

McRae, R. H., Sharples, J. J., & Fromm, M. (2015). Linking local wildfire dynamics to pyroCb development. *Natural Hazards and Earth System Sciences*, *15*, 417–428.

Metz, M. R., Varner, J. M., Frangioso, K. M., Meentemeyer, R. K., & Rizzo, D. M. (2013). Unexpected redwood mortality from synergies between wildfire and an emerging infectious disease. *Ecology*, *94*, 2152–2159. http://dx.doi.org/10.1890/13-0915.1

Naiman, R. J., & Decamps, H. (1997). The ecology of interfaces: Riparian zones. *Annual Review of Ecology and Systematics*, *28*, 621–658.

Norman, L. M., Lal, R., Wohl, E., Fairfax, E., Gellis, A. C., & Pollock, M. M. (2022). Natural infrastructure in dryland streams (NIDS) can establish regenerative wetland sinks that reverse desertification and strengthen climate resilience. *Science of the Total Environment*, *849*, 157738. http://dx.doi.org/10.1016/j.scitotenv.2022.157738

Oakley, N. S. (2021). A warming climate adds complexity to post-fire hydrologic hazard planning. *Earth's Future*, *9*, e2021EF002149.

O'Neil, S. T., Coates, P. S., Brussee, B. E., Ricca, M. A., Espinosa, S. P., Gardner, S. C., et al. (2020). Wildfire and the ecological niche: Diminishing habitat suitability for an indicator species within semi-arid ecosystems. *Global Change Biology*, *26*, 6296–6312.

Park, C. Y., Takahashi, K., Takakura, J., Li, F., Fujimori, S., Hasegawa, T., et al. (2021). How will deforestation and vegetation degradation affect global fire activity? *Earth's Future*, *9*, e2020EF001786. http://dx.doi.org/10.1029/2020EF001786

Payette, S., Couillard, P. L., Fregeau, M., Laflamme, J., & Lavoie, M. (2022). The velocity of post-glacial migration of fire-adapted species in eastern North America. *Proceedings of the National Academy of Sciences*, *119*, e2210496119. http://dx.doi.org/10.1073/pnas.2210496119

Peipoch, M., Brauns, M., Hauer, F. R., Weitere, M., & Valett, H. M. (2014). Ecological simplification: Human influences on riverscape complexity. *BioScience*, *65*, 1057–1065.

Perry, M. C., Vanvyve, E., Betts, R. A., & Palin, E. J. (2022). Past and future trends in fire weather for the UK. *Natural Hazards and Earth System Sciences*, *22*, 559–575. http://dx.doi.org/10.5194/nhess-22-559-2022

Poff, N. L., Olden, J. D., Merritt, D. M., & Pepin, D. M. (2007). Homogenization of regional river dynamics by dams and global biodiversity implications. *Proceedings of the National Academy of Sciences*, *104*, 5732–5737.

Preisler, H. K., Riley, K. L., Stonesifer, C. S., Calkin, D. E., & Jolly, W. M. (2016). Near-term probabilistic forecast of significant wildfire events for the western United States. *International Journal of Wildland Fire*, *25*, 1169–1180.

Pugh, B. E., Colley, M., Dugdale, S. J., Edwards, P., Flitcroft, R., Holz, A., et al. (2022). A possible role for river restoration enhancing biodiversity through interaction with wildfire. *Global Ecology and Biogeography*, *31*, 1990–2004.

Pyne, S. J. (2020). From Pleistocene to Pyrocene: Fire replaces ice. *Earth's Future*, *7*, e2020EF001722. http://dx.doi.org/10.1029/2020EF001722

Rahel, F. J., & Lynch, A. J. (2022). Managing fisheries within a RAD framework: Concepts and applications. *Fisheries Management and Ecology*, *29*, 323–328.

Rathburn, S. L., Shahverdian, S. M., & Ryan, S. E. (2018). Post-disturbance sediment recovery: Implications for watershed resilience. *Geomorphology*, *305*, 61–75.

Rego, F. C., Morgan, P., Fernandes, P., & Hoffman, C. (2021). *Fire Science: From Chemistry to Landscape Management*. Cham, Switzerland: Springer Nature Switzerland.

Rengers, F. K., & McGuire, L. A. (2021). Wildfire and earth surface processes. *Earth Surface Processes and Landforms*, *46*, 1099–1102.

Rideout, N. K., Wegscheider, B., Kattilakoski, M., McGee, K. M., Monk, W. A., & Baird, D. J. (2021). Rewilding watersheds: Using nature's algorithms to fix our broken rivers. *Marine and Freshwater Research*, *72*, 1118–1124. http://dx.doi.org/10.1071/MF20335

Robotham, J., Old, G., Rameshwaran, P., Sear, D., Trill, E., Bishop, J., et al. (2023). Nature-based solutions enhance sediment and nutrient storage in an agricultural lowland catchment. *Earth Surface Processes and Landforms*, *48*, 243–258. http://dx.doi.org/10.1002/esp.5483

Ryan, K. C., Knapp, E. E., & Varner, J. M. (2013). Prescribed fire in North American forests and woodlands: History, current practice, and challenges. *Frontiers in Ecology and the Environment*, *11*, e15–e24.

Selwood, K. E., & Zimmer, H. C. (2020). Refuges for biodiversity conservation: A review of the evidence. *Biological Conservation*, *245*, 108502.

Shuman, J. K., Balch, J. K., Barnes, R. T., Higuera, P. E., Roos, C. I., Schwilk, D. W., et al. (2022). Reimagine fire science for the Anthropocene. *PNAS Nexus*, *1*, 1–14. https://doi.org/10.1093/pnasnexus/pgac115

Sierra-Hernández, M. R., Beaudon, E., Porter, S. E., Mosley-Thompson, E., & Thompson, L. G. (2022). Increased fire activity in Alaska since the 1980s: Evidence from an ice core-derived black carbon record. *Journal of Geophysical Research: Atmospheres*, *127*, e2021JD035668. http://dx.doi.org/10.1029/2021JD035668

Southwell, D., Legge, S., Woinarski, J., Lindenmayer, D., Lavery, T., & Wintle, B. (2022). Design considerations for rapid biodiversity reconnaissance surveys and long-term monitoring to assess the impact of wildfire. *Diversity and Distributions*, *28*, 559–570.

Stanford, J. A., Ward, J. V., Liss, W. J., Frissell, C. A., Williams, R. N., Lichatowich, J. A., et al. (1996). A general protocol for restoration of regulated rivers. *Regulated Rivers*, *12*, 391–411.

Stevens-Rumann, C. S., Kemp, K. B., Higuera, P. E., Harvey, B. J., Rother, M. T., Donato, D. C., et al. (2018). Evidence for declining forest resilience to wildfires under climate change. *Ecology Letters*, *21*, 243–252.

Stevens-Rumann, C. S., & Morgan, P. (2019). Tree regeneration following wildfires in the western US: A review. *Fire Ecology*, *15*, 15. http://dx.doi.org/10.1186/s42408-019-0032-1

Stralberg, D., Wang, X., Parisien, M. A., Robinne, F. N., Sólymos, P., Mahon, C. L., et al. (2018). Wildfire-mediated vegetation change in boreal forests of Alberta, Canada. *Ecosphere*, *9*, e02156. http://dx.doi.org/10.1002/ecs2.2156

Sullivan, A., Baker, E., & Kurvits, T. (Eds.) (2022). *Spreading Like Wildfire: The Rising Threat of Extraordinary Landscape Fires*. United Nations Environment Programme.

Swain, D. L. (2021). A shorter, sharper rainy season amplifies California's wildfire risk. *Geophysical Research Letters*, *48*, e2021GL092843.

Swanson, F., Brodie, N., & Goodrich, C. (2012). Engaging the Humanities to Address Wildland Fire Issues. *JFSP Research Project Reports 61*. U.S. Joint Fire Science Program. http://digitalcommons.unl.edu/jfspresearch/61

Thompson, L. M., Lynch, A. J., Beever, E. A., Engman, A. C., Falke, J. A., Jackson, S. T., et al. (2021). Responding to ecosystem transformation: Resist, accept, or direct? *Fisheries*, *46*, 8–21.

Touma, D., Stevenson, S., Lehner, F., & Coats, S. (2021). Human-driven greenhouse gas and aerosol emissions cause distinct regional impacts on extreme fire weather. *Nature Communications*, *12*, 212. http://dx.doi.org/10.1038/s41467-020-20570-w

Triepke, F. J., Muldavin, E. H., & Wahlberg, M. M. (2019). Using climate projections to assess ecosystem vulnerability at scales relevant to managers. *Ecosphere*, *10*, e02854. https://doi.org/10.1002/ecs2.2854

Triska, F. J. (1984). Role of woody debris in modifying channel geomorphology and riparian areas of a large lowland river under pristine conditions: A historical case study. *Verhandlungen Internationale Vereinigung Limnologie*, *22*, 1876–1892.

Trnka, M., Mozny, M., Jurecka, F., Balek, J., Semeradov, D., Hlavinka, P., et al. (2021). Observed and estimated consequences of climate change for the fire weather regime in the moist-temperate climate of the Czech Republic. *Agricultural and Forest Meteorology*, *310*, 108583. http://dx.doi.org/10.1016/j.agrformet.2021.108583

Úbeda, X., & Sarricolea, P. (2016). Wildfires in Chile: A review. *Global and Planetary Change*, *146*, 152–161.

Wang, J., Stern, M. A., King, V. M., Alpers, C. N., Quinn, N. W. T., Flint, A. L., et al. (2020). PFHydro: A new watershed-scale model for post-fire runoff simulation. *Environmental Modelling and Software*, *12*, 104555.

Wang, S. S. C., Qian, Y., Leung, L. R., & Zhang, Y. (2021). Identifying key drivers of wildfires in the contiguous US using machine learning and game theory interpretation. *Earth's Future*, *9*, e2020EF001910. http://dx.doi.org/10.1029/2020EF001910

Whitman, E., Parisien, M. A., Thompson, D. K., & Flannigan, M. D. (2019). Short-interval wildfire and drought overwhelm boreal forest resilience. *Scientific Reports*, *9*, 18796. http://dx.doi.org/10.1038/s4159

Wiener, K. D., Schlegel, P. K., Grenfell, S. E., & van der Waal, B. (2022). Contextualising sediment trapping and phosphorus removal regulating services: A critical review of the influence of spatial and temporal variability in geomorphic processes in alluvial wetlands in drylands. *Wetlands Ecology and Management*, *30*, 737–770.

Wine, M. L., & Cadol, D. (2016). Hydrologic effects of large southwestern USA wildfires significantly increase regional water supply: Fact or fiction? *Environmental Research Letters*, *11*, 085006.

Wohl, E. (2013a). Landscape-scale carbon storage associated with beaver dams. *Geophysical Research Letters*, *40*, 1–6.

Wohl, E. (2013b). Redistribution of forest carbon caused by patch blowdowns in subalpine forests of the southern Rocky Mountains, USA. *Global Biogeochemical Cycles*, *27*, 1205–1213.

Wohl, E. (2014). A legacy of absence: Wood removal in U.S. rivers. *Progress in Physical Geography*, *38*, 637–663.

Wohl, E. (2016). Spatial heterogeneity as a component of river geomorphic complexity. *Progress in Physical Geography*, *40*, 598–615.

Wohl, E., Castro, J., Cluer, B., Merritts, D., Powers, P., Staab, B., et al. (2021). Rediscovering, reevaluating, and restoring lost river-wetland corridors. *Frontiers in Earth Science*, *9*, 653623.

Wohl, E., Greenbaum, N., Schick, A. P., & Baker, V. R. (1994). Controls on bedrock channel incision along Nahal Paran, Israel. *Earth Surface Processes and Landforms*, *19*, 1–13.

Wohl, E., Lininger, K. B., & Baron, J. (2017). Land before water: The relative temporal sequence of human alteration of freshwater ecosystems in the conterminous United States. *Anthropocene*, *18*, 27–46.
Wohl, E., Lininger, K. B., & Scott, D. N. (2018). River beads as a conceptual framework for building carbon storage and resilience to extreme climate events into river management. *Biogeochemistry*, *141*, 365–3803.
Wohl, E., Marshall, A. E., Scamardo, J., White, D., & Morrison, R. R. (2022). Biogeomorphic influences on river corridor resilience to wildfire disturbances in a mountain stream of the southern Rockies, USA. *Science of the Total Environment*, *820*, 153321.
Yu, K., Okin, G. S., Ravi, S., & D'Odorico, P. (2016). Potential of grass invasions in desert shrublands to create novel ecosystem states under variable climate. *Ecohydrology*, *9*, 1496–1506. http://dx.doi.org/10.1002/eco.1742
Zhuang, Y., Fu, R., Santer, B. D., Dickinson, R. E., & Hall, A. (2021). Quantifying contributions of natural variability and anthropogenic forcings on increased fire weather risk over the western United States. *Proceedings of the National Academy of Sciences*, *118*, e2111875118. http://dx.doi.org/10.1073/pnas.2111875118

INDEX

Landscapes on Fire: Impacts on Uplands, Rivers, and Communities, Special Publications 80, First Edition. Ellen Wohl.

DOI: 10.1002/9781394235186.index

L

M

N

O

P

Q

R

S